T0210636

STRUCTURAL ANALYSIS
&
SELECTED TOPICS

Dr. Mohammed Bin Salem
University of Qatar
College of Engineering
Civil Engineering Dept.
P.O.Box 2713
Doha – Qatar
e-mail m.alansari@qu.edu.qa

Order this book online at www.trafford.com
or email orders@trafford.com

Most Trafford titles are also available at major online book retailers.

Print information available on the last page.

ISBN: 978-1-4120-8608-0 (sc)
ISBN: 978-1-4269-2879-6 (e)

Because of the dynamic nature of the Internet, any web addresses or links contained in this book may have changed since publication and may no longer be valid. The views expressed in this work are solely those of the author and do not necessarily reflect the views of the publisher, and the publisher hereby disclaims any responsibility for them.

Any people depicted in stock imagery provided by Thinkstock are models, and such images are being used for illustrative purposes only.
Certain stock imagery © Thinkstock.

Trafford rev. 10/24/2016

www.trafford.com
North America & international
toll-free: 1 888 232 4444 (USA & Canada)
fax: 812 355 4082

Preface

The main goal of this book is to help students understand structural analysis topics through numerical examples that explain the theory behind these topics. An equally important goal is to have students recognize and understand the link between structural analysis and design topics. In other words, students are helped to know how to use structural analysis results for design purpose.

This book minimizes theoretical derivations and maximizes numerical analyses through a large number of illustrated examples. The book is divided into sixteen chapters:

- Chapter 1 is an introduction to some structural concepts such as types of structures, loads, structural modeling, structural analysis, and design links.
- Chapters 2,3, and 4 cover basic structural analysis techniques to determine the reactions, the shear and the moment diagrams of determinate beams and frames as well as the reactions and the bar forces of determinate trusses.
- Chapter 5 covers the deflection analysis of determinate structures using the following methods: 1) double integration, 2) moment area, 3) conjugate beam, and 4) energy methods. These methods will be used in analyzing indeterminate structures by the consistent deformation method in chapter 9.
- Chapter six covers influence lines for the structural analysis of determinate structures under moving loads.
- Chapter 7 covers the analysis of three-hinged arches and cables under a number of loading types.
- Chapters 8 through 11 covers the following analysis methods of indeterminate structures: 1) force or consistent deformation method, 2) slope deflection method, 3) and moment distribution method.
- Chapters 12 through 15 introduce the following matrix analysis methods of indeterminate structures: 1) flexibility method, 2) stiffness method, and 3) direct stiffness method.
- Chapter 16 covers the following topics related to structural analysis: 1) concrete design , 2) steel design, 3) analysis of non-prismatic member, and 4) analysis of other types of structural members. A number of MathCad programs are introduced as learning and teaching tools that could be used by instructors and students to minimize hand calculations and to maximize the understanding of structural analysis topics.

Acknowledgements

The author is particularly indebted to the following contributors who gave time, effort, and assistance in developing this book:

- Dr. Ahmed Senouci (University of Qatar) for his review and discussion of the book.
- Eng. Fadi Bahjet (GHD Consulting office) for his help in documenting, editing, and preparing some of the book drawings.
- Mr. Irfan (Draftsman) for preparing most of the book drawings.
- Mr. Ibrahim Al-Ansari (Manager of Dar Al-Thagafa book stores and Press.) for providing the book cover picture.

CONTENTS

CHAPTER - 1
INTRODUCTION

Abu Dhabi - UAE

Chapter 1. Introduction

1.1   STRUCTURES

Structures, such as bridges, buildings, towers, dams, airports, high ways, high-pressure vessels, are composed of a number of interconnected structural systems which in turn are composed of a number of interconnected structural members and/or elements. Beams, columns, and ties are common examples of structural members, Fig. 1-1. Shells, trusses, arches, cables, and frames are common examples of structural systems, Fig1-2.

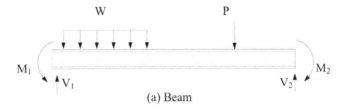

(a) Beam

(b) Simply-Supported Beam

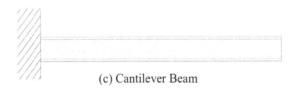

(c) Cantilever Beam

(d) Continuous Beam

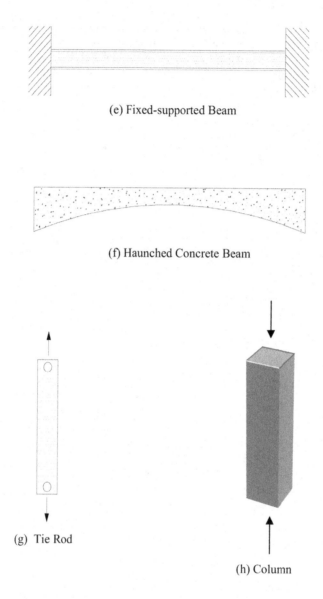

(e) Fixed-supported Beam

(f) Haunched Concrete Beam

(g) Tie Rod

(h) Column

Figure 1-1. Common Structural Elements

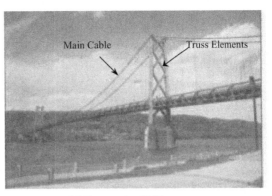

(b) Suspension Bridge

(a) Shell Structural System

(c) Circular Structure

(d) Cable

(e) Fort-Pitt Bridge

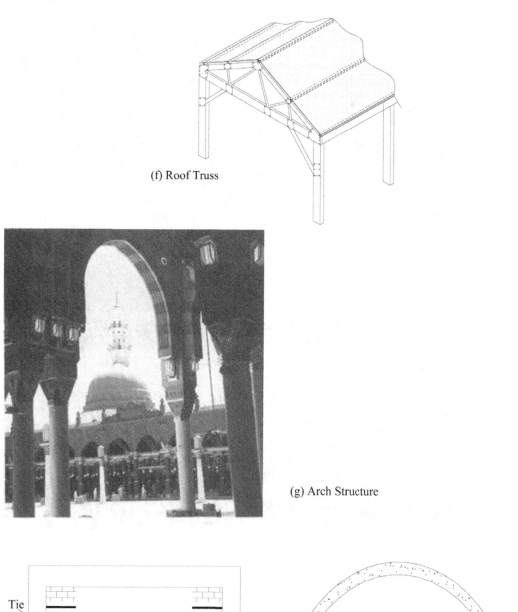

(f) Roof Truss

(g) Arch Structure

(i) Masonry Wall Frame

(h) Concrete Arch

Tie

Figure 1-2. Common Structural Systems

1.2   LOADS

Loads are due to gravity, earthquake, wind, temperature change, water pressure, blast impact, vibration, gas pressure, shrinkage, friction, creep, and support moment. Loads are subdivided into two groups: *dead loads and live loads*. Dead loads are assumed to remain constant over life span of the structure. The structure weight and any permanent loads fixed on the structure are examples of dead loads. Live loads can be removed or replaced on a structure. Occupancy loads, snow loads, wind loads, earthquake loads, vehicle moment, and blast loads are all examples of live loads. Tables 1-1, 1-2, 1-3 list minimum design dead loads, live loads, and construction material densities , respectively. National and local design codes must always be checked for the minimum required design loads.

Table 1-1 Minimum Live Loads

Type of use	lb / ft^2	kN / m^2
Apartment buildings:		
Private units	40	1.92
Public rooms	100	4.80
Corridors	80	3.84
Office buildings:		
Offices	50	2.40
Lobbies	100	4.80
Corridors above first floor	80	3.84
Garages (cars only)	50	2.4
Stores:		
First floor	100	4.8
Upper floors	75	3.6
Warehouse:		
Light storage	125	6.0
Heavy storage	250	12.0
Libraries:		
Reading rooms	60	2.9
Stack rooms, books and shelving	150	7.2
Corridors above first floor	80	3.8
Schools:		
Classrooms	40	1.9
Corridors	80	3.8
Shop with light equipment	60	2.9
Panel institutions:		
Cell blocks	40	1.9
Corridors	100	4.8

Table 1-2 Minimum Dead Loads

Type of use	Ib / ft^2	kN / m^2
Masonry:		
Cast-stone masonry	144	6.9
Concrete, stone aggregate, reinforced	150	7.2
Ashlar:		
Granite	165	7.9
Limestone, crystalline	165	7.9
Limestone, olitic	135	6.5
Marble	173	8.3
Sandstone	144	6.9
Walls:		
Concrete brick:		
4-in, with heavy aggregate	46	2.2
4-in, with light aggregate	33	1.6
Concrete block, hollow:		
8-in, with heavy aggregate	55	2.6
8-in, with light aggregate	35	1.7
12-in, with heavy aggregate	85	4.1
12-in, with light aggregate	55	2.6
Clay brick:		
High-absorption, per 4-in. wythe.	34	1.6
Med-absorption, per 4-in. wythe.	39	1.9
Low-absorption, per 4-in. wythe.	46	2.2
Sand-lime brick, per 4-in. wythe.	38	1.8
Clay tile, load-bearing:		
4-in.	24	1.1
8-in.	42	2.0
12-in.	58	2.8
Clay tile, non-load-bearing:		
2-in.	11	0.5
4-in.	18	0.9
8-in.	34	1.6
Furring tile:		
1.5- in.	8	0.4
2-in.	10	0.5
Glass block, 4-in.	18	0.9
Gypsum block, hollow:		
2-in.	9.5	0.5
4-in.	12.5	0.6
6-in.	18.5	0.9
Ceiling		
Plaster (on tile or concrete)	5	0.2
Suspended metal lath and gypsum plaster	10	0.5
Suspended metal lath and cement plaster	15	0.77

Table 1-3 Minimum densities.

Type of use	lb / ft^3	kN / m^3
Aluminum, cast	165	25.9
Earth (not submerged):		
Clay, dry	63	9.9
Clay, damp	110	17.3
Clay and gravel, dry	100	15.7
Silt, moist, loose	78	12.3
Silt, moist, packed	96	15.1
Sand and gravel, dry, loose	100	15.7
Sand and gravel, dry, packed	110	17.3
Sand and gravel, wet	120	18.9
Gold, solid	1.2	0.2
Gravel, dry	104	16.3
Gypsum, loose	70	11
Ice	57	9
Iron, cast	450	70.7
Lead	710	111.5
Lime, hydrate, loose	32	5
Lime, hydrate, compacted	45	7.1
Magnesium alloys	112	17.6
Mortar, hardened:		
Cement	130	20.4
Lime	110	17.3

Example 1-1

Determine the beam loading per meter length for the beams shown in Fig. 1-3. The normal- weight concrete slab, which has a thickness of 200 mm, is covered with ashlars marble. All of the beams carry a 4-m high, 200 mm-thick heavy aggregate hollow block wall.

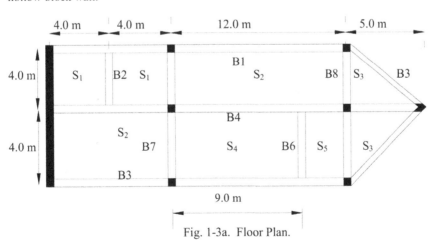

Fig. 1-3a. Floor Plan.

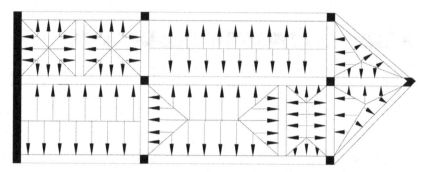

Fig. 1-3b. Load Distribution

To determine the loading on beam B2 the Live and Dead load must be calculated

Min. Live load for residential building = 1.9 kN/m^2
Min. Design load for marble = 8.3 kN/m^2
Min. Density for concrete = 25 kN/m^3
Min. Design load for concrete block = 2.6 kN/m^2

Slab = Thickness (density)
    = 0.2(25) = 5 kN/m^2

Wall = Height(concrete block)
    = 4(2.6) = 10.4 kN/m

Loads :

W_{Slab} = 2(5) + 2(5) = 20 kN/m
W_{Marble} = 2(8.3) + 2(8.3) = 33.2 kN/m
W_{LL} =2(1.9) + 2(1.9) = 7.6 kN/m

W_u = 1.4 DL + 1.7 LL
    = 1.4(20+33.2) + 1.7 (7.6) = 87.5

For wall calculation
W_{Wall} = 1.4(10.4) = 14.56 kN/m

1.2.1 Wind Loads

Wind is due to the natural horizontal motion of the atmosphere. Extreme winds, such as thunderstorms, typhoons, tornadoes, and hurricanes, impose on structures loads that are many times more than those normally considered in the design of these structures. Building codes and standards such as, the Uniform Building Code, the American National Standard specify the minimum load requirements for buildings and other structures which provide accepted design and construction practices. The design wind loads according to UBC code (Uniform Building Code) is given by the following equation:

$$P = C_e\, C_q\, q_s\, I \qquad\qquad [1\text{-}1]$$

Where

P	Design wind load
C_e	Combined height, exposure, and gust factor. Table 1-1.
C_q	Dimensionless pressure coefficient. Table 1-2.
q_s	Wind velocity. Table 1-3.
I	Important factor. Table 1-4

For more details on the parameters and the equation derivation, one should refer to the UBC codes.

Table 1-1. Combined height, exposure, and gust factor, C_e

Height above average level of adjoining ground (feet) (×304.8 for mm)	Exposure D	Exposure C	Exposure B
0-15	1.39	1.06	0.62
20	1.45	1.13	0.67
25	1.50	1.19	0.72
30	1.54	1.23	0.76
40	1.62	1.31	0.84
60	1.73	1.43	0.95
80	1.81	1.53	1.04
100	1.88	1.61	1.13
120	1.93	1.67	1.20
160	2.02	1.79	1.31
200	2.10	1.87	1.42
300	2.23	2.05	1.63
400	2.34	2.19	1.80

Table 1-2. Pressure Coefficient C_q

Description	C_q
Method 1 (Normal Force Method)	
Walls:	
Windward wall	0.8 inward
Leeward wall	0.5 outward
Roofs:	
Wind perpendicular to ridge	
Leeward roof or flat ridge	0.7 outward
Windward roof	
Less than 2:12 (16.7%)	0.7 outward
Slope 2:12 (16.7%) to less	
than 9:12 (75%)	0.9 outward
	or 0.3 inward
Slope 9:12 (75%) to 2:12 (100%)	0.4 inward
Slope > (12:12) (100%)	0.7 inward
Wind parallel to ridge and flat roofs	0.7 outward
Method 2 (Projected Area Method)	1.3 horizontal any
On vertical projected area	direction
Structures 40 feet (12,192 mm) or less in height	1.4 horizontal any
Structures over 40 feet (12,192 mm) in height	direction
On horizontal projected area	0.7 upward

$$q = 0.00256\ V^2 \qquad V \rightarrow \textbf{\textit{miles/hr}} \qquad\qquad [1\text{-}2]$$

$$q = 0.047\ V^2 \qquad \textbf{\textit{V}} \rightarrow \text{km/hour} \qquad\qquad [1\text{-}3]$$

Table 1-3. Wind Pressure q_s at 33 ft (10 m) For Various Basic Wind Speeds

Basic wind speed (mph)*(× 1.61 for km/h)	70	80	90	100	110	120	130
Pressure q_s (psf) (× 0.0479 for kN/m²)	12.6	16.4	20.8	25.6	31.0	36.9	43.3

Table 1-4. Importance Factor

Occupancy Category	Importance Factor
Essential facilities	1.15
Hazardous facilities	1.15
Special occupancy structures	1.00
Standard occupancy structures	1.00

1.2.2   Earthquake loads

Earthquakes generate ground motions that can affect everything supported above or below the ground within a quite large distance from the epicenter. Of the three basic seismic waves that cause damage during an earthquake, only two propagate with in a body of solid rock. The faster wave is called the primary or P wave while the slower one is called the secondary or S wave. The third wave is called a surface wave, because its motion is near the ground surface.

The structure responses to seismic motions is interdependently affected by the characteristics of the input motion at the structure base (motion frequencies, amplitude, phase relationship, and duration), the stiffness characteristics of the foundation materials, and the dynamic response characteristics of the structure. Forces shears and moment develop because of the deformation that resulted as a structure response to the ground motion, Fig. 1-4. The seismic shear force required by the uniform building code for earthquake design shell be at least equal to hat specified by equation [1-4].

$$V = \frac{ZIC}{R_w} W \qquad\qquad [1\text{-}4]$$

Where, R_w = response modification factor, Z = seismic zone factor, W = total seismic dead load, and   C = Coefficient related to the fundamental period of vibration of the structure, and including the site structure response factor.

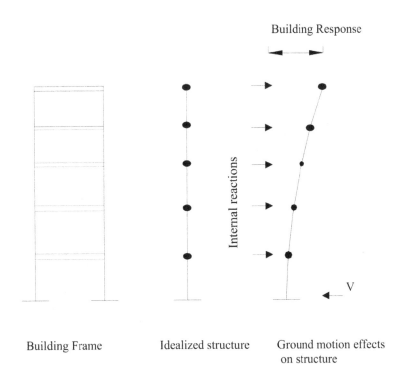

Building Response

Internal reactions

V

Building Frame   Idealized structure   Ground motion effects
                                                          on structure

Fig. 1-4. Ground Motion Effects on Structures

1.2.3 Other Loads

Loads due to temperature change. When the structure or structural elements are subjected to differential temperature change, displacements will occur. Constraining theses displacements induces internal forces, shear, and moments within the structural member.

Loads due to support settlement. The additional axial forces, shears, and moments due to support settlements may exceed the original design loads. These settlements are the result of lack of attention to design details, inadequate supervision of soil compaction, and sudden or long time settlement of the structure.

Bridge loads. Bridges must be designed to support the vehicular loads associated with their functional use. The minimum design loads are prescribed by the American Association of State Highway and Transportation Officials (AASHTO) and other codes. The prescribed loadings should be so placed that they will produce the maximum design effects for each of the examined conditions. All bridge live loads to be increased to account for the impact factor. For more details see AASHTO.

Snow and ice loads. Depend mainly on the relative moisture content, the roof shape, the wind velocity, and the type of heating and insulation adjacent to the roof. For example, the design snow load in the USA is about 3 kN/m^2 not including the state of Alaska.

1.3   Structural Modeling

Once the loads are determined, the structure model is established by joining the structure members together through various types of joints called supports as shown in Fig. 1-5. Roller supports resist forces in one direction only, pinned or hinged supports resist forces in two mutually perpendicular directions, and fixed supports resist forces in two mutually perpendicular directions as well as a bending moment. These three supports are the most commonly used type of supports. Real structures, which are three dimensional structures (3-D). However, it is desirable to reduce the 3-D structures to (2-D) structures (2-D) or (1-D) structures, provided that the reduced structure will give acceptable structure analysis results.

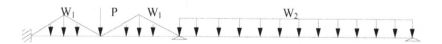

Fig. 1-5. Continuous Beam Structural Model

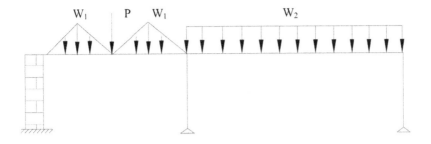

Fig. 1-6 2-D Frame Structural Model

1.4   Structural Analysis

Structural analysis is the determination of the structure (or member) internal forces as a response to the external loads acting upon it. Internal forces include bending moments, shear, axial compression or tension, and torsional moments. Structural elements are classified as beams (members subject to bending moment and shear), columns (members subject to axial forces), and beam-columns (members subject to bending moments and axial forces), Fig.1-8. A very important part of structural analysis is the determination of the displacements of the structural joints and members. The equilibrium concept, which is the most important concept of structural analysis, represents the balance condition between the applied loads and the internal resisting forces. At any structural joint, the external loads and the internal forces has to satisfy the following laws of static equilibrium:

$$
\begin{array}{|c|c|c|}
\hline
\sum F_X = 0 & \sum F_Y = 0 & \sum F_Z = 0 \\
\hline
\sum M_X = 0 & \sum M_Y = 0 & \sum M_Z = 0 \\
\hline
\end{array}
\qquad [1\text{-}5]
$$

The summation of all forces acting in any direction, and summation of all moments about any axis must be equal to zero.

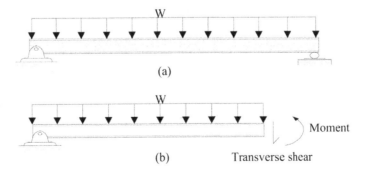

Figure 1-8. (a) Beam Element, (b) Beam Internal Forces

1.5 Design Link

Economical, safe, and efficient structural designs are the aim and goal of the structural engineer. Structural analysis and structural design are interlocked subjects. Structural analysis is an important link to efficient designs.

Summary

- The characteristics of the common structural elements commonly found in buildings and bridges such as beams, trusses, arches, frames with rigid joints, cables, and shells were reviewed.
- The loads that engineers must consider in the design of buildings, bridges, dams, water tanks, and foundations were also reviewed. These loads includes dead load, live load, wind, earthquake, snow, rain, fluid pressure, and soil pressure.
- The loads that govern the design of structures are specified by national and local building codes. Structural codes also specify additional loading provisions that apply specifically to construction materials such as steel, reinforced concrete, aluminum, and wood.
- Wind velocities increase with height above the ground. The parameters C_e, C_q, q_s, and I used for wind load calculation according to UBC code are tabulated in Tables 1-1 through 1-4, respectively.
- The ground motions produced by earthquakes cause buildings, bridges, and other structures to sway. In buildings this motion creates lateral inertia forces that are assumed to be concentrated at each floor. The inertia forces are greatest at the top of buildings where the displacements are greatest.
- The magnitude of the earthquake shear forces depends on the size of the earthquake, the weight of the building, the natural period of the building, the stiffness and ductility of the structural frame, and the soil type. The earthquake shear force according to the UBC is given by Eq. 1-4.
- Buildings with a ductile frame (that can undergo large deformations without collapsing) may be designed for much smaller seismic forces than structures that depend on a brittle structural system (for example, un-reinforced masonry).
- Although they are most of the time three-dimensional, structures can often be divided into a series of simpler planar structures for analysis. A simplified and idealized model that accurately represents the essentials of the real structure can be selected.
- In zones where wind or earthquake forces are small, low-rise buildings are initially proportioned for live and dead load, and then checked for wind or earthquake, or both, depending on the region; the design can be easily modified as needed.
  On the other hand, for high-rise buildings located in regions where large earthquakes or high winds are common, designers must give high priority in the preliminary design phase to select structural systems (for example, shear walls or braced frames) that resist lateral loads efficiently.

CHAPTER - 2
REACTIONS

Dubai -UAE

Chapter 2 Reactions

2.1   INTRODUCTION

Reaction forces are developed at the structure supports to equilibrate the effects of the applied forces.
The main steps in the analysis of structures are as follows.

1)   Draw a free body diagram. This is a simplified representation of the structure that shows the external loads and the reactions acting on the structure.

2)   Apply equations of equilibrium $\sum F_X = 0, \sum F_Y = 0, \sum M_0 = 0$.

3)   Draw the free body diagram to show the magnitude of loads and reactions.

The support types and restraints are shown in Table 2-1. The equations of equilibrium should be applied in a way that provide a short and fast solution of the problem.

A negative value for the unknown force or moment means that the correct direction of the force or moment is opposite to the original assumption made on the free body diagram.

2.2   STABILITY, DETERMINACY AND INDETERMINACY

The structure is statically determinant if the number of reactions (unknown) is equal to the number of available static equations **(R = E)**. The structure is externally indeterminate if the number of unknowns is greater than the number of available static equations **(R > E)**. The structure is unstable externally if the number of unknowns is less than the number of available static equations **(R < E)**. An extra static equation is added for each hinge in the structure, Tables 2-2, 2-3, and 2-4. Indeterminate structures require additional equations, which can be obtained by relating the applied loads and reactions to the displacements and slopes at different points on the structure.

Table 2-1 Support Types

Type of Connection	Reaction	Number of Unknowns
1) Roller	F_y	One unknown.
2) Pin	F_y F_x	Two unknowns.
3) Fixed	F_y M F_x	Three unknowns.
4) Translation spring	Δ K_{spr} $F_y = K_{Spr}(\Delta)$	One unknown.
5) Rotational spring	K_θ θ θ $M = K_\theta(\theta)$	One unknown.

Table 2-2. Frame Determinacy and Stability

Frame	Reactions	Equations	Conditions
	6	3	R > E Indeterminate 3rd degree
	10	3	R > E Indeterminate 7th degree
	7	3	R > E Indeterminate 4th degree
	24	3	R > E Indeterminate 21th degree
	30	3	R > E Indeterminate 27th degree
	6	4	R > E Indeterminate 2th degree

* R = Reactions, E = Equations of equilibrium.

Table 2-3. Beam Determinacy and Stability

Beam	Reactions	Equations	Conditions
	3	3	R = E Determinate
	2	3	R < E Unstable
	3	3	R = E Determinate
Hinge	4	4	R = E Determinate
	5	3	R > E Indeterminate to 2nd degree
	4	4	R = E Determinate
Cable	5	3	R > E Indeterminate to 2nd degree
	2	3	R < E Unstable
	5	5	R = E Determinate

Table 2-3 Beam Determinacy and Stability (Continued)

Beam	Reactions	Equations	Conditions
	3	3	Parallel Reaction Unstable
	8	5	$R > E$ Indeterminate 3rd degree

Table 2-4. Truss Determinacy and Stability

Truss	E	R	B	J	Conditions
	3	3	6	4	$B + R = 9 > 2J = 8$ Indeterminate 1st degree internally
	3	3	6	4	$B + R = 9 > 2J = 8$ Indeterminate 1st degree internally
	3	3	8	5	$B + R > 2J$ Indeterminate 1st degree internally
	3	3	13	8	$B + R = 2J$ Stable
	3	3	19	11	$B + R = 2J$ Stable
	4	4	11	7	$B + R > 2J$ Indeterminate 1st degree internally
	4	4	9	6	$B + R > 2J$ Indeterminate 1st degree internally

* E = Equations , R = Reactions, B = Bars, J = Joints

Example: 2-1

Determine the reactions for the beam shown in Fig. 2-1a.

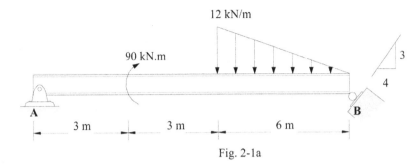

Fig. 2-1a

SOLUTION

Free Body Diagram.
As shown in Fig. 2-1b, the support (roller) at B exerts a normal force on the beam at its point of contact. The line of action of this force is defined by the triangle (3-4-5).

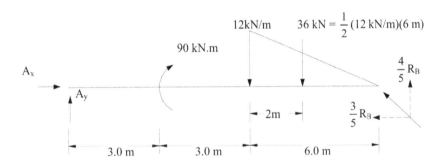

Fig. 2-1b.

The application of the equations of equilibrium (Eq. 1-5) yields the following:

$$\sum M_A = 0 \; + \circlearrowleft \qquad -R_B\left(\frac{4}{5}\right)(12) + 36(8) + 90 = 0 \qquad \Rightarrow R_B = 39.375 \text{ kN} \; \nwarrow$$

$$\sum F_y = 0 \; + \uparrow \qquad 39.375\left(\frac{4}{5}\right) - 36 + A_y = 0 \qquad \Rightarrow A_y = 4.5 \text{ kN} \; \uparrow$$

$$\sum F_x = 0 \; + \rightarrow \qquad -39.375\left(\frac{3}{5}\right) + A_x = 0 \qquad \Rightarrow A_x = 23.625 \text{ kN} \; \rightarrow$$

To check our answer, it is a very good practice to show the magnitudes of all loads and forces on the free-body diagram as shown in Fig. 2-1c.

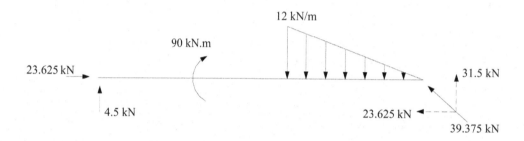

Fig. 2-1c

Example: 2-2

Determine the reactions for the beam shown in Fig. 2-2a.

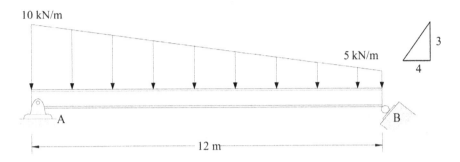

Fig. 2-2a

SOLUTION

Free Body Diagram.
As shown in Fig. 2-2b, the support (roller) at B exerts a normal force on the beam at its point
of contact and the line of action of this force is defined by the triangle (3-4-5).

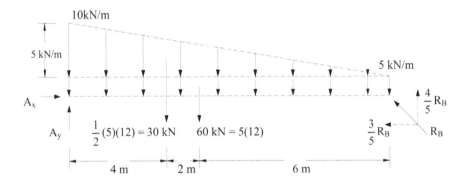

Fig. 2-2b.

The application of the equations of equilibrium (Eq. 1-5) yields the following:

$$\sum M_A = 0 \ + \circlearrowright \qquad -R_B \left(\frac{4}{5}\right)(12) + 5(12)(6) + 30(4) = 0 \ \Rightarrow \ R_B = 50 \text{ kN} \ \nwarrow$$

27

$$\sum F_y = 0 \quad +\uparrow \qquad A_y - 60 - 30 + \left(\frac{4}{5}\right)(50) = 0 \qquad \Rightarrow A_y = 50 \text{ kN} \quad \uparrow$$

$$\sum F_x = 0 \quad + \rightarrow \qquad A_x - 50\left(\frac{3}{5}\right) = 0 \qquad \Rightarrow A_x = 30 \text{ kN} \quad \rightarrow$$

Final
Free Body
Diagram

To check our answer, it is a very good practice to show the magnitudes of all loads and forces on the free-body diagram as shown in Fig. 2-2c.

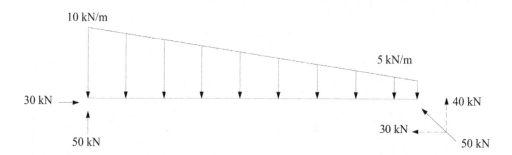

Fig. 2-2c

Example: 2-3

Determine the reactions for the beam shown in Fig. 2-3a.

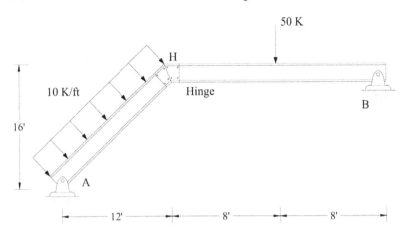

Fig. 2-3a

SOLUTION

Free Body Diagram. The free body diagram of member HB and the frame AHB are shown in Figs. 2-3b and Fig. 2-3c, respectively.

SECTION (HB)

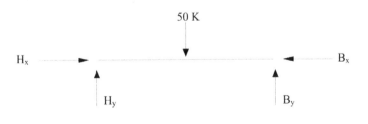

Fig. 2-3b

The application of the equations of equilibrium (Eq. 1-5) yields the following:

$$\sum M_H = 0 \quad + \circlearrowleft \qquad -B_y(16) + 8(50) = 0 \qquad \Rightarrow B_y = 25^{\,k} \uparrow$$

FRAME (AHB)

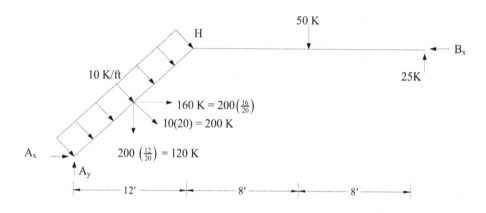

Fig. 2-3c.

$$\sum M_A = 0 \quad + \circlearrowleft \qquad -B_x(16) - 25(28) + 50(20) + 160(8) + 120(6) = 0 \Rightarrow B_x = 143.75^{\,k} \leftarrow$$

Or

$$\sum M_A = 0 \quad + \circlearrowleft \qquad -B_x(16) - 25(28) + 50(20) + 200(10) = 0 \qquad \Rightarrow B_x = 143.75^{\,k} \leftarrow$$

$$\sum F_y = 0 \quad + \uparrow \qquad A_y - 120 - 50 + 25 = 0 \qquad\qquad \Rightarrow A_y = 145^{\,k} \uparrow$$

$$\sum F_x = 0 \quad + \rightarrow \qquad A_x - 143.75 + 160 = 0 \;\Rightarrow A_x = -16.25 \text{ k} \qquad \Rightarrow A_x = 16.25^{k} \leftarrow$$

To check our answer, it is a very good practice to show the magnitudes of all loads and forces on the free-body diagram as shown in Fig. 2-3d.

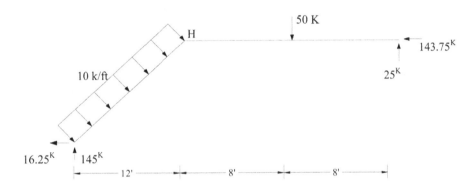

Fig. 2-3d

Example: 2-4

Determine the reactions for the beam shown in Fig. 2-4a. A hinge is located at point H.

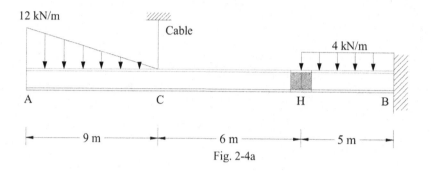

Fig. 2-4a

SOLUTION

Free Body Diagram
The free body diagram of member ACH, the structure ACHB, and section HB are, respectively, shown in Figs. 2-4b, 2-4c, 2-4d, and 2-4e.

SECTION (ACH)

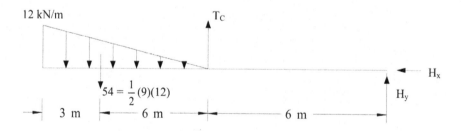

Fig. 2-4b

The application of the equations of equilibrium (Eq. 1-5) yields the following:

$\sum M_H = 0 \quad +\circlearrowright \qquad -54\,(6+6) + T_C\,(6) = 0 \qquad\qquad \Rightarrow T_C = 108 \text{ kN} \uparrow$

$\sum F_y = 0 \quad +\uparrow \qquad H_y - 54 + 108 = 0 \qquad\qquad\qquad \Rightarrow H_y = 54 \text{ kN} \downarrow$

$\sum F_x = 0 \quad +\rightarrow \qquad 0 - H_x = 0 \qquad\qquad\qquad\qquad \Rightarrow H_x = 0$

STRUCTURE (ACHB)

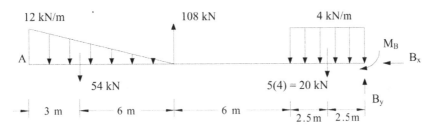

Fig. 2-4c.

$\sum M_B = 0 \quad +\circlearrowright \qquad -54\,(6+6+5) + 108(6+5) - 20(2.5) + M_B = 0 \Rightarrow M_B = 220 \text{ kN.m} \ \circlearrowleft$

$\sum F_y = 0 \quad +\uparrow \qquad B_y - 54 + 108 - 20 = 0 \qquad\qquad \Rightarrow B_y = 34 \text{ kN} \downarrow$

$\sum F_x = 0 \quad +\rightarrow \qquad -B_x + 0 = 0 \qquad\qquad\qquad \Rightarrow B_x = 0$

OR

SECTION (HB)

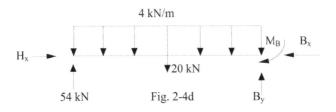

Fig. 2-4d

$$\sum M_B = 0 \quad +\circlearrowright \qquad 54\,(5) - 20(2.5) + M_B = 0 \qquad\qquad \Rightarrow M_B = 220^{\text{ kN.m}}$$

$$\sum F_y = 0 \quad +\uparrow \qquad B_y + 54 - 20 = 0 \qquad\qquad \Rightarrow B_y = 34^{\text{ kN}} \quad \downarrow$$

$$\sum F_x = 0 \quad +\rightarrow \qquad -B_x + 0 = 0 \qquad\qquad \Rightarrow B_x = 0$$

Final
Free Body
Diagram

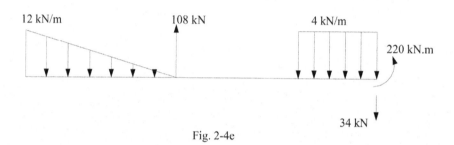

12 kN/m   108 kN   4 kN/m

220 kN.m

34 kN

Fig. 2-4e

Example: 2-5

Determine the reactions for the beam shown in Fig. 2-5a.

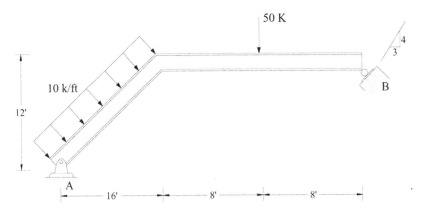

Fig. 2-5a

SOLUTION

Free-Body Diagram.
As shown in Fig. 2-5b, the support (roller) at B exerts a normal force on the beam at its point of contact and the line of action of this force is defined by the triangle (3-4-5).

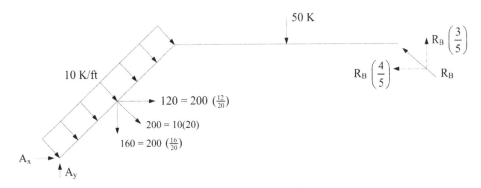

Fig. 2-5b.

The application of the equations of equilibrium (Eq. 1-5) yields the following:

$$\sum M_A = 0 \quad +\circlearrowleft \qquad -R_B\left(\frac{3}{5}\right)(8+8+16) - R_B\left(\frac{4}{5}\right)(12) + 50(8+16) + 200(10) = 0$$

$$R_B = 111.111 \text{ k} \quad \nwarrow$$

$$\sum F_y = 0 \quad +\uparrow \qquad -160 - 50 + 111.111\left(\frac{3}{5}\right) + A_y = 0 \quad \Rightarrow \quad A_y = 143.333 \text{ k} \quad \uparrow$$

$$\sum F_x = 0 \quad + \rightarrow \qquad 120 - 111.111\left(\frac{4}{5}\right) + A_x = 0 \qquad \Rightarrow \quad A_x = 31.111 \text{ k} \quad \leftarrow$$

Final Free Body Diagram

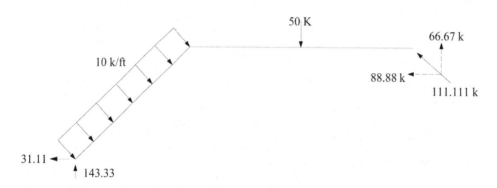

Fig. 2-5c

Example: 2-6

Determine the reactions for the beam shown in Fig. 2-6a. A hinge is located at point H and a cable is located at point C.

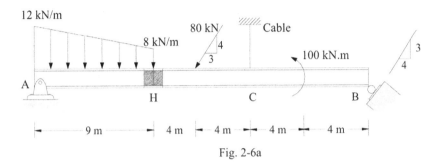

Fig. 2-6a

SOLUTION

Free Body Diagram.
The free body diagrams of member AH, and the structure ACHB are, respectively, shown in Figs. 2-6b and 2-6c.

SECTION (AH)

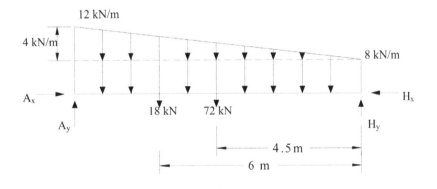

Fig. 2-6b

The application of the equations of equilibrium (Eq. 1-5) yields the following:

$$\sum M_H = 0 \quad \circlearrowleft \qquad A_y\,(9) - 18(6) - 72(4.5) = 0 \qquad\qquad \Rightarrow A_y = 48 \text{ kN} \uparrow$$

STRUCTURE (AHCB)

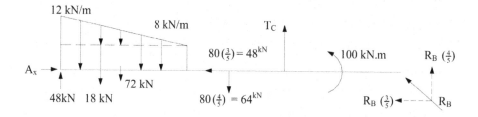

Fig. 2-6c.

$$\sum M_B = 0 \quad \circlearrowleft \qquad 48\,(25) - 18(6+16) - 72(4.5+16) - 64(12) + T_c(8) - 100 = 0$$

$$T_C = 192.5 \text{ kN} \uparrow$$

$$\sum F_y = 0 \quad +\uparrow \qquad 48 - 18 - 72 - 64 + 192.5 + R_B\left(\frac{4}{5}\right) = 0 \qquad R_B = -108.125 \text{ kN}$$

$$R_B = 108.125 \text{ kN} \searrow$$

$$\sum F_x = 0 \quad + \rightarrow \qquad A_x - 48 + 108.125\left(\frac{3}{5}\right) = 0 \qquad\qquad A_x = -16.875 \text{ kN}$$

$$A_x = 16.875 \text{ kN} \leftarrow$$

Final
Free Body
Diagram

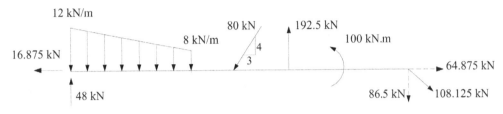

Fig. 2-6d

Summary

- Since most loaded structures are at rest and restrained against displacements by their supports, their behavior is governed by the laws of statics, which for planar structures can be stated as follows:
$$\sum F_x = 0$$
$$\sum F_y = 0$$
$$\sum M_o = 0$$
- Designers use a variety of symbols to represent actual supports as summarized in Table 2-1. These symbols represent the primary action of a particular support. For example, a pin support is assumed to apply restraint against displacement in any direction but to provide no rotational restraint.
- Planar structures whose reactions and internal forces can be determined by applying the three statics equations are called determinate structures. Highly restrained structures that cannot be analyzed by the three statics equations are termed indeterminate structures. These structures require additional equations based on the geometry of the deflected shape. If the statics equations cannot be satisfied for a structure or any part of a structure, the structure is considered unstable. Tables 2-2, 2-3, and 2-4 show the determinacy and stability conditions of frames, beams, and trusses, respectively.
- Because indeterminate structures have more supports or members than the minimum required to produce a stable determinate structure, they are therefore generally stiffer than determinate structures and less likely to collapse if a single support or member fails.

Problems

For the problems 1 to 14, determine the reactions and show the Free Body Diagrams.

1-

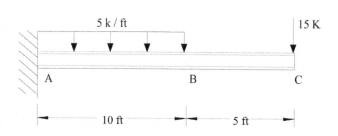

2-

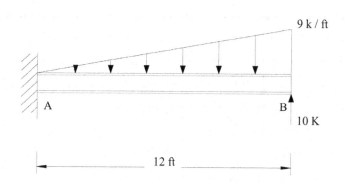

3-

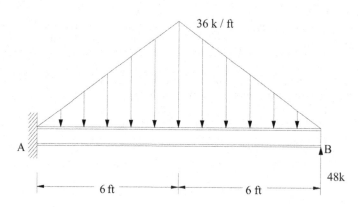

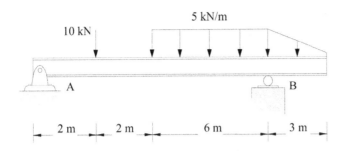

4-

5-

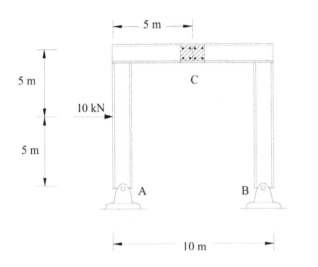

6-

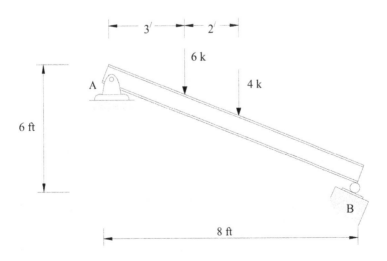

7-

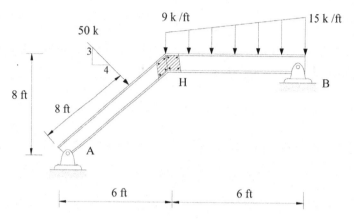

8-

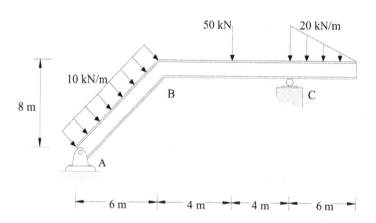

9-

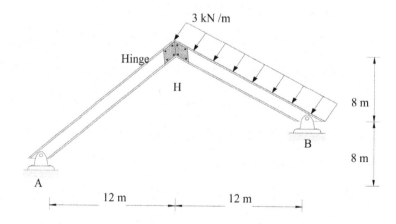

10-

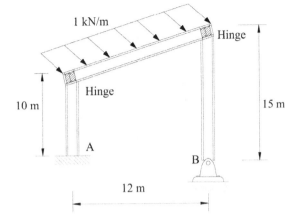

11-

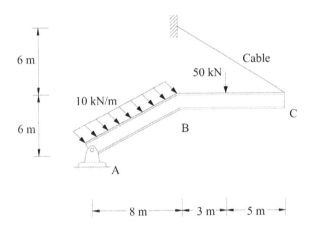

12-

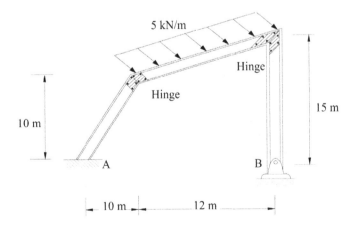

13-

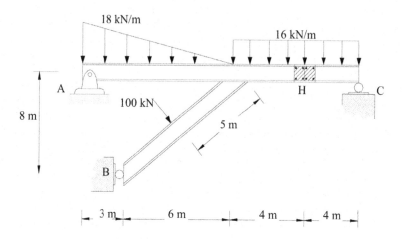

14-

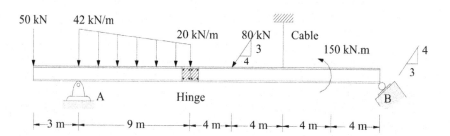

CHAPTER - 3
PLANE TRUSSES

Chapter 3 Plane Trusses

3.1 INTRODUCTION

A truss is a structure made of slender members joined together at their ends. Trusses are generally made of steel, but some times of timber, concrete or light alloy. The truss members are subjected to axial loads only (compression or tension). In other words, they do not carry any bending moments.

The analysis of trusses is based on the following assumptions:

        1) Pin-connected members.
        2) Perfectly hinged member ends
        3) Loads acting at the joints only.

Some of commonly-used truss types are shown in Fig. 3-1.

The unknown member forces will have negative values when their correct directions are opposite to the original assumptions made on the free–body diagram

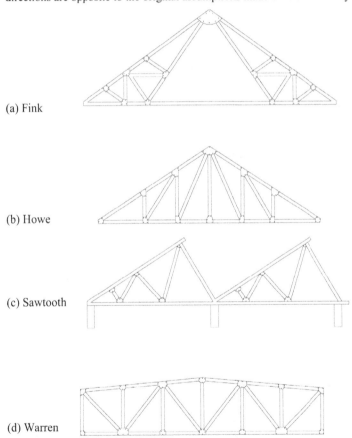

(a) Fink

(b) Howe

(c) Sawtooth

(d) Warren

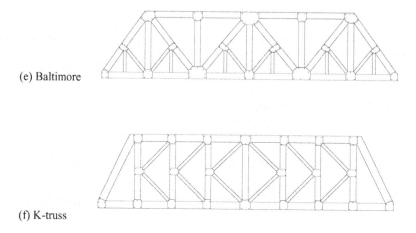

(e) Baltimore

(f) K-truss

Fig. 3-1. Different Truss Types

3.2 JOINT METHOD

All of the truss joints are in equilibrium when the truss is in equilibrium. The method of joints is based on satisfying the equations of equilibrium $\sum F_x = 0$ and $\sum F_y = 0$ for the forces acting on the truss joints.

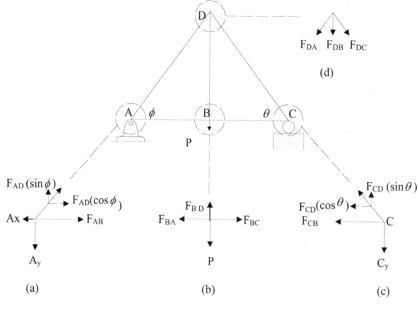

Figure 3-2

3.3 METHOD OF SECTIONS

The method of sections is used if certain member forces are to be found, or if some member forces cannot be determined by the method of joints alone. Hence, the method of sections provides an additional technique to determine truss member forces. A smart choice has to be made to cut the truss through the members where forces are to be determined. In the method of sections, a section is passed completely through the truss, cutting it into two parts. Each of these two parts is subjected to the forces acting on the cut members and the externally applied loads, as shown in Figs. 3-3a and 3-3b.

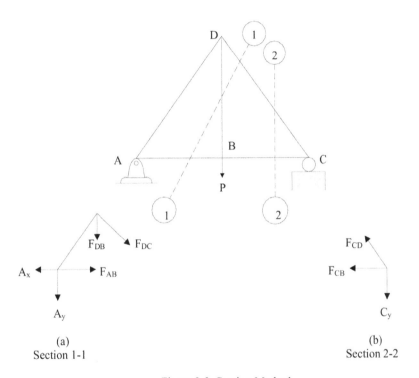

(a)
Section 1-1

(b)
Section 2-2

Figure 3-3. Section Method

Example: 3-1

For the truss shown in Fig. 3-4a, determine the following:
  a) Bar forces in JI, BH, and BC.
  b) Two zero members of the truss.

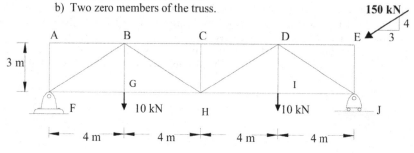

Fig. 3-4a

SOLUTION

The Free Body Diagram of the truss, the required joints, sections, and reactions are shown in Figs. 3-4b, 3-4c, 3-4d, 3-4e, and 3-4f, respectively.

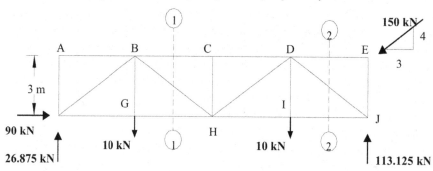

Fig. 3-4b

SECTION 1-1

To determine F_{BC} and F_{BH}

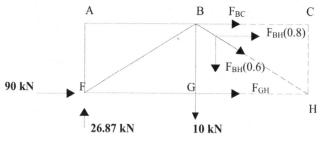

Fig. 3-4c

$$\sum M_H = 0 \quad \text{↻} \qquad -10(4) + 26.875(8) + F_{BC}(3) = 0 \qquad \Rightarrow F_{BC} = 58.33 \text{ kN } \underline{\textbf{C}}$$

(Compression)

$$\sum F_y = 0 \quad +\uparrow \qquad 26.87 - 10 - F_{BH}(0.6) = 0 \qquad \Rightarrow F_{BH} = 28.12 \text{ kN } \underline{\textbf{T}}$$

(Tension)

SECTION 2-2

To determine F_{JI}

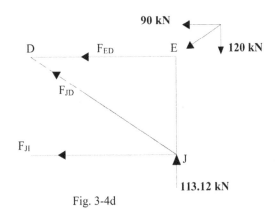

Fig. 3-4d

$$\sum M_D = 0 \quad \text{↻} \qquad -113.12(4) + 120(4) + F_{JI}(3) = 0 \qquad \Rightarrow F_{JI} = 9.167 \text{ kN } \underline{\textbf{C}}$$

Zero-load members

Joint A has two zero-load members, because the joint has no external load applied to it and is connected to two perpendicular members.

$$\sum F_y = 0 \quad +\uparrow \qquad -F_{AF} + 0 = 0 \quad \Rightarrow F_{AF} = 0$$
$$\sum F_x = 0 \quad +\rightarrow \qquad F_{AB} + 0 = 0 \quad \Rightarrow F_{AB} = 0$$

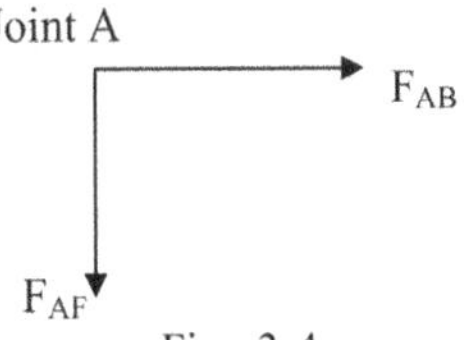

Joint A

F_{AB}

F_{AF}

Fig. 3-4e

51

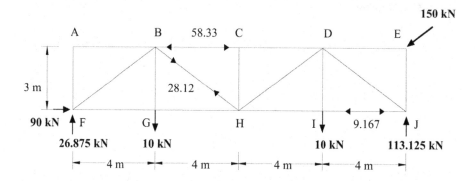

Final
Free Body
Diagram

Fig. 3-4f

Example: 3-2

Determine the forces in bars AB, IC, and LC for the truss shown in Fig. 3-5a.

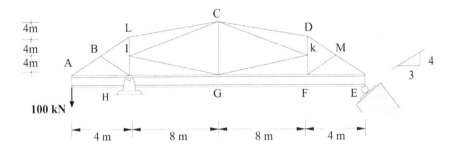

Fig. 3-5a

SOLUTION

The Free Body Diagram of the truss, the required joints, sections, and reactions are shown in Fig. 3-5b, 3-5c, 3-5d, 3-5e, and 3-5f, respectively.

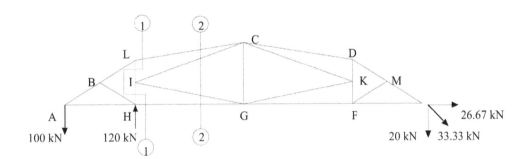

Fig. 3-5b

JOINT (A)

To determine F_{AB}

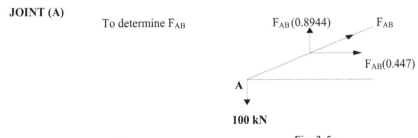

100 kN

Fig. 3-5c

53

$$\sum F_y = 0 \quad +\uparrow \qquad -100 + (0.8944) \, F_{AB} = 0 \qquad \Rightarrow F_{AB} = 111.8 \text{ kN } \underline{T}$$

SECTION 1-1

To determine F_{LC}

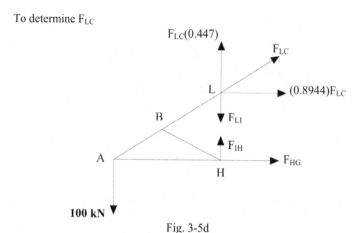

Fig. 3-5d

$$\sum M_H = 0 \quad +\curvearrowright \qquad -100(4) + F_{LC}\,(0.8944)(8) = 0 \qquad \Rightarrow F_{LC} = 55.9 \text{ kN } \underline{T}$$

SECTION 2-2

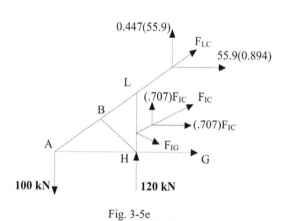

Fig. 3-5e

$$\sum M_G = 0 \quad +\curvearrowleft \qquad -100(12) + 120\,(8) + 0.477(55.9)(8) + 0.8944(55.9)(8) +$$
$$F_{IC}\,(\,0.707)(4+8) = 0 \qquad \Rightarrow \; F_{IC} = 42.42 \text{ kN } \underline{C}$$

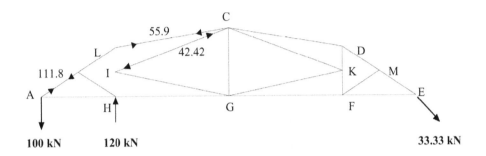

Final
Free Body
Diagram

55.9

42.42

111.8

100 kN   **120 kN**

33.33 kN

Fig. 3-5f

Example: 3-3

Determine the bar forces in AB, EF, and CK of the truss shown in Fig. 3-6a.

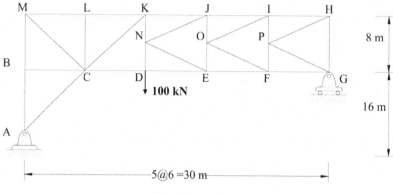

Fig. 3-6a

SOLUTION

The Free Body Diagram of the truss, the required joints, sections, and reactions are shown in Fig. 3-6b, 3-6c, 3-6d, 3-6e, and 3-6f, respectively.

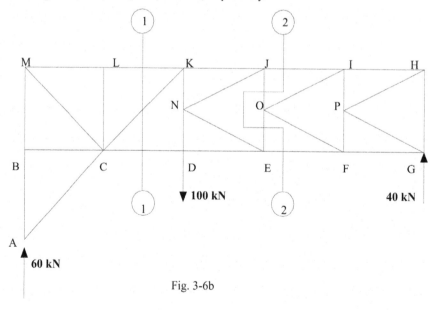

Fig. 3-6b

SECTION 1-1

To determine F_{CK}

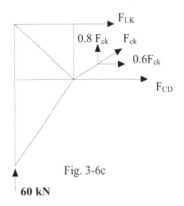

Fig. 3-6c

60 kN

$\sum F_y = 0 \quad +\uparrow \qquad 60 + F_{ck}(0.8) = 0 \qquad \Rightarrow F_{ck} = -75 \text{ kN} = 75 \text{ kN } \underline{\textbf{C}}$

JOINT (A)

To determine F_{AB}

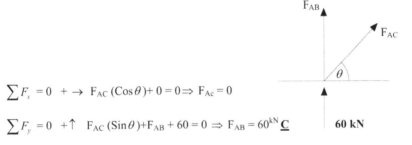

$\sum F_x = 0 \quad + \rightarrow \quad F_{AC}(\cos\theta) + 0 = 0 \Rightarrow F_{Ac} = 0$

$\sum F_y = 0 \quad +\uparrow \quad F_{AC}(\sin\theta) + F_{AB} + 60 = 0 \Rightarrow F_{AB} = 60^{kN} \underline{\textbf{C}}$

60 kN

Fig. 3-6d

SECTION 2-2

To determine F_{EF}

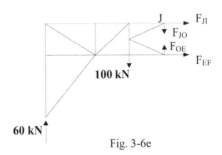

100 kN

60 kN

Fig. 3-6e

57

$$\sum M_J = 0 \quad +\circlearrowleft \qquad 60(18) - 100(6) - F_{EF}(8) = 0 \;\Rightarrow\; F_{EF} = 60\ kN\ \underline{\textbf{T}}$$

Final
Free Body
Diagram

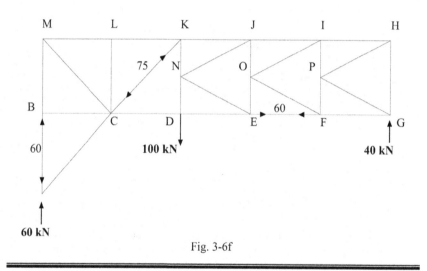

Fig. 3-6f

Summary

- Trusses are composed of slender bars that are assumed to carry only axial force. Joints in large trusses are formed by welding or bolting members to gusset plates. If members are relatively small and lightly stressed, joints are often formed by welding the ends of vertical and diagonal members to the top and bottom chords. Figure 3-1 shows different types of structures.
- Determinate trusses can be analyzed either by the method of joints or by the method of sections. The method of sections is used when the force in one or two bars is required. The method of joints is used when all bar forces are required.
- If the analysis of a truss results in inconsistent force values, that is, one or more joints are not in equilibrium, then the truss is unstable.

Problems

1- Determine the following:
  a) The forces in bars OF, EF, CK. (Section and joint method).
  b) Two zero-load members.

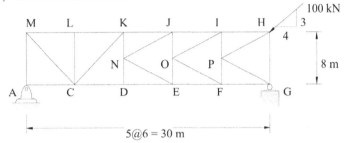

5@6 = 30 m

2- Determine the following:
  a) The forces in BC and FG, use joint method.
  b) One zero-load member.

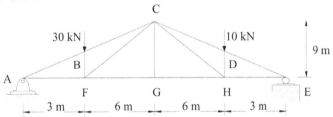

3- Determine the following:
  a) The forces in bars AB, AE, and EC.
  b) Two zero-load members.

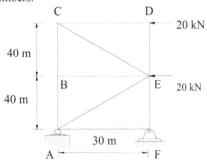

Chapter 3 Plane Trusses

4- Determine the following:

    a) The forces in bars BC and GK.
    b) One zero-load member.

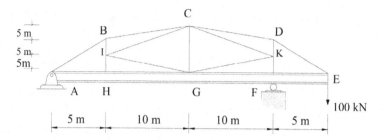

5- Determine the following:
    a) The forces in bars CD and IK
    b) Three zero-load members.

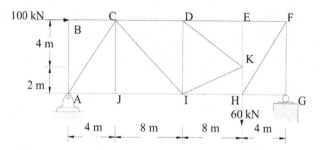

6- Determine the truss reactions and show the free body diagram

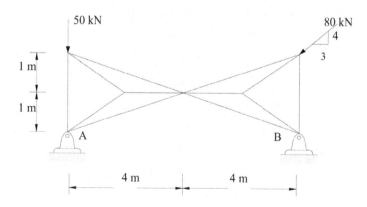

60

7- Determine the following:

a) The forces in bars BC and IC (section method).
b) One zero-load member.

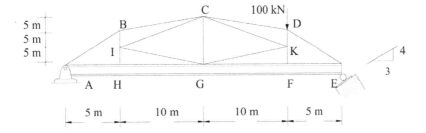

8- Determine the following:

a) The forces in bars IG and HG (section method).
b) Two zero-load members.

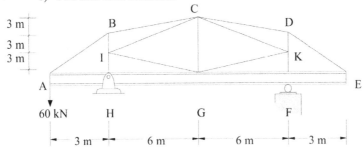

9- Determine the reactions and bar forces and show the free body diagram.

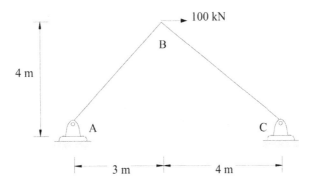

10- Determine the reactions and the force in bar BH and show the free body diagram.

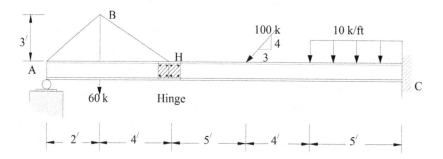

11- Determine the reactions and force in bar FG and show the free body diagram.

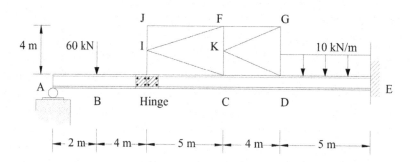

12- Determine the reactions and forces in bars FG and HD and show the free body diagram.

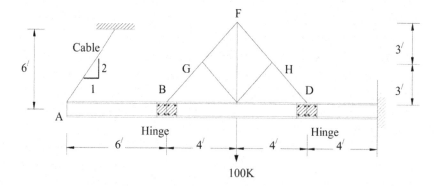

CHAPTER - 4
SHEAR AND BENDING MOMENT DIAGRAMS

Kuwait – Kuwait

Chapter 4.   Shear and Moment Diagrams

4.1   INTRODUCTION

The construction of shear and bending moment diagrams is a basic and essential tool that must be mastered   by   civil engineering students. Shear and moment diagrams, which are graphical plots of the variation of shear and moment in the structure, are used in the design of structural members. Structural members such as beams and frames have significant shear and bending forces-unlike trusses- due to their rigid connections and load applications.

4.2   INTERNAL FORCES

The sign conventions are used to specify the direction of internal forces and moments in a structural member. Positive axial forces elongate members, positive shear forces produce a clockwise couple, while positive bending moments compress the top fibers and tend the bottom fibers (see Fig. 4-1)

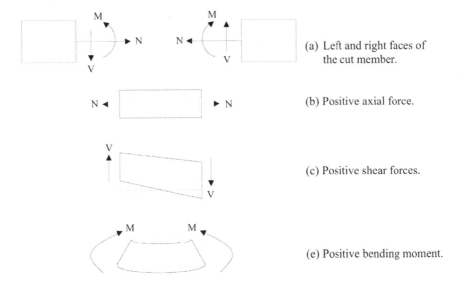

(a) Left and right faces of the cut member.

(b) Positive axial force.

(c) Positive shear forces.

(e) Positive bending moment.

Figure 4-1. Internal Force Sign Conventions

4.3   SHEAR AND BENDING MOMENT DIAGRAMS

Shear and moment diagrams are graphical plots of the variation of the shear and bending moments along the member span. In constructing the shear diagram the following points should be taken into consideration:

1)   The shear at any point is equal to the algebraic sum of the loads and reactions from the left end to the point of interest. The slope of the moment diagram at any point is equal to the shear at the point, $V = \dfrac{\Delta M}{\Delta x}$.

2)   On the shear diagrams:

- Concentrated loads produces straight lines (Fig. 4-2a).
- Uniform loads produces sloping lines (Fig. 4-2b).
- Linear loads produces curved lines (Fig. 4-2c).

3)   Concentrated loads will cause an abrupt change in the shear diagram.

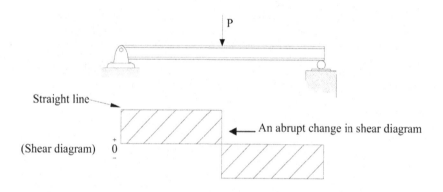

Fig. 4-2a

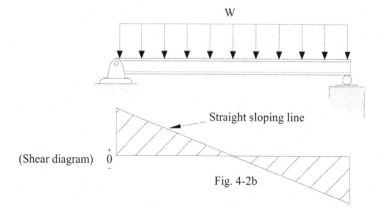

Fig. 4-2b

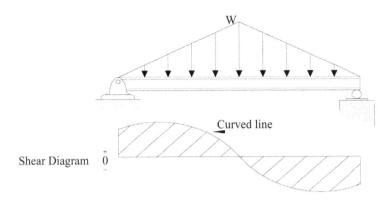

Fig. 4-2c

In the construction of the moment diagram, the following points should be taken into consideration:

    1)   Moment should start from the left end to the point of interest.

    2)   Clockwise moments are positive moments.

    3)   On the moment diagram:
- A concentrated load produces linear lines (Fig. 4-3a).
- A uniform load produces parabolic lines (Fig. 4-3b).
- Linear loads produces curved or parabolic lines of higher degree (Fig. 4-3c).

    4)   The moment at any point is equal to the area under the shear diagram up to the point $M = \int V dx$ [4.2]

    5)   The moment is zero at the hinge or the free end.

    6)   Concentrated moments cause abrupt changes in the moment diagram.

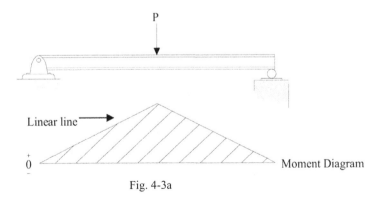

Fig. 4-3a

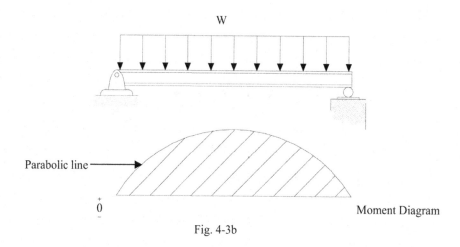

Parabolic line

0

Moment Diagram

Fig. 4-3b

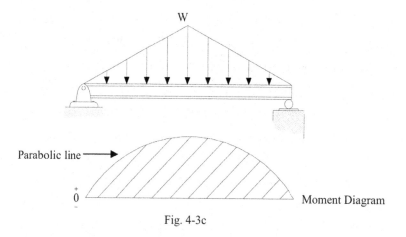

Parabolic line

0

Moment Diagram

Fig. 4-3c

The construction of shear and moment diagrams is performed using he following steps:

1)   Determine the reaction.
2)   Draw the free body diagram, showing all loads and reactions.
3)   Draw the shear diagram showing all straight, sloping, curved lines above and below the zero lines.
4)   Draw the moment diagram showing all straight, sloping, curved lines above and below the zero lines.

The following examples provide a clear picture of applying the steps needed to draw the shear and moment diagrams. No commentary is given unless it is needed to explain a certain idea of solving the examples.

Example: 4-1

Draw the shear and moment diagrams for the beam shown in Fig. 4-4a.

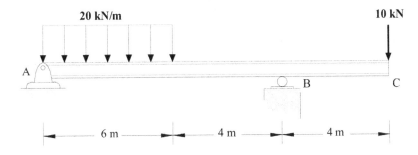

Fig. 4-4a

SOLUTION
        The free body diagram of the beam ABC is shown in Fig. 4-4b. The shear and moment diagrams are shown in Figs 4-4c and 4-4d, respectively. The reactions are shown on the free body diagram of Fig. 4-4b.

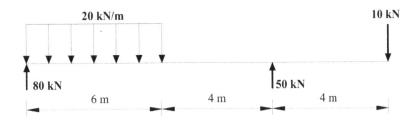

Fig. 4-4b. Applied Loads and Reactions

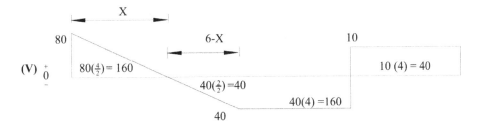

Fig. 4-4c. Shear Diagram (in kN)

69

To determine X, one of the following approaches may be used:

First approach

$$X = \frac{\text{Shear magnitude}}{\text{Load intensity}}$$

$$X = \frac{v}{w} = \frac{80}{20} = 4 \text{ m}$$

Second Approach: Similar Triangles

$$\frac{80}{X} = \frac{40}{6-X} \implies X = 4 \text{ m}$$

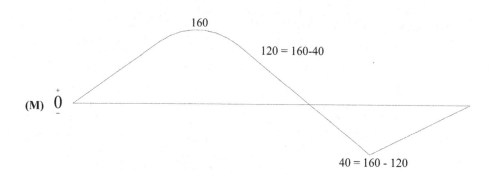

(M)

160

120 = 160-40

40 = 160 - 120

Fig. 4-4d. Moment Diagram (in kN.m)

Example: 4-2

Draw the shear and moment diagrams for the beam shown in Fig. 4-5a.

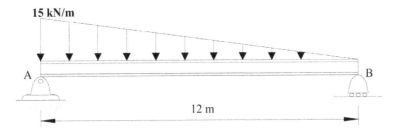

15 kN/m

A

B

12 m

Fig. 4-5a

SOLUTION

The shear and moment diagrams are shown in Figs 4-5b and 4-5d, respectively.

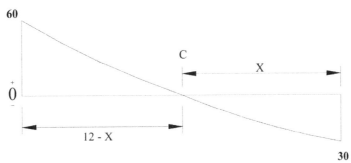

60

C

X

0

12 - X

30

Fig. 4-5b. Shear Diagram (in kN)

The segment BC is considered for the determination of the value of X, shown in Fig. 4-5b, because it gives the simplest solution (See Fig. 4-5c).

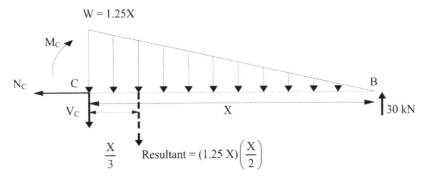

$W = 1.25X$

M_C

N_C

C

B

V_C

X

30 kN

$\dfrac{X}{3}$

$\text{Resultant} = (1.25\,X)\left(\dfrac{X}{2}\right)$

Fig. 4-5c. Segment BC

1) Load slope $= \dfrac{w(X)}{L} = \dfrac{15(X)}{12} = 1.25X$

2) Since the load resultant is a function of $X \Rightarrow$

$$\sum F_y = 0 \uparrow + \qquad -V_C - (1.25\,X)\!\left(\frac{X}{2}\right) + 30 = 0, \qquad \text{Since } V_C @\ x\ = 0$$

Then $-(1.25\,X)\!\left(\dfrac{X}{2}\right) + 30 = 0 \Rightarrow X = 6.93 \text{ m}$

3) $\sum M_C = 0 + \;\circlearrowright\;\; -30(6.93) + 1.25(6.93)\!\left(\dfrac{6.93}{2}\right)\!\left(\dfrac{6.93}{3}\right) + M_C \Rightarrow M_C = 138.56 \text{ kN.m} \;\circlearrowright$

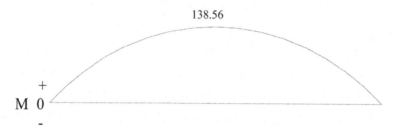

Fig. 4-5d. Moment Diagram (in kN.m)

Example: 4-3

Draw the shear and moment diagrams for the beam shown Fig. 4-6a, and derive the shear and moment equations for the segment CD. The beam has a hinge at C.

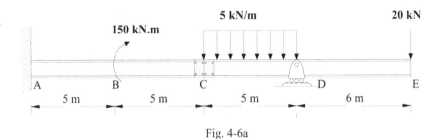

5 kN/m   **20 kN**

150 kN.m

A   B   C   D   E

5 m   5 m   5 m   6 m

Fig. 4-6a

SOLUTION

The free body diagram of the beam ABC is shown in Fig. 4-6b. The shear and moment diagrams of the beam ABC are shown in Figs. 4-6c and 4-6d, respectively. The reactions have been calculated and are shown in the free body diagram of Fig. 4-6b.

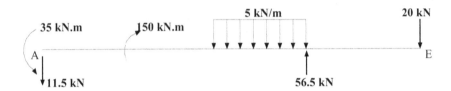

5 kN/m   **20 kN**

35 kN.m   **150 kN.m**

A   E

11.5 kN   **56.5 kN**

Fig. 4-6b. Applied Loads and Reactions

20

120

(V)   0

57.5   57.5

120

11.5

36.5

Fig. 4-6c. Shear Diagram (in kN)

Fig. 4-6d. Moment Diagram (in kN.m)

The shear and moment equations for the segment CD can be derived using either side of the beam as shown in Figs. 4-6e, 4-6f, and 4-6g.

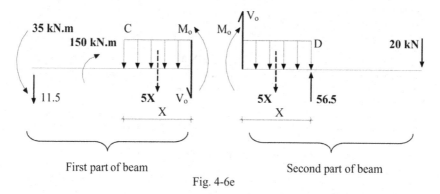

First part of beam   Second part of beam

Fig. 4-6e

First Part of the beam

$$\sum F_y = 0\uparrow + \quad -11.5 - 5X - V_o = 0$$
$$V_0 = -11.5 - 5X$$

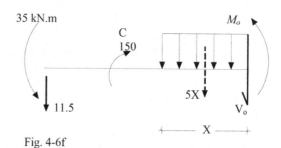

Fig. 4-6f

$$\sum M_o = 0 \circlearrowleft \quad + 11.5(10+X) - 150 + \frac{5X^2}{2} + 35 + M_o = 0$$

$$M_o = -(2.5X^2 + 11.5X)$$

Second Part of the beam

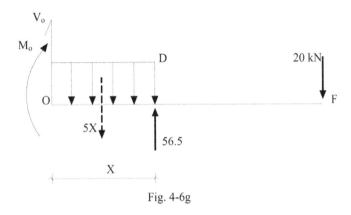

Fig. 4-6g

$$\sum F_y = 0 \uparrow + \quad V + 56.5 - 20 - 5X = 0$$

$$V_o = 5X - 36.5$$

$$\sum M_o = 0 \circlearrowleft \quad + -20(6+X) + 56.5X - \frac{5X^2}{2} - M_o = 0$$

$$M_o = -2.5X^2 + 36.5X - 120$$

Chapter 4.   Shear and Moment Diagrams

Example: 4-4

Determine the shear and moment diagrams for the frame shown in Fig. 4-7a.

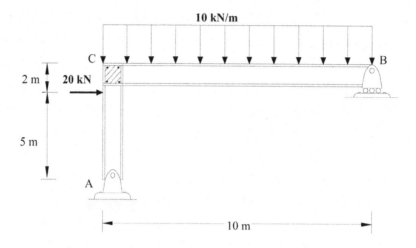

Fig. 4-7a

SOLUTION

The free body of the frame ACB is shown in Fig. 4-7b. The shear and moment diagrams of the frame ACB are shown in Figs. 4-7c and 4-7d, respectively. The reactions are shown on the free body diagram of Fig. 4-7b.

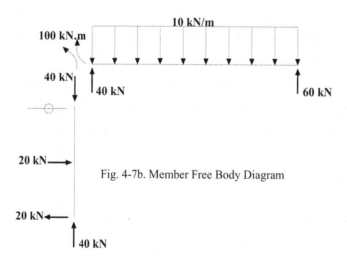

Fig. 4-7b. Member Free Body Diagram

The internal moment at C (M_C) can be calculated by summing the moment about C in either member AC or BC of the frame. For example, let us consider the member AC, we have:-

$$\sum M_C = 0 \quad + \quad 20(7) - 20(2) - M_C = 0 \;\Rightarrow\; M_C = 100 \text{ kN.m}$$

The shear diagram is constructed by following the directions of the forces shown in the free body diagram of Fig. 4-7b.

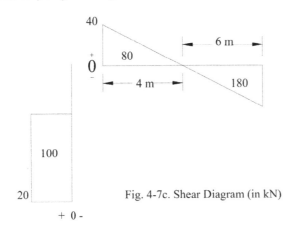

Fig. 4-7c. Shear Diagram (in kN)

The construction of frame moment diagrams is similar to that of beams. Keep in mind the effect of M_C as a constant applied moment.

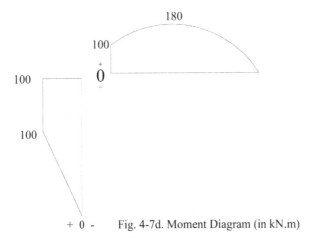

Fig. 4-7d. Moment Diagram (in kN.m)

Example: 4-5

Determine the shear and moment diagrams for the frame shown in Fig. 4-8a. The frame has a hinge at joint B.

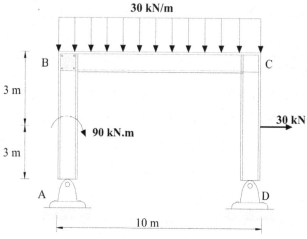

Fig. 4-8a

SOLUTION

The free body of the frame ABCD is shown in Fig. 4-8b. The shear and moment diagrams of the frame ACB are shown in Figs. 4-8c and 4-8d, respectively. The reactions are shown on the free body diagram of Fig. 4-8b.

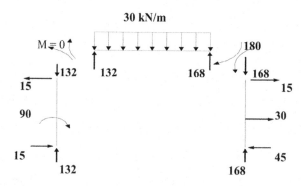

Fig. 4-8b. Member Free Body Diagram

The internal moments at B (M_B) and C (M_C) can be calculated by summing the moment about B and C in members AB and CD of the frame. Considering the members AB and DC, we have:-

$$\sum M_B = 0 \circlearrowleft + \qquad M_B = 0 \qquad \text{(Pin or hinge)}$$

$$\sum M_C = 0 \circlearrowright + \qquad M_C = 180 \text{ kN.m} \quad \circlearrowleft$$

The shear diagram is constructed by following the directions of the forces shown in the free body diagram of Fig. 4-8b.

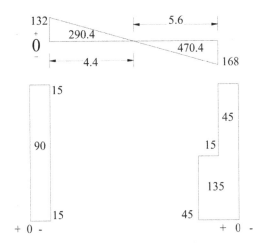

Fig. 4-8c. Shear Diagram (in kN)

The construction of frame moment diagrams is similar to that of beams.

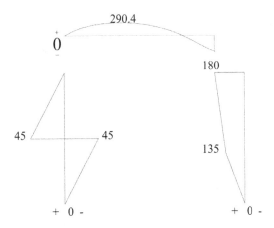

Fig. 4-8d Moment Diagram (kN.m)

Example: 4-6

Draw the shear and moment diagrams for the frame shown in Fig. 4-9a.

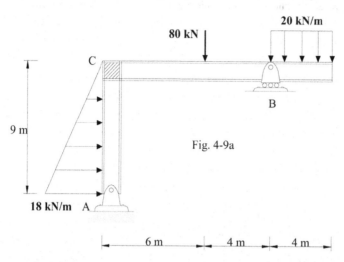

Fig. 4-9a

SOLUTION

The free body of the frame ACB is shown in Fig. 4-9b. The shear and moment diagrams of the frame ACB are shown in Figs. 4-9c and 4-9d, respectively. The reactions are shown on the free body diagram of Fig. 4-9b.

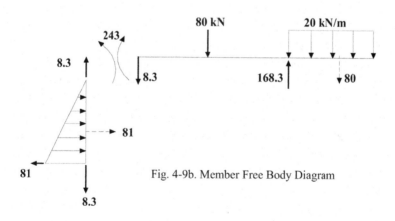

Fig. 4-9b. Member Free Body Diagram

The internal moment at C (M_C) can be calculated by summing the mome Examples either member AC or BC of the frame. For example let us consider the member AC, we have:-

$$\sum M_C = 0 \, \circlearrowleft + \qquad M_C = 243 \text{ kN.m} \, \circlearrowleft$$

The shear diagram is constructed by following the directions of the forces shown in the free body diagram of Fig. 4-9b.

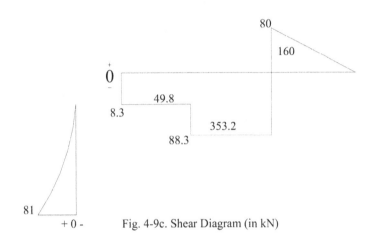

Fig. 4-9c. Shear Diagram (in kN)

The construction of frame moment diagrams is similar to that of beams.

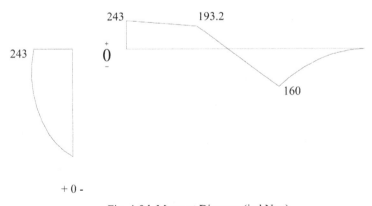

Fig. 4-9d. Moment Diagram (in kN.m)

Example: 4-7

Draw the shear and moment diagrams for the frame shown in Fig. 4-10a.

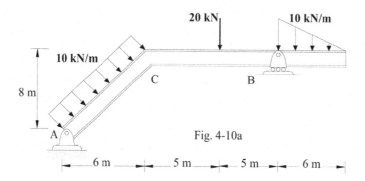

Fig. 4-10a

SOLUTION

The free body of the frame ACB is shown in Fig. 4-10b. The shear and moment diagrams of the frame ACB are shown in Figs. 4-10c and 4-10d, respectively. The reactions are shown on the free body diagram of Fig. 4-10b.

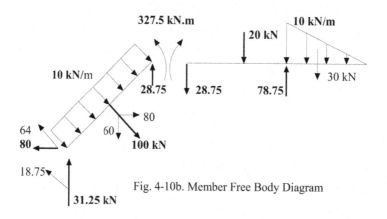

Fig. 4-10b. Member Free Body Diagram

The reaction components for the inclined member AC must be determined to draw the shear and moment diagrams for that member.

$$A_x \text{ (component)} \Rightarrow 80 \, (\text{Sin} \, \phi) = 64 \text{ kN}$$
$$A_y \text{ (component)} \Rightarrow 31.25 \, (\text{Cos} \, \phi) = 18.75 \text{ kN}$$

The internal moment at C (M_C) are computed by summing the moment about C in either member AC or BC of the frame. For example let us consider the member AC, we have:-

$$\sum M_C = 0 \quad + \qquad M_C = 327.5 \text{ kN.m}$$

The shear diagram is constructed by following the directions of the forces shown in the free body diagram of Fig. 4-10b.

$$V_A = 64 + 18.75 = 82.75 \text{ kN.m}$$
$$V_B = 82.75 - 100 = -17.25 \text{ kN.m}$$

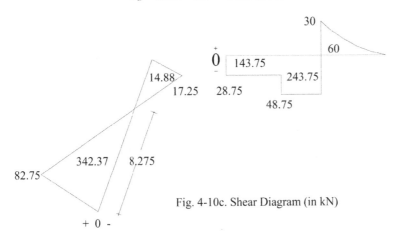

Fig. 4-10c. Shear Diagram (in kN)

The construction of frame moment diagrams is similar to that of beams. Keep in mind the effect of M_C as a constant applied moment.

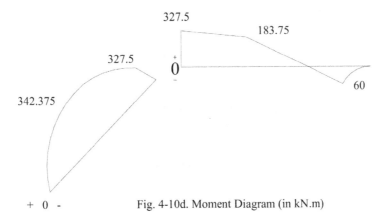

Fig. 4-10d. Moment Diagram (in kN.m)

Example: 4-8

Draw the shear and moment diagrams for the frame shown in Fig. 4-11a.

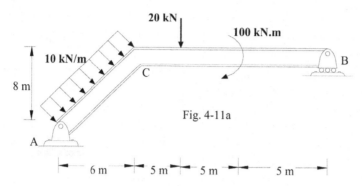

Fig. 4-11a

SOLUTION

The free body of the frame ACB is shown in Fig. 4-11b. The shear and moment diagrams of the frame ACB are shown in Figs. 4-11c and 4-11d, respectively. The reactions are shown on the free body diagram of Fig. 4-11b.

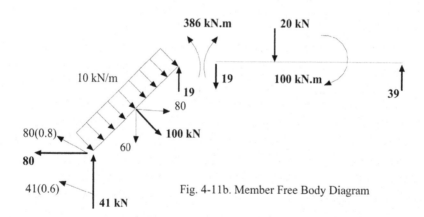

Fig. 4-11b. Member Free Body Diagram

The reaction components for the inclined member AC must be determined to draw the shear and moment diagrams for that member.

The internal moment at C (M_C) is computed by summing the moment about C in either member AC or BC of the frame. For example considering the member AC, we have:-

$$\sum M_C = 0 \qquad \qquad M_C = 386 \text{ kN.m}$$

The shear diagram is constructed by following the directions of the forces shown in the free body diagram of Fig. 4-11b.

$$V_A = 41(0.6) + 80(0.8) = 88.6 \text{ kN.m}$$
$$V_B = 88.6 - 100 = -11.4 \text{ kN.m}$$

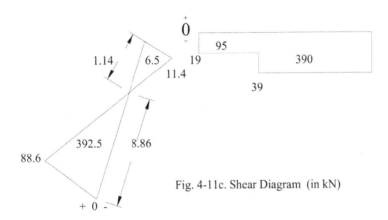

Fig. 4-11c. Shear Diagram (in kN)

The construction of frame moment diagrams is similar to that of beams. Keep in mind the effect of M_C as a constant applied moment.

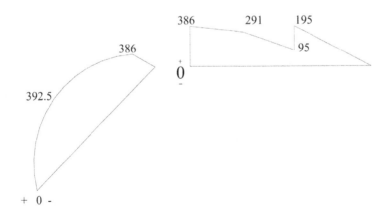

Fig. 4-11d. Moment Diagram (in kN.m)

Summary

- In this chapter, the beams and frames members are loaded primarily by forces (or component of forces) acting perpendicular to the member's longitudinal axis. These forces bend the member and produce internal forces of shear and moment on sections normal to the longitudinal axis.
- We compute the magnitude of the moment on a section by summing moments of all external forces on a free body to either side of the section. Moments of forces are computed about a horizontal axis passing through the centroid of the cross-section. The summation must include any reactions acting on the free-body. For horizontal members, we assume moments are positive when they produce curvature that is concave up and negative when curvature is concave down.
- Shear is the resultant force acting parallel to the surface of a section through the beam. We compute its magnitude by summing forces or components of forces that are parallel to the section,
- The following four relationships among load, shear, and moment were established to facilitate the construction of shear and moment diagrams:
  1. The change in shear ΔV between two points equals the area under the load curve between the two points.
  2. The slope of the shear curve at a given point equals the ordinate of the load curve at that point.
  3. The change in moment ΔM between two points equals the area under the shear curve between the two points.
  4. The slope of the moment curve at a given point equals the ordinate of the shear curve at that point.
- We also established that points of inflection (where curvature changes from positive to negative) in a beam's deflected shape occur where values of moment equal zero.

Problems

For the problems 1 to 20, show the free body diagram and draw the shear and moment diagrams.

1-

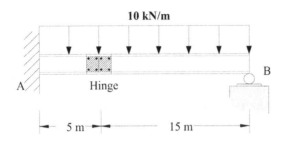

2-

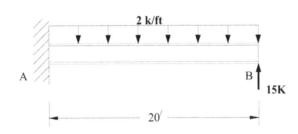

3-

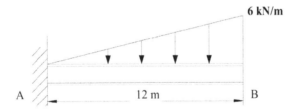

4-

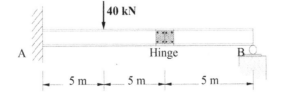

5-

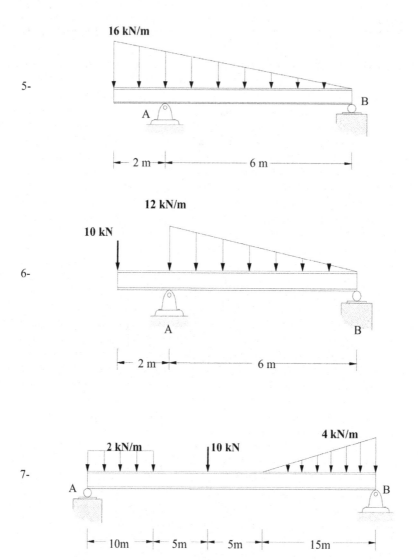

16 kN/m

A

B

2 m

6 m

6-

12 kN/m

10 kN

A

B

2 m

6 m

7-

2 kN/m

10 kN

4 kN/m

A

B

10m

5m

5m

15m

8-

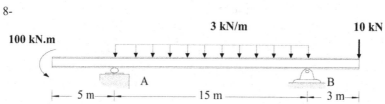

3 kN/m

10 kN

100 kN.m

A

B

5 m

15 m

3 m

9-

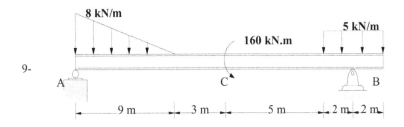

10-

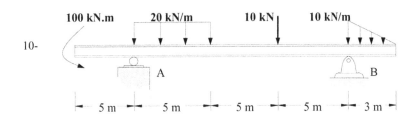

11-

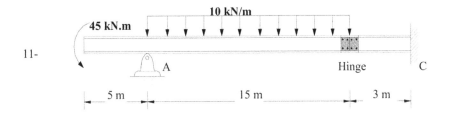

12-

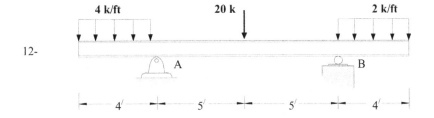

13-

14-

15-

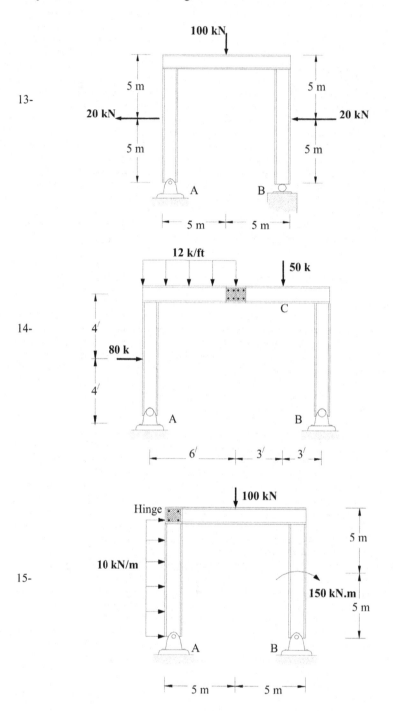

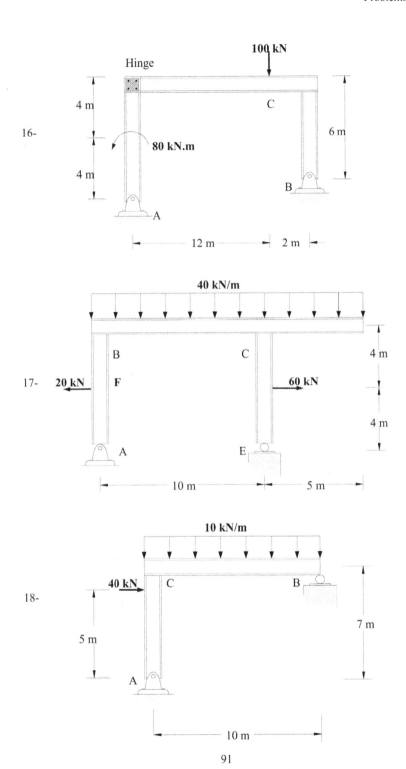

16-

17-

18-

19-

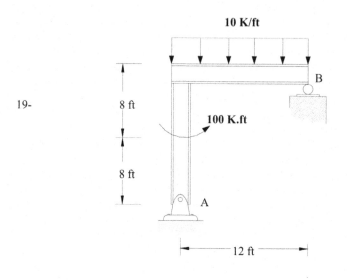

10 K/ft

B

8 ft

100 K.ft

8 ft

A

12 ft

20-

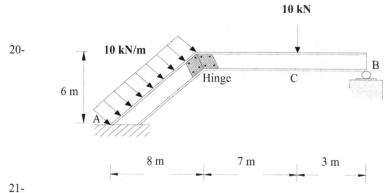

10 kN

10 kN/m

6 m

Hinge

C

B

A

8 m   7 m   3 m

21-

    i)   Show the free body diagram and draw shear and moment diagrams.

    ii)   Derive the shear and moment equations for **segment CD**

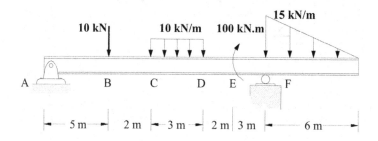

15 kN/m

10 kN   10 kN/m   100 kN.m

A   B   C   D   E   F

5 m   2 m   3 m   2 m   3 m   6 m

22-

   i)   Show the free body diagram and draw shear and moment diagrams.

   ii)   Derive the shear and moment equations for **DE**. There is a hinge at D.

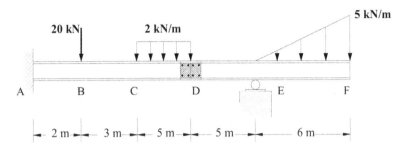

23-

   i)   Show the free body diagram and draw shear and moment diagrams.

   ii)   Derive the shear and moment equations for **CD.** There is a hinge at C.

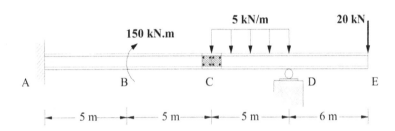

24-

   i)   Show the free body diagram and draw shear and moment diagrams.

   ii)   Derive the shear and moment equations for **BC**. There is a hinge at D.

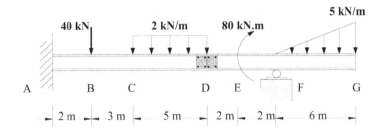

25-

i) Show the free body diagram and draw shear and moment diagrams.
ii) Derive the shear and moment equations for **DB**. There are hinges at D & E.

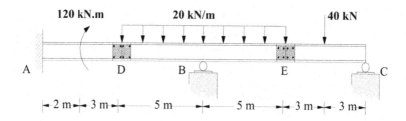

CHAPTER - 5
DEFLECTIONS

Doha – Qatar

Chapter 5   Deflections

5.1   INTRODUCTION

Deflection is the elastic movement of structures due to external loads such as gravity, earthquake, wind , vibrations, temperature change, and moving/cyclic loads. The deformation, which is a more general term than deflection, refers to both elastic and plastic deflections. The deflection of structural members must be less than required limits that are prescribed by design codes to ensure adequate serviceability.

Deflection analysis is performed for the following reasons:
> (1) Ensure adequate and safe structures.
> (2) Ensure acceptable and attractive aesthetic view.
> (3) Ensure satisfactory use of mechanical machines.
> (4) Ensure satisfactory users' comfort especially in tall buildings.
> (5) Analyze indeterminate structures.

The following methods are used to find the deflection of determinate structures:
> (1) Double Integration Method.
> (2) Moment Area Method.
> (3) Conjugate Beam Method.
> (4) Energy Methods.

5.2   DOUBLE INTEGRATION METHOD

When a beam deforms, it bends into an arc as shown in Fig. 5-1. The moment curvature relation for a deformed member is

$$\frac{1}{R} = \frac{M}{EI} \qquad\qquad [5\text{-}1]$$

where M = bending moment, E= modulus of elasticity, I = moment of inertia, R= radius of curvature..

$$\frac{1}{R} = \frac{d\theta}{ds} = \frac{d^2 y}{dx^2} \qquad\qquad [5\text{-}2]$$

Therefore,

$$\frac{d\theta}{ds} = \frac{M}{EI} \qquad\qquad [5\text{-}3]$$

Or

$$\frac{d^2 y}{dx^2} = \frac{M}{EI} \qquad\qquad [5\text{-}4]$$

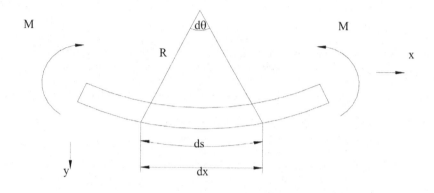

Fig. 5-1

Equation [5-3] is integrated once to obtain member slope.   Equation [5-4] is integrated twice to obtain member deflection.

$$\int \frac{M}{EI} = \theta \quad = \text{Slope} \tag{5-5}$$

$$\iint \frac{M}{EI} = y = \delta = \text{deflection} \tag{5-6}$$

The integration method consists of the following steps:
(1) Express the beam bending moment as a function of x because the method considers only deflection due to bending.
(2) Obtain an expression for the beam deflection using Equation [5-6]. The two constants of integration are determined using boundary conditions.
(3) The boundary conditions represent the beam locations where the values of the slope and deflection are known. Table 5-1 summarizes the boundary conditions for a beam with different support conditions.
(4) The slope and deflection at any beam location can be calculated by substituting back the integration constants into the slope and deflection expressions given by Equations [5-5] and [5-6], respectively.

Table 5-1. Beam Boundary Conditions

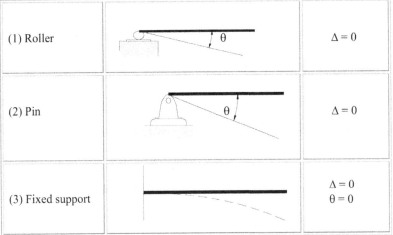

(1) Roller		$\Delta = 0$
(2) Pin		$\Delta = 0$
(3) Fixed support		$\Delta = 0$ $\theta = 0$

5.3   MOMENT AREA METHOD

The moment area method, which is used to determine slopes and deflections in simple beams, is based on the following two theorems (Fig. 5-2):

Theorem 1

The change in slope between two points on the beam deflected curve is equal to the area of the $\frac{M}{EI}$ diagram between these points.

$$\theta_{AB} = \int_A^B \frac{M}{EI} dx \qquad [5\text{-}7]$$

Where M = bending moment, E = modulus of elasticity, and I = moment of inertia.

Theorem 2

The vertical deflection δ, from a point B on the deflected curve of a beam to a tangent drawn to point A, is equal to the moment of the $\frac{M}{EI}$ diagram between A and B about B.

$$\delta_{AB} = \int_A^B \frac{M}{EI} x \, dx \qquad [5\text{-}8]$$

The main advantages of the moment area method over the integration method are as follows.

> (1) The moment area uses the moment diagram instead of the analytical moment expression.
> (2) The moment area can be applied even for moment diagram with discontinuities.

The moment area method consists of the following steps:

> (1) Determine support reactions, draw shear, and moment diagrams.
> (2) Sketch an exaggerated deflected shape of the member.
> (3) Draw the $\dfrac{M}{EI}$ diagram and show the required areas and centroids. Table 5-2 summarizes common areas and their centroids.
> (4) Apply **Theorem 1** to determine the slope between two points A and B on the deflected beam curve (Fig. 5-2)
> (5) Apply **Theorem 2** to determine the vertical deflection from a point A on the deflected curve of a beam to a tangent drawn to another point B. This deflection is equal to the moment of the $\dfrac{M}{EI}$ diagram between A and B about B (Fig. 5-2)

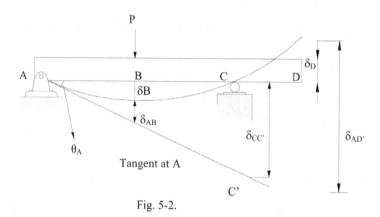

Fig. 5-2.

Table 5-2

1) Triangle		$A = \dfrac{1}{2}bh$ $\bar{x} = \dfrac{1}{3}b$
2) Trapezoid		$A = \dfrac{1}{2}b(h_1 + h_2)$ $\bar{x} = \dfrac{b(2h_2 + h_1)}{3(h_1 + h_2)}$
3) Semi Parabola		$A = \dfrac{2}{3}bh$ $\bar{x} = \dfrac{3}{8}b$
4) Parabolic spandrel		$A = \dfrac{1}{3}bh$ $\bar{x} = \dfrac{1}{4}b$

5.4   CONJUGATE BEAM METHOD

The conjugate beam method uses of the principles of statics to determine slopes and deflections in beams. A conjugate beam is an imaginary beam of same span as the original beam loaded with $\dfrac{M}{EI}$ of the original beam. The conjugate beam consists of transforming the problem of computing beam slopes and deflections to a problem of computing shears and moments in the corresponding conjugate beam. Table 5-3 summarizes he original beams and their corresponding conjugate beams.

The conjugate beam method is based on the following theorems:

Theorem 1

The slope at a point in a beam is equal to the shear force at the corresponding point in the conjugate beam.

$$\frac{dV}{dx} = w \qquad\qquad\qquad [5\text{-}9]$$

Where w = original load on the actual beam.

Or $V = \int w dx$ \qquad\qquad\qquad [5-10]

And $\theta = \int \frac{M}{EI} dx$ \qquad\qquad\qquad [5-11]

<u>Theorem 2</u>

The deflection at a point is equal to the bending moment at the corresponding point in the conjugate beam.

$$\frac{d^2 M}{dx^2} = w \qquad\qquad\qquad [5\text{-}12]$$

Or $M = \int\int w dx dx$ \qquad\qquad\qquad [5-12]

And $\delta = \int\int \frac{M}{EI} dx dx$ \qquad\qquad\qquad [5-13]

The computational steps of the conjugate beam method are:

    (1) Determine support reactions, draw shear and moment diagram.

    (2) Draw the $\frac{M}{EI}$ diagram.

    (3) Transform the original beam into the conjugate beam.

    (4) Load the conjugate beam with the original beam $\frac{M}{EI}$ diagram.

    (5) Determine the conjugate beam reactions.

    (6) Choose the section that will give the simplest solution to obtain the required slope and deflection at a specific point.

    (7) Use the equilibrium equations to determine V' and M' which are equal slope θ and deflection δ, respectively.

Table 5-3

Real Beam	Conjugate Beam
1) θ $\Delta = 0$ Pin	Pin
2) θ $\Delta = 0$ Roller	Roller
3) $\theta = 0$ $\Delta = 0$ Fixed	Free
4) θ Δ Free	Fixed
5) θ $\Delta = 0$ Internal pin	Hinge
6) θ $\Delta = 0$ Internal Roller	Hinge
7) θ Δ Hinge	Internal Roller

5.5   VIRTUAL WORK METHOD

The double integration method, the moment area method, and the conjugate beam method are not efficient for finding slopes and deflections for frames, trusses, and beams with complicated loadings. To compute the deflections at any point in any direction for any type of structural system, energy methods do provide an analysis procedure that meets these requirements.

Energy methods such as the strain energy, the virtual work or the unit load method, and Castiglianos' method, are based on the principle of conservation of energy, which states that the total work of external forces equals the total work of the internal forces (Fig. 5-3).

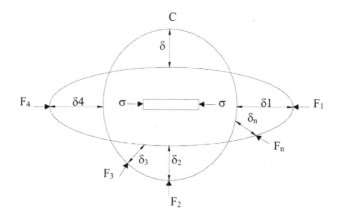

Fig. 5-3. Deformed structure under Applied loads

External work

The structure external work is equal to (Fig. 5-4)

$$W_E = \frac{1}{2}.F.\delta \qquad \text{External work due to force} \qquad [5\text{-}15]$$

$$W_E = \frac{1}{2}.M.\theta \qquad \text{External work due to moment} \qquad [5\text{-}16]$$

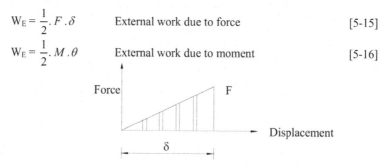

Fig. 5-4. Total External Work (Shaded Area)

Internal work

To resist the external forces, the internal forces increase gradually from an initial value of zero to their final value and the work is done and stored as energy in the structure to bring back the structure to its original shape and size upon the removal of the external loads to satisfy the elastic limit. The stored energy is due to the material straining. Thus, it is called **Strain Energy**.

$$W_I = \int \frac{1}{2}.\sigma.\varepsilon\,dv \qquad \text{Strain energy stored} \qquad [5\text{-}17]$$

where σ = element stress due to actual loads, ε = element strain due to applied loads, and dv = volume change .

Virtual work

Virtual work is the work done by real forces due to imaginary displacements or the work done by imaginary forces such as unit loads or unit moments during real displacements (See Fig. 5-5).

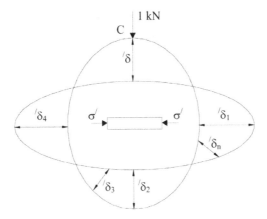

Fig. 5-5 Deformed Structure under a Unit Load

Virtual Work for Beams and Frames:

For a unit load,

$$W_E = \frac{1}{2}.1.\delta \qquad \text{External work of a unit load} \qquad [5\text{-}18]$$

$$W_E = \frac{1}{2}.1.\theta \qquad \text{External work of a unit moment couple} \quad [5\text{-}19]$$

$$W_I = \int \frac{1}{2}.\sigma'.\varepsilon\,dv \qquad \text{Internal work of a unit load} \qquad [5\text{-}20]$$

where σ' = element stress due to a unit load.

The principle of conservation of energy, which states the equality of external and internal work, can be expressed using the following equation:

$$W_E = W_I \qquad \text{Principle of conservation of energy} \qquad [5\text{-}21]$$

From equations [5-18], [5-20] and [5-21]

$$\frac{1}{2}.1.\delta = \int \frac{1}{2}.\sigma'.\varepsilon\, dv \qquad\qquad [5\text{-}22]$$

(Virtual / Applied)

The stress σ' is given by the following equation:

$$\sigma' = \frac{m\, y}{I}$$

where m = bending moment due to a unit load, y = distance from the element centeroid to the neutral axis, I = moment of inertia.

The strain ε is given by the following equation:

$$\varepsilon = \frac{\sigma}{E} = \frac{My}{EI}$$

where E = modulus of elasticity.

Substituting the stress and strain equations into Eq. [5-22] gives the following equation:

$$\delta = \int_0^L \frac{Mm}{EI^2}\left(\int_0^A y^2\, dA\right) dx \qquad\qquad [5\text{-}23]$$

but

$$I = \int_0^A y^2\, dA$$

Therefore the deflection is given by the following equation:

$$\delta = \int_0^L \frac{M\, m_\delta}{E\, I}\, dx \qquad\qquad [5\text{-}24]$$

In a similar manner, the rotation is given by the following equation:

$$\theta = \int_0^L \frac{M\, m_\theta}{E\, I}\, dx \qquad\qquad [5\text{-}25]$$

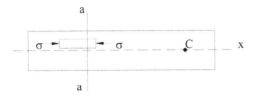

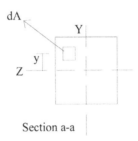

Section a-a

Fig. 5-6

The steps for determining the slope and deflection of a beam using the virtual work method are:

    (1) Assume a positive direction of the moment.

    (2) Choose the section with the appropriate x-coordinates and the simplest solution to determine the required displacement.

    (3) Determine the internal moment M due to applied loads.

    (4) Determine m_δ or m_θ or both due to a unit load or unit couple moment applied at the place of interest.

    (5) Construct the energy table.

    (6) Apply the virtual work equation to determine the required displacements.

The following illustrative examples will cover all methods and their procedures of analysis.

Example: 5-1

1. Determine the deflection equation of the elastic curve for the cantilever beam shown in Fig. 5-7a.
2. Determine the deflection at B (δ_B).

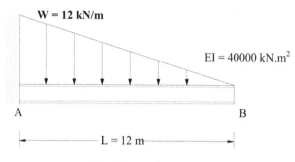

$W = 12$ kN/m

$EI = 40000$ kN.m^2

A

B

L = 12 m

Fig. 5-7a

SOLUTION

The moment M is expressed as a function of x as follows (See Fig. 5-7b)

$$M = \left(\frac{Wx^2}{2L}\right)\left(\frac{x}{3}\right) = \frac{Wx^3}{6L}$$

The deflection equation is obtained by integrating the moment twice as follows.

$$\delta = \int_0^L\int_0^L \frac{M}{EI}dx\,dx = \frac{1}{EI}\int\int\frac{Wx^3}{6L}dx\,dx$$

$$= \frac{1}{EI}\int\left[\frac{Wx^4}{24L}+C_1\right] \Rightarrow \theta_B = \frac{1}{EI}\left[\frac{Wx^4}{24L}+C_1\right]$$

$$\delta = \frac{1}{EI}\left[\frac{Wx^5}{120L}+C_1x+C_2\right]$$

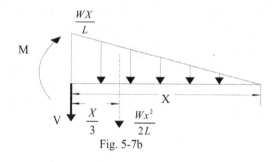

$\frac{WX}{L}$

M

X

V $\frac{X}{3}$ $\frac{Wx^2}{2L}$

Fig. 5-7b

The constants of integration are determined as to satisfy the boundary conditions.

$$\theta_{@x=L} = 0 \Rightarrow 0 = \left[\frac{WL^4}{24L} + C_1 \right]$$

$$C_1 = -\frac{WL^3}{24}$$

and

$$\delta_{@x=L} = 0 \Rightarrow 0 = \frac{1}{EI} \left[\frac{WL^5}{120L} + \left(-\frac{WL^3}{24} \right) L + C_2 \right]$$

$$C_2 = \frac{4WL^5}{120L} = \frac{WL^4}{30}$$

The elastic curve equation is equal to:

$$\delta = \frac{1}{EI} \left[\frac{Wx^5}{120L} - \frac{WL^3}{24}x + \frac{WL^4}{30} \right]$$

The deflection δ_B at point B is equal to

$$\delta_B = \delta_{@x=0} = \frac{1}{EI} \left[0 - 0 + \frac{WL^4}{30} \right]$$

$$\delta_B = \frac{1}{40000} \left[0 - 0 + \frac{12(12)^4}{30} \right]$$

$$\delta_B = 0.207 \text{ m} \downarrow$$

Is this deflection large? if so what to do about it?

Chapter 5   Deflections

Determine the slope (θ_B) at B and the deflection (δ_C) at C of the cantilever beam shown in Fig.5-8a.

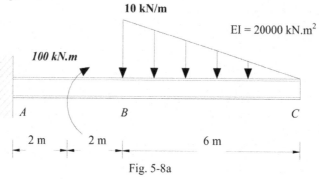

Fig. 5-8a

SOLUTION

The reactions are shown in Fig. 5-8b .

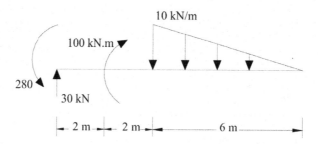

Fig. 5-8b. Applied Loads & Reactions

The shear diagram is shown in Fig. 5-2c.

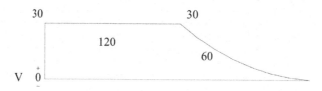

Fig. 5-8c. Shear Diagram (kN)

The bending diagram is shown in Fig. 5-8d.

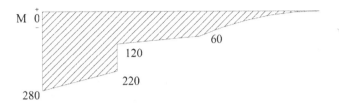

Fig. 5-8d. Moment Diagram (kN.m)

The M/EI diagram is shown in Fig. 5-8e.

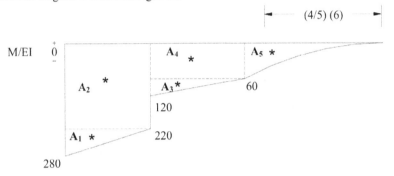

Fig. 5-8e. M/EI Diagram

Note:

$[A_1 = 60, A_2 = 440, A_3 = 60, A_4 = 120, A_5 = 90, \quad * => \text{centeroid}]$

$$\theta_B = \frac{1}{EI}[60 + 440 + 60 + 120] = \frac{680\,kN\,.m^2}{EI}$$

$\theta_B = 0.034$ **rad** ↷

$$\delta_C = \frac{1}{EI}\left[60\left(\left(\frac{2}{3}\right)(2) + 8\right) + 440(1 + 8) + 60\left(\frac{2}{3} \times 2 + 6\right) + 120(7) + 90\left(\frac{4}{5}\right)6 \right]$$

$\delta_C = 0.3116$ **m** ↓

Note:

The deflection is large. To reduce the deflection, the beam depth is to be increased. This way, the beam moment of inertia is increased which results in a reduction of the deflection.

Example: 5-3

Determine the deflection and slope at C of the beam shown in Fig.5-9a

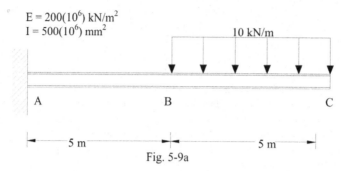

$E = 200(10^6) \text{ kN/m}^2$
$I = 500(10^6) \text{ mm}^2$

10 kN/m

A   B   C

5 m   5 m

Fig. 5-9a

SOLUTION

The beam reactions are shown in Fig. 5-9b .

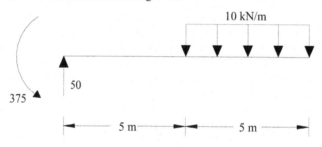

10 kN/m

375   50

5 m   5 m

Fig. 5-9b. Applied Loads & Reactions

The shear and moment diagrams are shown in Figs. 5-9c and 5-9d, respectively.

50

250

125

$V \overset{+}{\underset{-}{0}}$

Fig. 5-9c. Shear Diagram (kN)

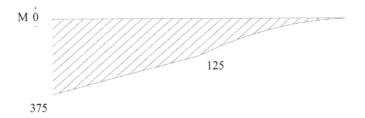

Fig. 5-9d. Moment Diagram (kN.m)

The deflected curve is shown in Fig. 5-9e.

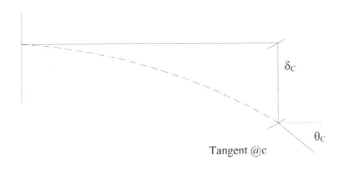

Fig. 5-9e. Deflected Curve

The M/EI diagram is shown in Fig. 5-9g.

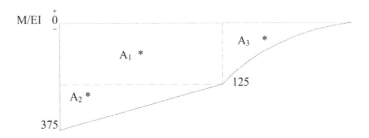

Fig. 5-9f. M/EI Diagram

To determine the slope at C, the area of the M/EI diagram between A and C is computed as follows.

$\theta_C = [A_1 + A_2 + A_3] / EI$

113

$$\theta_C = \frac{[625 + 625 + 208.33]}{EI} = \frac{1458.33\,kN\,.m^2}{EI}$$

$\theta_C = 0.0146\ \text{rad}$ ↷

$$\delta_c = \frac{1}{EI}\left[625\left(10 - \frac{5}{3}\right) + 625(10 - 2.5) + 208.33\left(5 - \frac{1}{4}(5)\right)\right] = \frac{10667\,kN\,.m^3}{E\,I}$$

$\delta_C = 0.107\ \text{m}$↓

Note:

The positive values of θ_C and δ_C indicate that the deflection and slope are as shown in the deflected curve of Fig. 5-9e.

Is this deflection acceptable? How do you define an acceptable deflection?

Example: 5-4

Determine the deflection at point B of the cantilever beam shown in Fig. 5-10a. Use the superposition method.

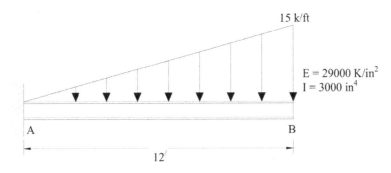

Fig. 5-10a

SOLUTION

In the super position method, the actual beam is replaced by a combination of two loads LOAD1 and LOAD2 as shown in Fig. 5-10b. Each of these loads is treated separately.

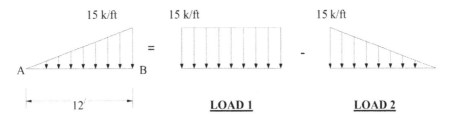

Fig. 5-10b

LOAD1
The reactions due to LOAD1 are shown in Fig. 5-10c.

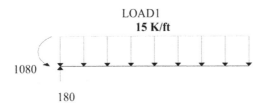

Fig. 5-10c. Applied Loads & Reactions

The shear, moment, and M/EI diagrams due to LOAD1 are shown in Figs. 5-10d, 5-10e, and 5-10f, respectively.

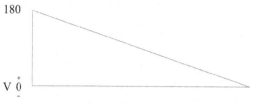

Fig. 5-10d. Shear Diagram (Kips)

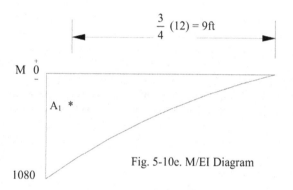

Fig. 5-10e. M/EI Diagram

$$A_1 = 1080 \ (12) \ \frac{1}{3} \frac{1}{EI} = \frac{4320}{EI} \ K \, ft^2$$

Deflection due to LOAD1

$$\delta_{B(LOAD1)} = \frac{1}{EI} \left[A_1 \left(\frac{3}{4} \right) 12 \right] = \frac{38880}{EI} \ K \, ft^3$$

LOAD 2

The reactions due to LOAD2 are shown in Fig. 5-10f.

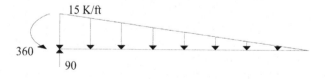

Fig. 5-10f. Applied Loads and Reactions

The shear, moment, and M/EI diagrams due to LOAD2 are shown in Figs. 5-10g and 5-10h, respectively.

Fig. 5-10g. Shear Diagram (Kips)

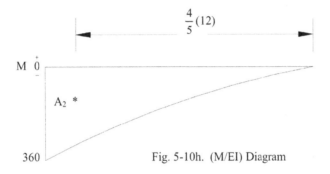

Fig. 5-10h. (M/EI) Diagram

$$A_2 = 360 \,(12) \,\frac{1}{4} = 1080 \ K\,ft^2$$

Deflection due to LOAD2

$$\delta_{B(LOAD2)} = \frac{1}{EI}\left[A_2 \left(\frac{4}{5}\right)12 \right] = \frac{10368}{EI} \ K\,ft^3$$

Total deflection at B $= \delta_B = \delta_{B(LOAD1)} - \delta_{B(LOAD2)} = \dfrac{28512}{EI} \ K\,ft^3$

$$\delta_B = \frac{28512(12^3)}{29000(3000)} = \textbf{0.566 in} \downarrow$$

Example: 5-5

Determine the deflection at point B (δ_B) of the beam shown in Fig. 5-11a. Use the moment area method.

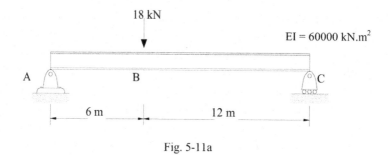

18 kN

EI = 60000 kN.m^2

A   B   C

6 m   12 m

Fig. 5-11a

SOLUTION

The reactions are shown in Fig. 5-11b.

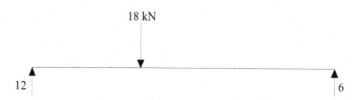

18 kN

12   6

Fig. 5-11b. Applied Loads and Reactions

The shear, moment, and M/EI diagrams are shown in Figs. 5-11c, 5-11d, and 5-11e, respectively.

V 0

72

72

6

Fig. 5-11c. Shear Diagram (kN)

72

M 0

Fig. 5-11d. Moment Diagram (kN.m)

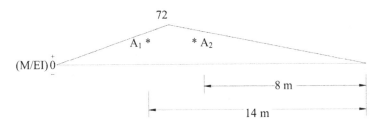

$A_1 = 216, A_2 = 432$

Fig. 5-11c. M/EI Diagram

The deflected shape of the beam is shown in Fig. 5-11f.

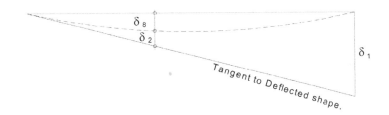

Fig. 5-11f. Deflected Shape

Making use of Fig. 5-11f and similar triangles

$$\frac{\delta_1}{18} = \frac{\delta_B + \delta_2}{6} \implies \delta_B = \frac{\delta_1}{3} - \delta_2$$

δ_1 and δ_2 need to be determined using the 2nd theorem as follows.

$$\delta_1 = \frac{1}{EI} [\, 216(14) + 432(8) \,] \quad = \frac{6480 \, kN.m^3}{EI}$$

$$\delta_2 = \frac{1}{EI} [\, 216(2)] \quad = \frac{432 \, kN.m^3}{EI}$$

$$\delta_B = \frac{1}{EI} \left(\frac{6480}{3} - 432 \right) \quad = \frac{1728 \, kN.m^3}{EI} = \textbf{0.0288 m} \downarrow$$

Example: 5-6

Determine the slope at point B (θ_B) and the deflection at C (δ_C) for the beam shown in Fig. 5-12a.

EI = 40000 kN.m^2

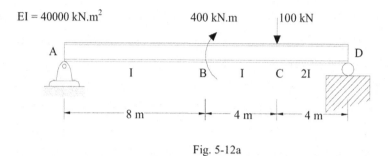

Fig. 5-12a

SOLUTION

The reactions are shown in Fig. 5-12b.

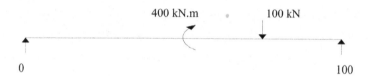

Fig. 5-12b. Applied Loads and Reactions

The shear, moment, and M/EI diagrams are shown in Figs. 5-12c, 5-12d, and 5-12e, respectively.

Fig. 5-12c. Shear Diagram (kN)

Fig. 5-12d. Moment Diagram (kN.m)

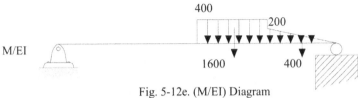

Fig. 5-12e. (M/EI) Diagram

The conjugate beam reactions, which are shown in Fig. 5-12f, are determined using the equilibrium equations.

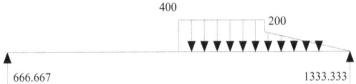

Fig. 5-12f. Conjugate Beam Reaction

To determine the slope at B (θ_B), the conjugate beam shear at B (V_B) is computed as shown in Fig.5-12g.

Fig. 5-12g

$$\Sigma F_y = 0 \uparrow + \; \frac{666.667}{EI} - V_B = 0$$

$$V_B = \frac{666.667 \, kN \, .m^2}{EI}$$

$$\theta_B = V_B = \mathbf{0.0167 \; rad} \; \curvearrowright$$

To determine the deflection at C (δ_C), the conjugate beam moment at C is determined as shown in Fig. 5-12h.

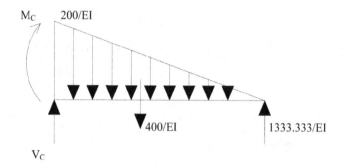

Fig. 5-12h

$$\sum M_C = 0 \circlearrowleft + \quad \frac{-1333.33}{EI}(4) + \frac{400\left(\frac{1}{3}\right)(4)}{EI} + M_C = 0$$

$M_C = \delta_C = \textbf{0.12 m} \downarrow$

Example: 5-7

For the beam shown in Fig. 5-13a, determine the slope θ at point B and the deflection δ at point D.

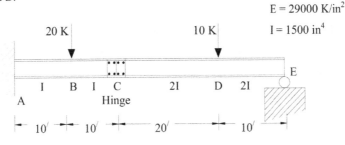

$E = 29000 \ K/in^2$

$I = 1500 \ in^4$

Fig. 5-13a

SOLUTION

The reactions are shown in Fig. 5-13b.

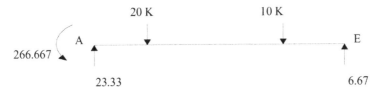

Fig. 5-13b. Applied Loads and Reactions

The shear, moment, and M/EI diagrams are shown in Figs. 5-13c, 5-13d, and 5-13e, respectively.

Fig. 5-13c. Shear Diagram, kN

Fig. 5-13d. Moment Diagram, kN.m

123

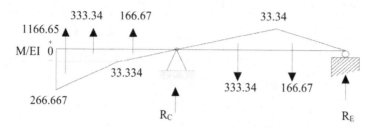

Fig. 5-13e. M/EI diagram

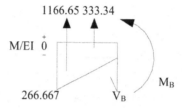

Fig. 5-13f

The slope at B (θ_B) is determined as follows (Fig. 5-13f).

$$V_B = \theta_B = \frac{144}{EI}[1166.665 + 333.34] = \frac{1500(144)}{EI} \Rightarrow \theta_B = \textbf{0.005 rad} \curvearrowright$$

The deflection at D (δ_D) is determined as follows (Fig. 5-13g)

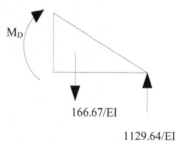

$$M_D = \delta_D = \left(1129.64(10) - 166.67\left(\tfrac{1}{3}\times10\right)\right)\frac{12^3}{EI}$$

$$\delta_D \cong \textbf{0.427 in} \downarrow$$

166.67/EI

1129.64/EI

Fig. 5-13g

Example: 5-8

Determine the deflection and slope at the points D and B for the beam shown in Fig. 5-14a.

$E = 200(10^6)$ kN/m^2
$I = 70(10^6)$ mm^4

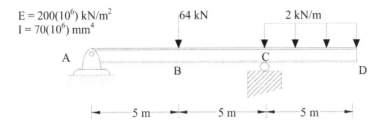

64 kN

2 kN/m

A

B

C

D

5 m   5 m   5 m

Fig. 5-14a

SOLUTION

The reactions are shown in Fig. 5-14b.

64 kN

2 kN/m

29.5

44.5

Fig. 5-14b. Applied Loads and Reactions

The shear and M/EI diagrams are shown in Figs. 5-14c and 5-14d, respectively.

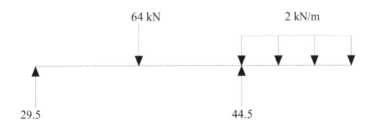

29.5

10

147.5

V   0

25

172.5

34.5

Fig. 5-14c. Shear Diagram, kN

125

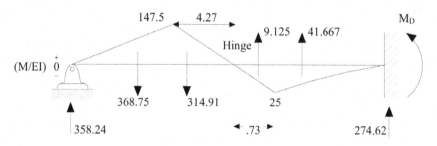

Fig. 5-14d. M/EI Diagram

SECTION AC

$$\Sigma M_C = 0 \;\circlearrowright + \qquad R_A = \frac{358.245}{EI}$$

Beam AD

$$\Sigma F_Y = 0 \uparrow + \quad R_D = \theta_D = \frac{274.6225}{EI} = \textbf{0.0196 rad.} \;\curvearrowleft$$

$$\Sigma M_D = 0 \;\circlearrowright +$$

$$M_D = \delta_D = -\frac{1425.2}{EI} = \textbf{0.102 m} \uparrow$$

SECTION AB

$$\Sigma M_B = 0 \;\circlearrowright +$$

$$M_B = \delta_B = \frac{1176.6}{EI} = \textbf{0.084 m} \downarrow$$

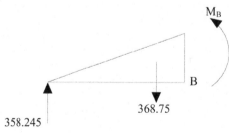

Fig. 5-14e

Example: 5-9

For the beam shown in Fig. 5-15a, determine the deflection at points A and C, and the slope at point E.

$E = 200.10^6 \text{ kN/m}^2$
$I = 200.10^6 \text{ mm}^4$

50 kN

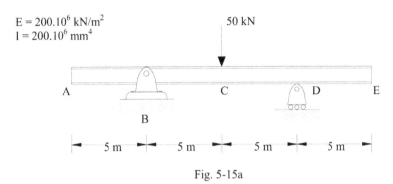

A   C   D   E

B

| 5 m | 5 m | 5 m | 5 m |

Fig. 5-15a

SOLUTION

The reactions are shown in Fig. 5-15b.

50 kN

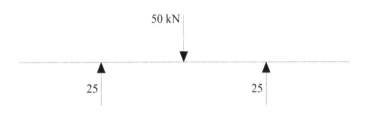

25   25

Fig. 5-15b. Applied Loads and Reactions

The shear and M/EI diagrams are shown in Figs. 5-15c and 5-15d, respectively.

25
125

V   0

125

25

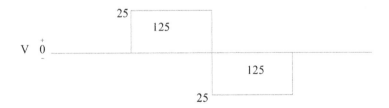

Fig. 5-15c. Shear Diagram, kN

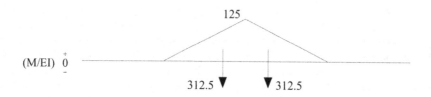

(M/EI) $\overset{+}{\underset{-}{0}}$

Fig. 5-15d. M/EI Diagram

Breaking up the beam at the hinges yields Fig. 5-15e.

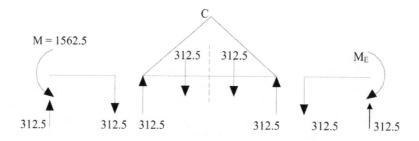

Fig. 5-15e

Applying the equilibrium equations to the beam yields the following results:

$$\delta_A = M_A = \left(\frac{1562.5}{EI}\right) = \textbf{0.039m} \uparrow$$

$$\theta_E = V_E = R_E = \frac{312.5}{EI} = \textbf{0.0078 rad} \curvearrowright$$

$$\delta_c = \frac{1041.667}{EI} = \textbf{0.0206 m} \downarrow$$

Example: 5-10

Determine the deflection at point B of the beam shown in Fig. 5-16a.
$E = 200(10^6) \, kN/m^2$
$I = 100(10^6) \, mm^4$

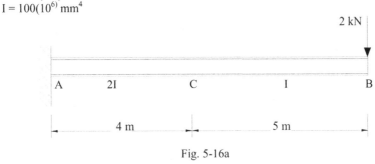

Fig. 5-16a

SOLUTION

The beam reactions due to applied loads are shown in Fig. 5-16b.

Fig. 5-16b. Applied Loads and Reactions

To determine the displacement δ_B at B a unit load is applied at B. The beam reactions due a unit load at B are shown in Fig. 5-16c.

Fig. 5-16c. Reactions due a Unit Load at B

Clockwise moments are assumed positive. Two sections are also chosen: 1) section 1-1 for member BC and 2) section 2-2 for member CA. Theses sections are selected because of the moment of inertia change along the beam AB (Figs. 5-16e through 5-16h).

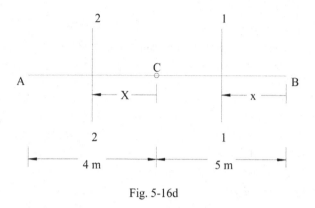

Fig. 5-16d

SECTION 1-1

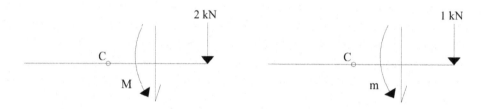

Fig. 5-16e. Section 1-1, M-Applied Load   Fig. 5-16f. Section 1-1, m-virtual load

SECTION 2-2

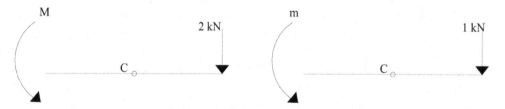

Fig. 5-16g. Section 2-2, M-applied load   Fig. 5-16h. Section 2-2, m-virtual load

The energy table is constructed after selecting the sections and the appropriate coordinates that will yield the simplest solution.

Member	EI	X=0	M	m_δ	$\int_a^b$
BC	1	B	2X	X	$\int_0^5$
CA	2	C	2(X+5)	1 (X+5)	$\int_0^4$

The required displacement is obtained by applying the virtual work equation as follows.

$$\delta_B = \int_0^L \frac{Mm}{EI}dx = \frac{1}{EI}\int_0^5 2x(x)\,dx + \frac{1}{2EI}\int_0^4 2(x+5)(x+5)\,dx = \mathbf{0.014\ m}$$

The positive sign indicate δ_B has the same direction as the selected unit load.

$\delta_B = \mathbf{0.014\ m}\downarrow$

Example: 5-11

Determine the slope θ and deflection δ at point D of the beam shown in Fig. 5-17a.

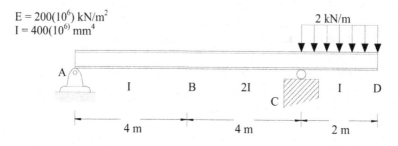

$E = 200(10^6) \text{ kN/m}^2$
$I = 400(10^6) \text{ mm}^4$

2 kN/m

A

I   B   2I   C   I   D

4 m   4 m   2 m

Fig. 5-17a

SOLUTION

The beam reactions due to applied loads are shown in Fig. 5-17b.

Fig. 5-17b. Applied Loads and Reactions

To determine the displacement δ at D and the slope θ at D, a unit load of 1 kN and a unit couple moment are applied at D. The beam reactions due to a unit load and a unit couple moment at D are shown in Fig. 5-17c and 5-17d, respectively.

1 kN

0.25   1.25

Fig. 5-17c. Reactions due to a unit load at D

132

Fig. 5-17d. Reactions due a unit couple moment at D

Clockwise moments are chosen to be positive. The energy table is constructed after choosing the sections and appropriate x coordinates that yield the simplest solution.

Member	EI	X=0	M	m_δ	m_θ	$\int_a^b$
AB	1	A	$-0.5X$	$-0.25X$	$-\dfrac{1}{8}X$	$\int_0^4$
BC	2	B	$-0.5(4+X)$	$-0.25(4+X)$	$\dfrac{1(4+X)}{8}$	$\int_0^4$
CD	1	D	$\dfrac{2X^2}{2}=X^2$	X	1	$\int_0^2$

The required displacements are obtained by applying the virtual work equation:

$$\delta_D = \int_0^4 \frac{Mm}{EI}\,dx = \frac{1}{EI}\int_0^4 (-0.5x)(-0.25x)\,dx + \frac{1}{2EI}\int_0^4 [-0.5(4+x)][-0.25(4+x)]\,dx + \frac{1}{EI}\int_0^2 x^2(x)\,dx$$

$\delta_D = \mathbf{0.00037\ m}\ \downarrow$

$$\theta_D = \frac{1}{EI}\int_0^4 (-0.5x)\left(\frac{-x}{8}\right)dx + \frac{1}{2EI}\int_0^4 \left[-0.5(4+x)\left(\frac{-(4+x)}{8}\right)\right]dx + \frac{1}{EI}\int_0^2 x^2(1)\,dx$$

$\theta_D = \mathbf{0.0002\ rad}\ \curvearrowright$

Example: 5-12

Determine the slope θ, the vertical deflection δ_V, and the horizontal deflection δ_H at point A of the frame shown in Fig. 5-18a.

EI = 20000 kN.m^2

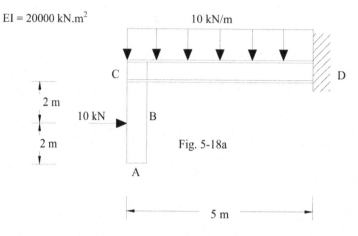

Fig. 5-18a

SOLUTION
The frame reactions due to applied loads are shown in Fig. 5-18b.

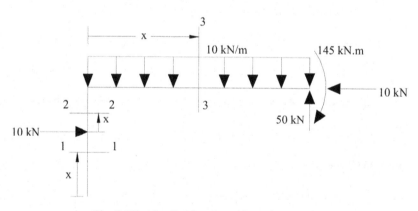

Fig. 5-18b. Applied Loads and Reactions

To determine the displacements δ_{VA}, δ_{HA}, and θ_A, a unit vertical load, a unit horizontal load , and a unit couple moment are applied at A as shown in Figs. 5-18c, 5-18d, and 5-18e.

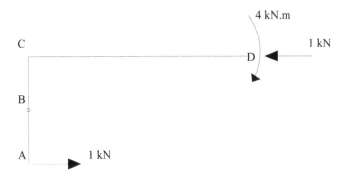

Fig. 5-18c. Reactions due to a unit horizontal load at A

Fig. 5-12d. Loads and Reactions due to a unit vertical load at A

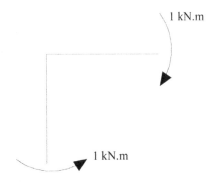

Fig. 5-12e. Loads and Reactions due to a unit couple moment at A

The energy table is constructed after choosing the sections and the appropriate x- coordinates that yields the simplest solution.

$\curvearrowleft$ +

Member	EI	X=o	M	$m_{\delta H}$	$m_{\delta V}$	m_θ	
AB	1	A	0	x	0	1	$\int_a^b$
BC	1	B	10x	1(x+2)	0	1	$\int_0^2$
CD	1	C	$\left[(10 \times 2) + \left(\dfrac{10 x^2}{2}\right)\right] = 20 + 5x^2$	4	x	1	$\int_0^2$ $\int_0^5$

The required displacements are obtained by applying the virtual work equation..

$$\delta_V = \int_0^L \frac{Mm_{\delta V}}{EI} = \frac{1}{EI}\left[\int_0^2 0 + \int_0^2 (10x)(0)dx + \int_0^5 (20 + 5x^2)(x)dx\right] = \mathbf{0.052\ m} \downarrow$$

$$\delta_H = \int_0^L \frac{Mm_{\delta H}}{EI} = \frac{1}{EI}\left[\int_0^2 0 + \int_0^2 (10x)(x+2)dx + \int_0^5 (20 + 5x^2)(4)dx\right] = \mathbf{0.065\ m} \rightarrow$$

$$\theta_A = \int_0^L \frac{Mm_\theta}{EI} = \frac{1}{EI}\left[\int_0^2 0 + \int_0^2 (10x)(1)dx + \int_0^5 (20 + 5x^2)(1)dx\right] = \mathbf{0.016\ rad} \curvearrowleft$$

Example: 5-13

Determine the slope θ and the horizontal deflection δ_H at point C of the frame shown in Fig. 5-19a.

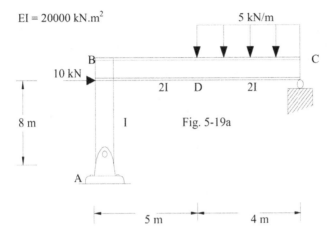

Fig. 5-19a

SOLUTION

The frame reactions due to the applied loads are shown in Fig. 5-19b.

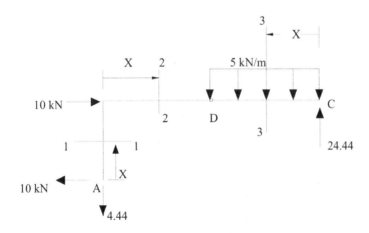

Fig. 5-19b. Applied Loads and Reactions

To determine the displacement δ_{HC} and the slope θ_C at point C, a horizontal unit load, and a unit couple moment are applied at C as shown in Fig. 5-19c and 5-19d.

137

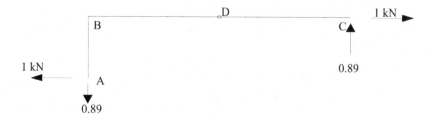

Fig. 5-19c. Loads and Reactions due to a Horizontal Unit Load at C.

Fig. 5-19d. Loads and Reactions due a Unit Couple Moment at C.

The energy table is constructed after selecting the sections and the appropriate x-coordinates that yields the simplest solution.

$\curvearrowleft$ +

Member	EI	X=0	M	$m_{\delta H}$	m_θ	$\int_a^b$
AB	1	A	$-10X$	$-X$	0	$\int_0^8$
BD	2	B	$-80+4.44X$	$-8 + 0.89X$	$-\dfrac{X}{9}$	$\int_0^5$
CD	2	C	$24.4X - \dfrac{5X^2}{2}$	$0.89x$	$1-\left(\dfrac{X}{9}\right)$	$\int_0^4$

138

The required displacement is obtained by applying the virtual work equation:

$$\delta_{HC} = \left[\frac{1}{EI} \int_0^8 (-10x)(-x) + \frac{1}{2EI} \int_0^5 (-80 + 4.44x)(-8 + 0.89x) + \frac{1}{2EI} \int_0^4 \left(24.4x - \frac{5x^2}{2} \right)(0.89x) \right]$$

$\delta_{HC} = \textbf{0.212m} \rightarrow$

$$\theta_C = \left[\frac{1}{EI} \int_0^8 (-10x)(0) + \frac{1}{2EI} \int_0^5 (-80 + 4.44x)\left(\frac{-x}{9} \right) + \frac{1}{2EI} \int_0^4 \left(24.4x - \frac{5x^2}{2} \right)\left(1 - \frac{x}{9} \right) \right]$$

$\theta_C = \textbf{0.011 rad} \curvearrowleft$

Example: 5-14

Determine the slope (θ_D) and the deflection (δ_D) at point D of the beam shown in Fig. 5-20a.

$E = 200(10^6) \text{ kN/m}^2$
$I = 400(10^6) \text{ mm}^4$

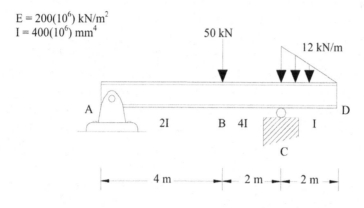

Fig. 5-20a

SOLUTION

The beam reactions due to applied loads are shown in Fig. 5-20b.

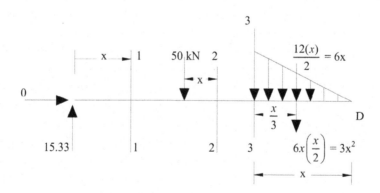

Fig. 5-20b. Applied Loads and Reactions

To determine the displacement δ_B and the slope θ_B at point D, a horizontal unit load, and a unit couple moment are applied at D as shown in Fig. 5-20c and 5-20d.

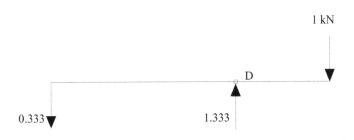

Fig. 5-20c. Loads and Reactions to a Horizontal Unit Load at D

Fig. 5-20d. Loads and Reactions due to a Unit Couple Moment at D

The energy table is constructed after selecting the sections and the appropriate x-coordinates that yields the simplest solution.

Member	EI	X=0	M	m_δ	m_θ	$\int_a^b$
AB	2	A	$15.33X$	$-0.333X$	$-0.167X$	$\int_0^4$
BC	4	B	$15.33(4+X) - 50X$	$-0.333(4+X)$	$-0.167(4+X)$	$\int_0^2$
CD	1	D	$X^3 = 3X^2\left(\dfrac{X}{3}\right)$	X	1	$\int_0^2$

141

The required displacement and slope at D are determined by applying the virtual work equation

$$\delta_D = \left[\frac{1}{2EI} \int_0^4 (15.33x)(-0.33x) + \frac{1}{4EI} \int_0^2 (15.33(4+x) - 50x)(-0.333(4+x)) + \frac{1}{EI} \int_0^2 x^3(x) \right]$$

$$= -0.00085$$

$\delta_D = 0.00085$ m ↑

$$\theta_D = \left[\frac{1}{2EI} \int_0^4 (15.33x)(-0.167x) + \frac{1}{4EI} \int_0^2 (15.33(4+x) - 50x)(-0.167(4+x)) + \frac{1}{EI} \int_0^2 x^3(1) \right]$$

$$= -0.00042$$

$\theta_D = 0.00042$ rad ↶

Example: 5-15

For the truss shown in Fig. 5-21a, determine the vertical deflection (δ_{VC}) and the horizontal deflection (δ_{HD}).
[$A_{BAR} = 400(10^{-6})$ m^2, E = 200.10^6 kN/m^2]

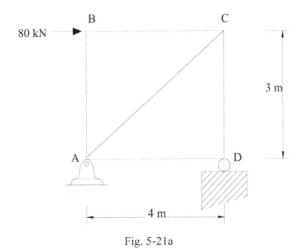

Fig. 5-21a

SOLUTION

The bar forces due to applied loads are shown in Fig. 5-21b. To determine the vertical displacement (δ_{VC}) at C and the horizontal displacement (δ_{HD}) at D, a vertical unit load is applied at C and a horizontal unit load is applied at D as shown in Figs. 5-21c and 5-21d.

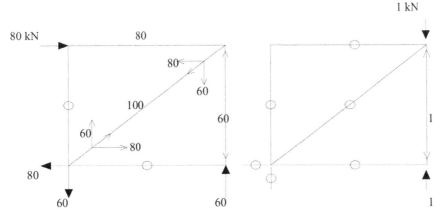

| Fig. 5-21b | Bar forces (F) due to Applied Loads | Fig. 5-21c. | Bar forces (f_v) due to a Vertical Unit Load at C |

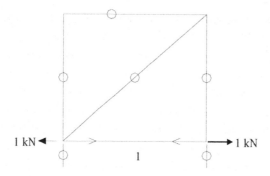

Fig. 5-21d. Bar forces (f_H) due to a Horizontal Unit Load at D

The energy table is as follows (Tension (+), Compression (-)).

Member	L/A	F	f_{VC}	$F(f_{VC})L/A$	f_{HD}	$F(f_{HD})L/A$
AB		0	0	0	0	0
BC		-80	0	0	0	0
CD	7500	-60	-1	450000	0	0
DA		0	0	0	1	0
AC		100	0	0	0	0
Σ				**450000**		**0**

The required displacements are obtained by Applying the virtual work equations.

$$\delta_{VC} = \sum \frac{F f_V L}{AE}$$

$$\delta_{VC} = \frac{450000}{E = 200.10^6} = \mathbf{0.00225 \ m} \downarrow$$

$$\delta_{HC} = \sum \frac{F f_H L}{AE}$$

$$\delta_{HD} = \frac{0}{E = 200.10^6} = \mathbf{0}$$

Example: 5-16

For the truss shown in Fig. 5-22a, determine the vertical deflection (δ_{VB}) at B and the horizontal deflection (δ_{HC}) at C.

$A_{BAR} = 500(10^{-6})$ m^2
$E = 200(10^6)$ kN/m^2

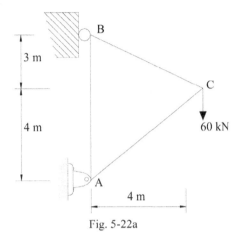

Fig. 5-22a

SOLUTION
The truss bar forces due to applied loads are shown in Fig. 5-22b. To determine the vertical displacement (δ_{VB}) at B and the horizontal displacement (δ_{HC}) at C, a vertical unit load is applied at B and a horizontal unit load is applied at C as shown in Fig. 5-22c and 5-22d.

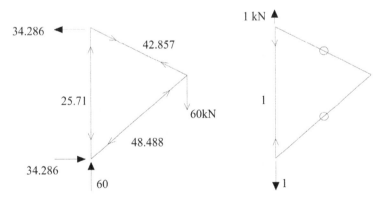

Fig. 5-22b   Bar forces (F) due to
            Applied Loads

Fig. 5-22c.   Bar forces (f_v) due to a
            Vertical Unit Load at B

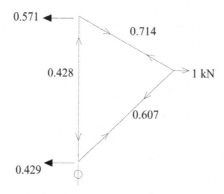

Fig. 5-22d. Bar forces (f$_H$) due to a Horizontal Unit Load at C.

The energy table is shown below (Tension (+), Compression (-))

Member	L/A	F	f$_v$	F(f$_v$)(L/A)	f$_H$	F(f$_H$)(L/A)
AB	14000	-25.715	1	-360010	-0.428	154084.28
BC	10000	42.8575	0	0	0.714	306002.55
AC	11313.7	-48.488	0	0	0.607	-332987.26
Σ				**-360010**		**127099.57**

The required displacements are obtained by applying the virtual work equation.

$$\delta_{VB} = \sum \frac{F f_V L}{AE}$$

$$\delta_{VB} = \frac{-360010}{E = 200.10^6}$$

$$\delta_{VB} = \textbf{0.0018 m} \downarrow$$

$$\delta_{HC} = \sum \frac{F f_H L}{AE}$$

$$\delta_{HC} = \frac{127099.57}{E = 200.10^6}$$

$$\delta_{HC} = \textbf{0.00064 m} \rightarrow$$

Example: 5-17

For the frame shown in Fig. 5-23a, determine the displacements δ_{vc}, δ_{Hc} and θ_C.

$A_{CABLE} = 50 \text{ mm}^2$
$E = 200(10^6) \text{ kN/m}^2$
$I = 150(10^6) \text{ mm}^4$

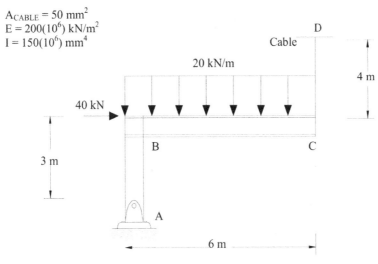

Fig. 5-23a

SOLUTION
The frame reactions due to applied loads are shown in Fig. 5-23b. To determine the vertical displacement (δ_{VC}), the horizontal displacement (δ_{HC}), and the slope (θ_C) at point C, vertical and horizontal unit loads as well as a unit couple moment are applied at C as shown in Figs. 5-23c, 5-23d, and 5-23e.

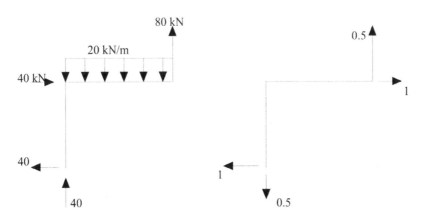

Fig. 5-23b.   Loads and Reactions due to   Fig. 5-23c. Loads and Reactions due to a
            Applied Loads   Horizontal Unit Load at C.

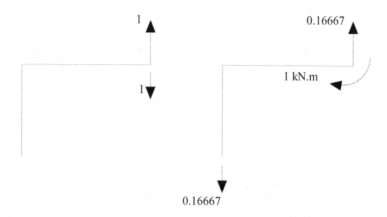

5-23d. Loads and Reactions due to a   Fig. 5-23e. Loads and Reactions due to a
       A Vertical Unit Load at C   Unit Couple Moment at C.

The energy table is constructed after selecting the sections and the appropriate
x- coordinates that yields the simplest solution.

Member	EI	X=0	M	$m_{\delta H}$	$m_{\delta V}$	m_θ	$\int_a^b$
AB	1	A	40X	X	0	0	$\int_0^3$
BC	1	B	$120+40X-10X^2$	$3 - 0.5X$	0	$-\dfrac{1}{6}X$	$\int_0^6$

The required displacements are determined by applying the virtual work equation.

$\delta = \delta_{Beam} + \delta_{Cable}$ = Flexural displacements + Axial displacements

$$= \int_0^6 \frac{Mm}{EI}\,dx + \sum \frac{FfL}{AE}$$

$$\delta_{HC} = \frac{1}{EI}\left[\int_0^3 (40x)(x) + \int_0^6 (120+40x-10x^2)(3-0.5x)\right] + \frac{0.5(80\times4)}{10000} = \textbf{0.07 m} \rightarrow$$

$$\delta_{VC} = \frac{1}{EI}\left[\int_0^3 0 + \int_0^6 0\right] + \frac{1(80\times4)}{EA=10000} = \textbf{0.032 m} \downarrow$$

$$\theta_c = \frac{1}{EI}\left[\int_0^3 (0) + \int_0^6 (120+40x-10x^2)\left(-\frac{1}{6}x\right)\right] + \frac{0.1667(80\times4)}{10000} = \textbf{-0.0047 rad}$$

$$= \textbf{0.0047 rad} \curvearrowleft$$

Example: 5-18

For the tapered beam shown in Fig. 5-24a, determine the slope θ_C and deflection δ_C at point C.

$E = 200(10^6)$ kN/m^2
$I = 100(10^{-6})$ m^4

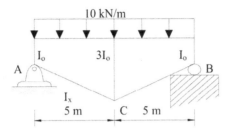

Fig. 5-24a

SOLUTION

A moment of inertia function, which defines the change of the moment of inertia along the beam, needs to be developed for the tapered beam AB.

The moment of inertia function (I_x) for the tapered beam is equal to

$$I_x = I_o\left(a + b\frac{x}{L}\right) \Rightarrow \begin{bmatrix} a = 1 \\ b = 2 \end{bmatrix}$$

$$I_x = I_o\left(1 + 2\frac{x}{L}\right) \Rightarrow \begin{bmatrix} @x = 0 \Rightarrow I_x = I_0 \text{ at A or B} \\ @x = L \Rightarrow I_x = 3I_0 \text{ at C} \end{bmatrix}$$

The beam reactions due to applied loads are shown in Fig. 5-24b. To determine the vertical displacement δ_C, a vertical unit load is applied at point C as shown in Fig. 5-24c.

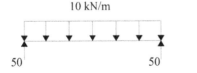

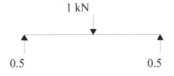

Fig. 5-24b.   Load and Reaction due
to Applied Loads

Fig. 5-24c   Loads and Reactions due
to a Vertical Unit Load at C

The energy table is constructed after selecting the sections and the appropriate x-coordinates that yields the simplest solution.

Member	EI	X=0	M	m$_\delta$	$\int_a^b$
AC	I_x	A	$50X - 5X^2$	0.5X	$\int_0^5$
BC	I_x	B	$5X^2 - 50X$	-0.5X	$\int_0^5$

The required displacements are obtained by applying the virtual work equation.

$$\delta_C = \frac{1}{E}\int_a^b \frac{Mm}{I_x}\ dx$$

$$\delta_C = \frac{1}{EI_o}\int_0^5 \frac{(50x - 5x^2)(0.5x)}{\left(1 + \frac{2x}{5}\right)} + \frac{1}{EI_o}\int_0^5 \frac{(5x^2 - 50x)(-0.5x)}{\left(1 + \frac{2x}{5}\right)}$$

$\delta_c = $ **0.0069 m** $\downarrow$

Example: 5-19

For the curved cantilever beam shown in Fig. 5-25a, determine the vertical and horizontal displacements (δ_V) and (δ_H) at the free end.

EI = Constant

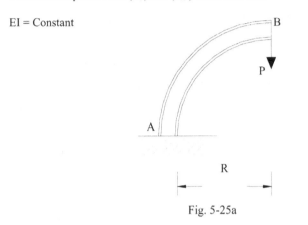

Fig. 5-25a

SOLUTION

The curved beam reactions due to applied loads are shown in Fig. 5-25b. To determine the vertical displacement δ_V at B, a vertical unit load is applied at point B as shown in Fig. 5-25c. To determine the horizontal displacement δ_H at B, a horizontal unit load is applied at point B as shown in Fig. 5-25d.

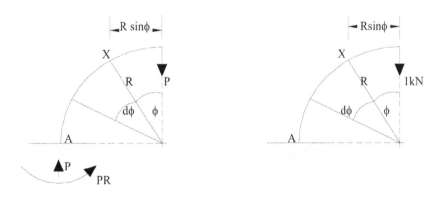

Fig. 5-25c   Loads and Reactions due to
               Applied Loads

Fig. 5-25d Loads and Reactions due to
               a Vertical Unit Load at B

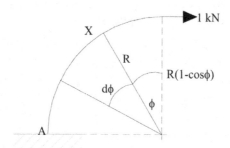

Fig. 5-25e.   Loads and Reactions due to a Horizontal Unit Load at B

The energy table is constructed after selecting the sections and the appropriate x-coordinates that yield the simplest solution.

Member	EI	X=0	M	$m_{\delta V}$	$m_{\delta H}$	$\int_a^b$
AB	EI	B	PRsinθ	(1) Rsinθ	(1)R(1-cosθ)	$\int_{\frac{\pi}{2}}^{a}$

The required displacements are obtained by applying the virtual work equation.

$$\delta_{VB} = \int_0^\theta \frac{Mm_{\delta v}}{EI} = \frac{1}{EI} \int_0^{\frac{\pi}{2}} (\text{PRsin}\theta)[(1)(\text{Rsin}\theta)]R\,d\theta$$

$$\delta_{VB} = \frac{\pi PR^3}{4EI} \downarrow$$

$$\delta_{HB} = \int_0^\theta \frac{Mm_{\delta H}}{EI} = \frac{1}{EI} \int_0^{\frac{\pi}{2}} (\text{PRsin}\theta)[(1)(R)(1\text{-Rsin}\theta)]R\,d\theta$$

$$\delta_{HB} = \frac{PR^3}{2EI} \rightarrow$$

Example: 5-20

For the beam shown in Fig. 5-26a, determine the vertical deflection δ_{VC} at point C.
$E = 200.10^6$ kN/m^2
$I = 80.10^6$ mm^4
$A_{cable} = 150$ mm^2

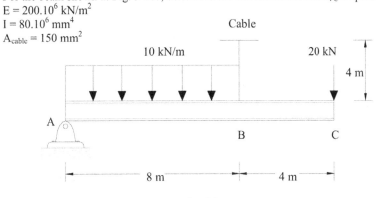

Fig. 5-26a

SOLUTION

The beam reactions due to applied loads are shown in Fig. 5-26b. To determine the vertical displacement δ_C, a vertical unit load is applied at C as shown in Fig 5-26c.

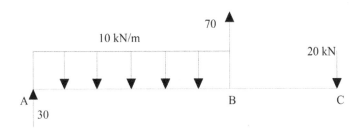

Fig. 5-26b.   Applied Loads and Reactions

Fig. 5-26c. Loads and Reactions due to a Vertical Unit Load at C

153

The energy table is constructed after selecting the sections and the appropriate x-coordinates that yield the simplest solution.

$$+ \circlearrowright$$

Member	EI	X=0	M	m_δ	$\int_a^b$
AB	1	A	$30X - \dfrac{10X^2}{2}$	-0.5X	$\int_0^8$
BC	1	C	$20X$	X	$\int_0^4$

The required displacements are obtained by applying the virtual work equation.

$\delta = \delta_{\text{Beam}} + \delta_{\text{Cable}} = $ Flexural displacements + Axial displacements

$$= \int_0^6 \frac{Mm}{EI}\,dx + \sum \frac{FfL}{AE}$$

$$\delta_{VC} = \frac{1}{EI}\left[\int_0^8 (30X - 5X^2)(-0.5X) + \int_0^4 (20X)(X) \right] + \frac{70(1.5)(4)}{EA}$$

$$\delta_{VC} = 0.041\text{m} \downarrow$$

Example: 5-21

For the tapered beam shown in Fig. 5-27a, determine the slope θ_A and the deflection δ_A at point A.

$E = 200.10^6$ kN/m^2
$I_0 = 400.10^{-6}$ m^4

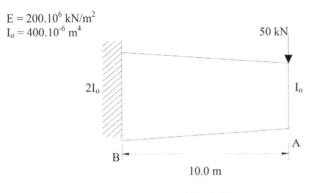

Fig. 5-27a

SOLUTION

A moment of inertia function, which defines the change of the moment of inertia along the beam, needs to be developed for the tapered beam AB.

The moment of inertia function (I_x) for the tapered beam is equal to

$$I_x = I_o\left(a + b\frac{x}{L}\right) \Rightarrow \begin{bmatrix} a = 1 \\ b = 1 \end{bmatrix}$$

$$I_x = I_o\left(1 + \frac{x}{L}\right) \Rightarrow \begin{bmatrix} @x = 0 \Rightarrow I_x = I_o \text{ at A} \\ @x = L \Rightarrow I_x = 2I_o \text{ at B} \end{bmatrix}$$

The beam reactions due to applied loads are shown in Fig. 5-27b. To determine the vertical displacement δ_A, a vertical unit load is applied at A as shown in Fig 5-27c. To determine the slope θ_A, a unit couple moment is applied at A as shown in Fig 5-27d.

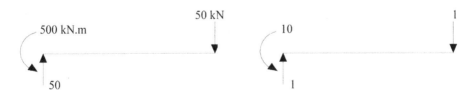

Fig. 5-27b. Loads and Reactions due to
Applied Loads

Fig. 5-27c. Loads and Reactions due
to a Vertical Unit Load at A

Fig. 5-27d   Loads and Reactions due to a Unit Couple Moment at A

The energy table is constructed after selecting the sections and the appropriate x- coordinates that yield the simplest solution.

$+ \circlearrowleft$

Member	EI	X=0	M	$m_{\delta V}$	m_θ	$\int_a^b$
AB	I_x	A	50X	X	1	$\int_0^{10}$

The required displacements are obtained by applying virtual work equation .

$$\delta = \frac{1}{E} \int_a^b \frac{M m_{\delta v}}{I_x} dx$$

$$\delta_A = \frac{1}{EI_o} \int_0^{10} \left(\frac{50x \, (x)}{\left(1+\dfrac{x}{10}\right)} \right) dx$$

$\delta_A = \textbf{0.12m} \downarrow$

$$\theta_A = \frac{1}{E} \int_a^b \frac{M m_\theta}{I_x} dx = \frac{1}{EI_o} \int_0^{10} \left(\frac{50x \, (1)}{\left(1+\dfrac{x}{10}\right)} \right) dx$$

$\theta_A = \textbf{0.019 rad} \curvearrowright$

Example: 5-22

For the beam shown in Fig. 5-28a, determine the deflection and the slope at point B.

$E = 29(10^6)$ psi
$I = 3000$ in^4

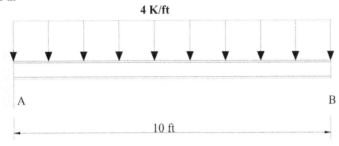

4 K/ft

A   B

|←——————————— 10 ft ———————————→|

Fig. 5-28a

SOLUTION
The beam reactions due to applied loads are shown in Fig. 5-28b.

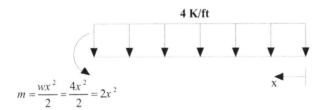

4 K/ft

$$m = \frac{wx^2}{2} = \frac{4x^2}{2} = 2x^2$$

x

Fig. 5-28b. Applied Loads and Reactions

To determine the vertical displacement (δ_B), an external vertical force P is applied at B as shown in Fig. 5-28c.

$Px = m_{\delta v}$ P

x

Fig. 5-28c. Loads and Reactions due to vertical Force P

To determine the slope (θ_B), an external couple moment $\overline{m}$ is applied at B as shown in Fig. 5-28d.

157

Fig. 5-28d. Loads and Reactions due to couple moment $\overline{m}$

The energy table is as follows.

Member	X=0	EI	M	$\dfrac{\partial m_{\delta_v}}{\partial P}$	$\dfrac{\partial m_\theta}{\partial \overline{m}}$	$\displaystyle\int_a^b$
AB	B	1	$2X^2$	$\dfrac{\partial (PX)}{\partial P} = X$	1	$\displaystyle\int_0^{10}$

The required displacements are obtained by applying virtual work equations [5-29] and [5-30].

$$\theta_B = \frac{1}{EI}\int_0^{10}\left(2x^2\right)(1)\,dx = \frac{2}{3EI}\left[x^3\right]_0^{10} = \frac{(10\times12)^3\left(\dfrac{2000}{12}\right)}{3\left(29\times10^6\right)(3000)}$$

$\theta_B = $ **0.001 rad** $\curvearrowright$

$$\delta_B = \frac{1}{EI}\int_0^{10}\left(2x^2\right)(x)\,dx = \frac{2}{4EI}\left[x^4\right]_0^{10} = \frac{(10\times12)^4\left(\dfrac{2000}{12}\right)}{3\left(29\times10^6\right)(3000)}$$

$\delta_B = $ **0.1 in** $\downarrow$

Example: 5-23

For the frame shown in Fig. 5-29a, determine the horizontal deflection δ_H and the vertical deflection δ_V at point C.
$E = 200.10^6 \text{ kN/m}^2$
$I = 200.10^6 \text{ mm}^4$

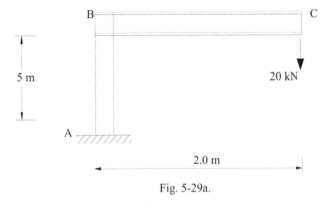

Fig. 5-29a.

SOLUTION

The beam reactions due to applied loads are shown in Fig. 5-29b. To determine the vertical and horizontal displacement δ_H, and δ_V, a horizontal and vertical loads are applied at C as shown in Figs. 5-29c and 5-29d.

Fig. 5-29b. Applied Loads and Reactions

To determine the horizontal displacement δ_H, a horizontal force P is applied at C as shown in Fig. 5-29c.

Fig. 5-29c. Loads and Reactions due to a Horizontal Force at C

To determine the vertical displacement δ_V, a vertical force (P) is applied at C as shown in Fig. 5-29d.

Fig. 5-29d. Loads and Reactions to a Vertical Force at C

The energy table is constructed after selecting the sections and the appropriate x-coordinates that yield the simplest solution.

Member	X=0	EI	M	$\dfrac{\partial m\,\delta_H}{\partial P}$	$\dfrac{\partial m\,\delta_V}{\partial P}$	$\displaystyle\int_a^b$
AB	B	1	40	$\dfrac{\partial(Px)}{\partial P} = x$	$\dfrac{\partial(2P)}{\partial P} = 2$	$\displaystyle\int_0^5$
BC	C	1	20x	0	$\dfrac{\partial(Px)}{\partial P} = x$	$\displaystyle\int_0^2$

$$\delta_H = \frac{1}{EI}\left[\int_0^5 (40)(x)\,dx + \int_0^2 (20x)(0)\,dx\right] = \mathbf{0.00125\ m} \rightarrow$$

$$\delta_V = \frac{1}{EI}\left[\int_0^5 (40)(2)\,dx + \int_0^2 (20x)(x)\,dx\right] = \mathbf{0.00113\ m} \downarrow$$

Example: 5-24

For the truss shown in Fig. 5-30a, determine the vertical deflection (δ_{VB}) and the horizontal deflection (δ_{HB}).

$A_{BAR} = 400(10^{-6})$ m^2
$E = 200(10^6)$ kN/m^2

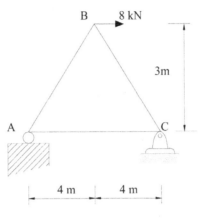

Fig. 5-30a

SOLUTION

The truss bar forces due to applied loads are shown in Fig. 5-30b. To determine the displacement (δ_V) at B a vertical unit load is applied at B as shown in Fig. 5-30c.

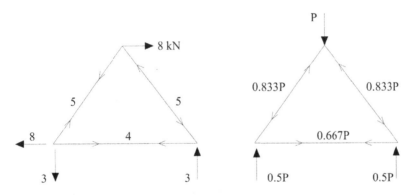

Fig. 5-30b. Bar Forces (F) due to   Fig. 5-30c.   Bar Forces (f_v) due to a
         Applied Loads   Vertical Force at B

To determine the horizontal displacement (δ_H) a horizontal unit load is applied at B as shown in Fig. 5-30d.

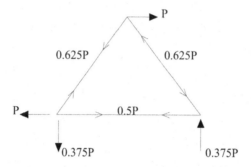

Fig. 5-30d. Bar Forces (f_h) due to Horizontal Force at B

The energy table is as follows (Tension (+), Compression (-))

Member	$\dfrac{L(m)}{A(m^2)}$	F	$\dfrac{\partial f_v}{\partial p}$	$F\left(\dfrac{\partial f_v}{\partial p}\right)\left(\dfrac{L}{A}\right)$	$\dfrac{\partial f_H}{\partial p}$	$F\left(\dfrac{\partial f_H}{\partial p}\right)\left(\dfrac{L}{A}\right)$
AB	12500	5	-0.833	-5206.5	0.625	39062.5
BC	12500	-5	-0.833	5206.5	-0.625	39062.5
CA	20000	4	0.667	53360	0.5	40000
$\sum$				53360		118125

The required displacements are obtained by applying virtual work equations.

$$\delta_V = \sum F\left(\frac{\partial f_v}{\partial p}\right)\left(\frac{L}{AE}\right)$$

$$\delta_{VB} = \frac{53360}{E}$$

$\delta_{VB} = $ **0.0002668 m** $\downarrow$

$$\delta_H = \sum F\left(\frac{\partial f_H}{\partial p}\right)\left(\frac{L}{AE}\right)$$

$$\delta_{HB} = \frac{118125}{E}$$

$\delta_{HB} = $ **0.00059 m** $\rightarrow$

Summary

- The maximum deflections (i.e., elastic movement) of beams and frames must be checked to ensure that structures are not excessively flexible. Large deflection of beams and frames can produce cracking of attached nonstructural elements (masonry, tiles windows, …) and excessive vibrations of floors and bridge decks under moving loads.

- The deflection of a beam or frame is a function of the bending moment M and the member's flexural stiffness, which is related member's moment of inertia / and modulus of elasticity E. Deflections due to shear are typically neglected unless members are very deep, shear stresses are high, and the shear modulus G is low.

- The following methods are used to find the deflection of determinate structures: 1) Double Integration Method, 2) Moment Area Method, 3) Conjugate Beam Method, and 4) Energy Methods.

- The Double Integration Method establishes equations for the beam slope and deflection by integrating the differential equations of the elastic curve (Eqs. 5-5 and 5-6). This method becomes cumbersome when loads vary in a complex manner.

- The Moment Area Method, which is described in Section 5-3, uses the M/EI diagram as a load to compute slopes and deflections at selected points along the beam's axis. The shear at any point is the slope, and the moment is the deflections. Points of maximum deflections occur where the shear is zero.

- The Conjugate Beam Method applies to members with a variety of boundary conditions. This method requires that actual supports be replaced by fictitious supports to impose boundary conditions that ensure that the values of shear and moment in the conjugate beam, loaded by the M/EI diagram, are equal at each point to the slope and deflection, respectively, of the real beam. Once equations for evaluating maximum deflections are established for a particular beam and loading, tables available in structural engineering reference books (see Table 9.1) supply all the important data required to analyze and design beams.

- The Virtual Work Method is used to compute a single component of deflection with each application of the method. To compute a component of deflection by the method of virtual work, we apply a dummy load to the structure at the point of, as well as in the direction of, the desired displacement. The force and its associated reactions are called the Q system. If a slope or angle change is required, the force is a moment. With the dummy load in place the actual loads —called the P system— are applied to the structure. The external virtual work W_Q is done by the dummy loads as they move through the real displacements produced by the P system. Simultaneously an equivalent quantity of virtual strain energy U_Q is stored in the structure. That is, $W_Q = U_Q$

- If a deflection has both vertical and horizontal components, two separate analyses by virtual work are required; the unit load is applied first in the vertical direction and then in the horizontal direction. The actual deflection is the vector sum of the two orthogonal components.

- The use of a unit load to establish a Q system is arbitrary. However, since deflections due to unit loads (called flexibility coefficients) are utilized in the analysis of indeterminate structures (see Chap. 11), use of unit loads is common practice among structural engineers.

Problems

1-   Determine δ_B using the integration method.

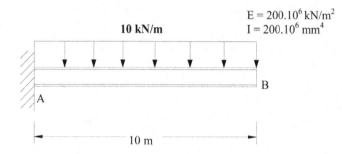

E = 200.10^6 kN/m^2
I = 200.10^6 mm^4

10 kN/m

A

B

10 m

2-   Determine δ_B , θ_B using the integration method

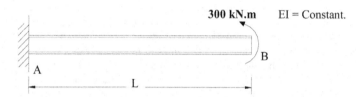

300 kN.m   EI = Constant.

A

B

L

3-   Determine δ_B using the integration method .

E = 200.10^6 kN/m^2
I = 200 10^6 mm^4

10 kN

A   B

200 kN.m

10 m

4-   Determine δ_B and θ_B using the integration method

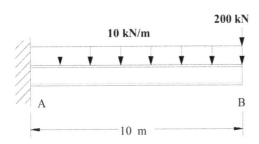

200 kN $E = 200.10^6 \, kN/m^2$

10 kN/m $I = 200.10^6 \, mm^4$

A   B

10 m

5-   Find the deflection (δ_B) at B using the moment area method.

90 kN $EI = 20000 \, kN.m^2$

A   B   C

3 m   6 m

6-   Determine θ_B and δ_B using the moment area method.

$E = 29000 \, ksi$ and $I = 120 \, in^4$

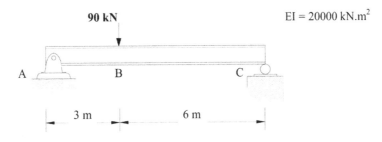

A   B

100 K.ft

15 ft

7-   Determine θ_C and δ_B using the moment area method.

E = 200.10^6 kN/m^2 and I = $100 .10^6$ mm^4

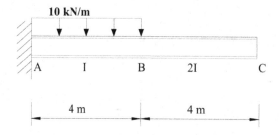

8-   Determine δ_B and δ_C using the moment area method.

EI = 60000 kN.m^2

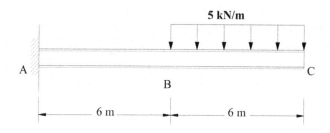

9-   Determine δ_B and θ_B using the moment area method.

EI = 140000 kN.m^2

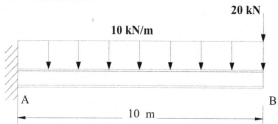

10- Determine θ_B and δ_C using the moment area method.

EI = 20000 kN.m^2

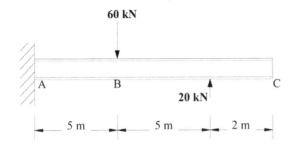

11- Determine θ_B and δ_C using the moment area method.

$I_1 = 6000$ in^4 and $I_2 = 2000$ in^4
E = 29000 ksi

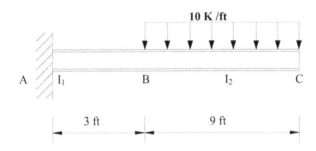

12- Determine θ_B and δ_C using the moment area method.
$I_1 = 3000$ in^4 and $I_2 = 1000$ in^4
E = 29000 ksi   150 K

Chapter 5   Deflections

13-   Determine θ_C and δ_D using the moment area method.

$I = 100.10^6 \, mm^4$
$E = 200.10^6 \, kN/m^2$

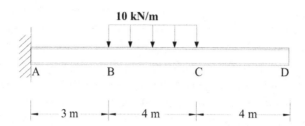

14-   Determine δ_B and δ_C using the moment area method.

$EI = 20000 \, kN.m^2$

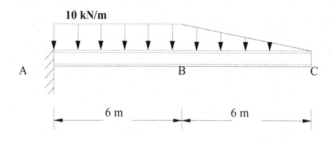

15-   Determine θ_B and δ_D using the moment area method.

$EI = 60000 \, kN.m^2$

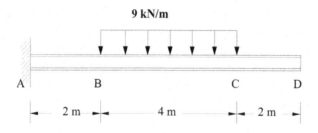

16-   Determine δ_B and δ_D using the moment area method.

$$EI = 20000 \text{ kN.m}^2$$

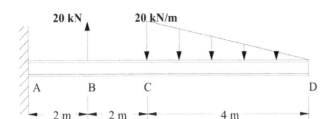

17-   Determine δ_B and θ_B using the moment area method.

$$EI = 60000 \text{ kN.m}^2$$

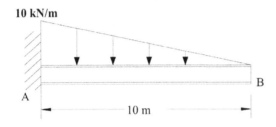

18-   Determine δ_A and θ_C using the moment area method.

$$E = 29.10^6 \text{ Ksi}$$
$$I = 1000 \text{ in}^4$$

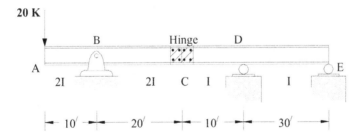

19-   Determine δ_A and θ_B using the moment area method.
$$I_1 = 800.10^6 \, mm^4$$
$$I_2 = 400.10^6 \, mm^4$$
$$E = 600.10^6 kN/m^2$$

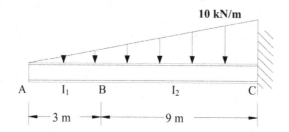

20-   Determine δ_B and θ_B using the conjugate beam method.
$$E = 40000 \, kN.m^2$$

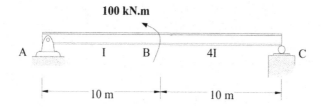

21-   Determine δ_C using the conjugate beam method.
$$EI = 20000 \, kN.m^2$$

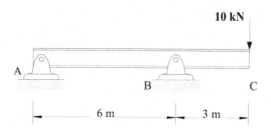

22- Determine δ_B and δ_D using the conjugate beam method.

E = 29000 Ksi
I = 120 in^4

100 K.ft

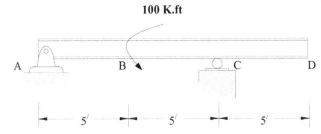

23- Determine δ_B and δ_D using the conjugate beam method.

E = 29000 Ksi
I = 120 in^4

200 K.ft

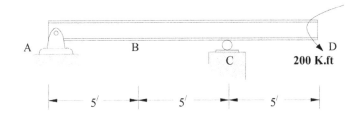

24- Determine δ_B and θ_B using the conjugate beam method.

E = 29000 Ksi
I = 120 in^4

10 K/ft

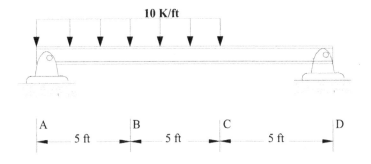

171

25-   Determine δ_A, δ_C and θ_E using the conjugate beam method.

E = 200.10^6 kN/m^2
I = 200.10^6 mm^4

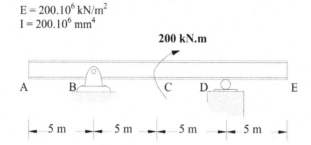

26-   Determine δ_B and δ_D using the conjugate beam method.

E = 29000 Ksi
I = 360 in^4

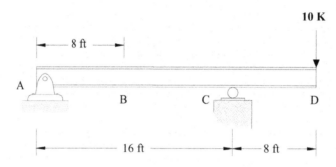

27-   Determine the δ_B, δ_D and θ_D using the conjugate beam method.
E = 200.10^6 kN/m^2
I = 350.10^6 mm^4

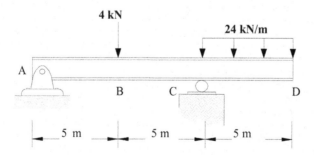

28-   Determine δ_C, δ_A and θ_E using the conjugate beam method
       EI = 20000 kN.m^2

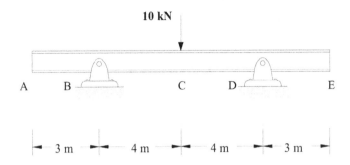

29-   Determine δ_B and θ_B using the conjugate beam method.
       EI = 120000 kN.m^2

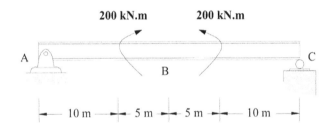

30-   Determine δ_B and θ_B using the conjugate beam method.
       EI = 40000 kN.m^2

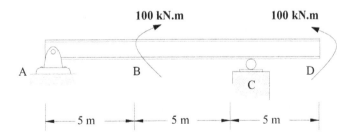

31-   Determine δ_B, δ_D and θ_D using the conjugate beam method.
       $E = 200.10^6$ kN/m^2
       $I = 400.10^6$ mm^4

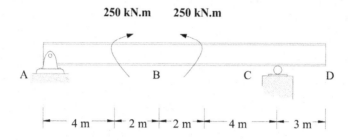

32-   Construct the energy table for θ_C and δ_C.
       EI = Constant.

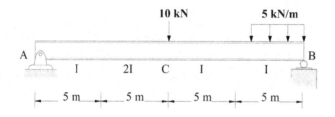

33-   Determine θ_D and δ_B using the energy method.
       $E = 200.10^6$ kN/m^2
       $I = 200.10^6$ mm^4

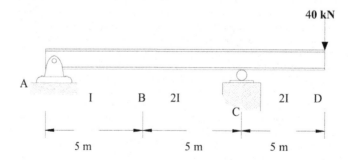

34-   Find the deflection and slope at point C using the energy method.
       EI = 20000kN.m^2

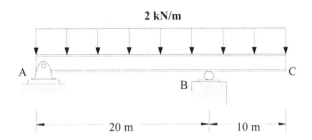

35-   Determine θ_D and δ_D using the virtual work method (Energy).
       E = 200.10^6 kN/m^2
       I = 400.10^6 mm^4

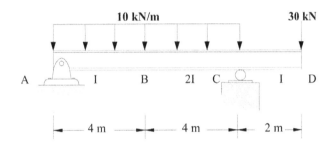

36-   Determine δ_D using the energy method.
       EI = 20000 kN.m^2

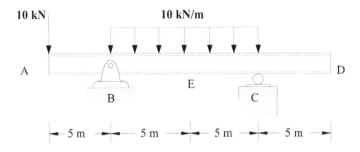

37-   Determine θ_D and δ_D using the virtual work method (Energy).
       $E = 200.10^6 \text{ kN/m}^2$
       $I = 400.10^6 \text{ mm}^4$

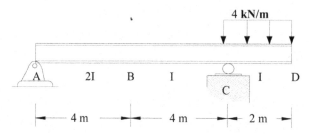

38-   Determine θ_B and δ_C using the virtual work method (Energy).
       $E = 30000 \text{ ksi}$
       $I = 9000 \text{ in}^4$

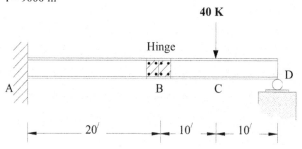

39-   Determine θ_C and δ_C using the virtual work method (Energy).
       $E = 200.10^6 \text{ kN/m}^2$
       $I = 80.10^6 \text{ mm}^4$
       $A_c = 150 \text{ mm}^2$

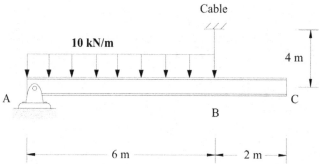

40-   Determine δ_V at C using the energy method.
        $E = 200.10^6 \text{ kN/m}^2$
        $I = 80.10^6 \text{ mm}^4$
        $A_c = 150 \text{ mm}^2$

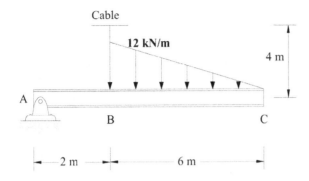

41-   Find the deflection at C .using the energy Method.
        $E = 200.10^6 \text{ kN/m}^2$
        $I = 80.10^6 \text{ mm}^4$
        $A_c = 150 \text{ mm}^2$

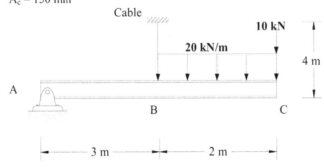

42-   Determine the vertical and horizontal deflections at B using the energy
        Method . E = 30000 ksi

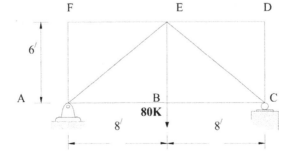

43-   Determine the vertical and horizontal deflections at B using the energy method.

$A_{Bar} = 400.10^{-6} \ m^2$

$E = 200.10^6 \ kN/m^2$

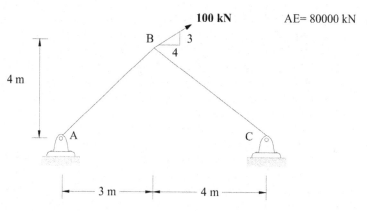

44-   Determine the displacements δ_H and δ_V at <u>B</u> using the energy method

100 kN   AE= 80000 kN

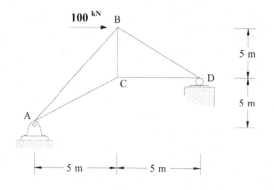

45-   Determine the vertical and horizontal deflections at C using the energy method

$L/A_{BAR} = 5000 \ m^{-1}$

$E = 200.10^6 \ kN/m^2$

178

46-   Determine the vertical deflection at B and the support motion at C using the
       energy method.

       E = 30000 Ksi
       L/A = 1

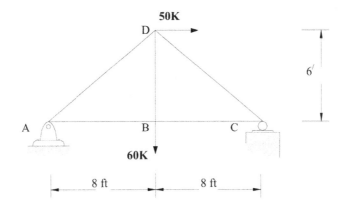

47-   Using the energy method, determine the following:
       1- δ_H, δ_V at F
       2- δ_H, δ_V at F due to rise in the temperature of 25^0c in bars <u>DE and EF.</u>

       A = 400 mm^2 for all bars
       E = 200.10^6 kN/m^2
       A$_t$ = 0.000012/^{0}C

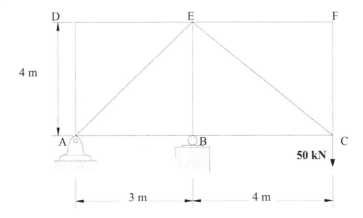

48-   Determine the vertical displacement δ_V at C using the energy method.
       EI = 20000 kN.m^2

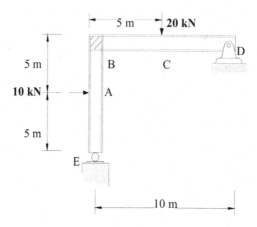

49-   Construct the energy table for the displacement δ_H, δ_V, δ_θ, at C.

                                                              EI = Constant.

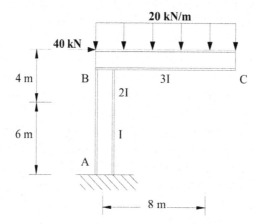

50-   Construct the energy table for the displacements δ_H, δ_V, θ at C.

EI = Constant

10 kN/m

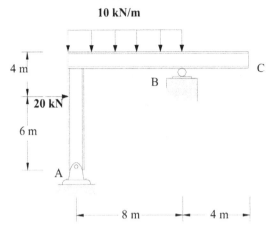

51- Construct the energy table for the displacements δ_H, δ_V, θ at C.

10 kN/m

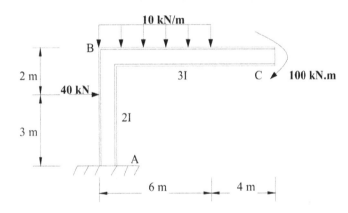

52-   Determine the vertical displacement δ_V at C using the energy method
       $E = 200.10^6 \, kN/m^2$
       $I = 80.10^6 \, mm^4$

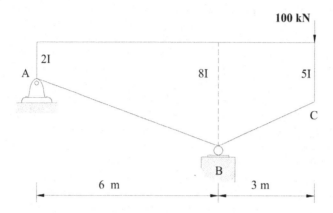

53-   1- Determine δ_B, θ_B using the double integration method.
       2- Is this beam adequate if the building code requires $\delta_{allowable} = L/480$?
       3- Design a square section for the cantilever beam if he building code requires
          an $h_{min} = L/10$. (L = cantilever span)

       $E = 200.10^6 \, kN/m^2$
       $I = 500.10^6 \, mm^4$

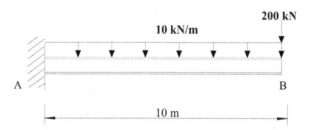

54-   Find the deflection and slope at point C by Castigliano's Method.
       $EI = 100000 kN.m^2$

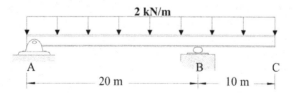

55-   Determine the vertical displacement at B and the support motion at C using
       Castigliano's method.
       E = 30000 ksi
       L/A = 1

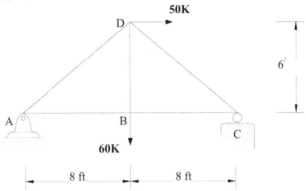

54-   Determine δ_H, δ_V, and θ at C using Castigliano's method
       EI = Constant.

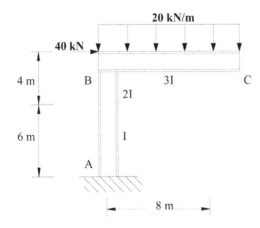

Chapter 5   Deflections

CHAPTER - 6
INFLUENCE LINES

Abu Dhabi-UAE

6.1   INTRODUCTION

The loads acting on structures are **dead** and **live** loads. Dead loads do not change their position during the structure service life, but live loads do, Fig. 6-1. As live loads change locations on a structure, the values of the reactions, shears, and bending moments also change. The critical (or design) load conditions are obtained by placing the live loads at the locations where they produce the maximum effects. Influence line diagrams are used to identify these critical load conditions.

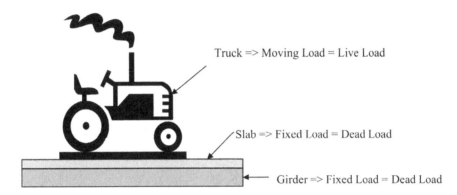

Truck => Moving Load = Live Load

Slab => Fixed Load = Dead Load

Girder => Fixed Load = Dead Load

Fig. 6-1   Dead and Live Loads

6.2   INFLUENCE LINE DIAGRAM

Influence line diagrams are used to determine the location of the live load as to produce maximum load effect functions (i.e., reaction, shear, or moment) and also to compute these load effect functions. They are constructed by placing a unit load at several locations on the structure and by computing the value of the desired load effect for each load position, Fig. 6-2.

The construction of the influence line diagram consists of the following steps:
1. Place a unit load at various locations and compute the values of the desired load effect function (i.e., reactions, shear, and moment) at a specified location.
2. For a concentrated load at a location, the actual value of the load effect function is equal to the product of the value of the load effect function obtained from the influence line diagram and the value of the concentrated load.

3. For a uniform load, the actual value of the load effect function is equal to the product of the area of the load effect function under the uniform load obtained from the influence line diagram and the intensity of the uniform load..

The following illustrative examples will employ this procedure of analysis.

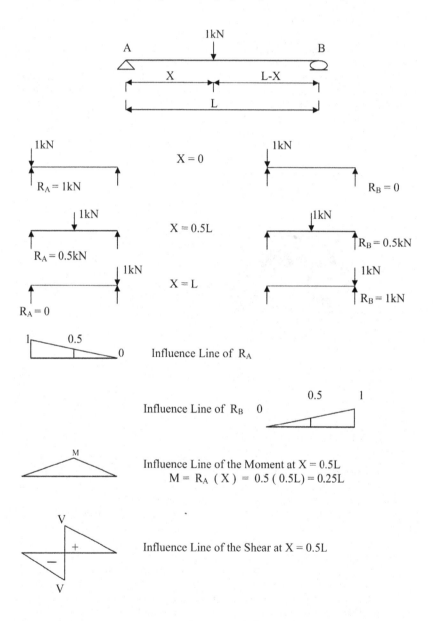

Fig. 6-2   Influence Lines Diagrams for Reaction, Moment, and Shear

Chapter 6.   Influence Lines

Example: 6-1

Construct the influence lines and determine M_C for the beam shown in Fig 6-3a.

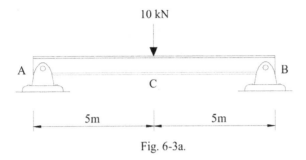

Fig. 6-3a.

SOLUTION

The influence lines for the reaction at A (R_A) and the moment at C (M_C) are shown in Fig. 6-3b and 6-3c, respectively.

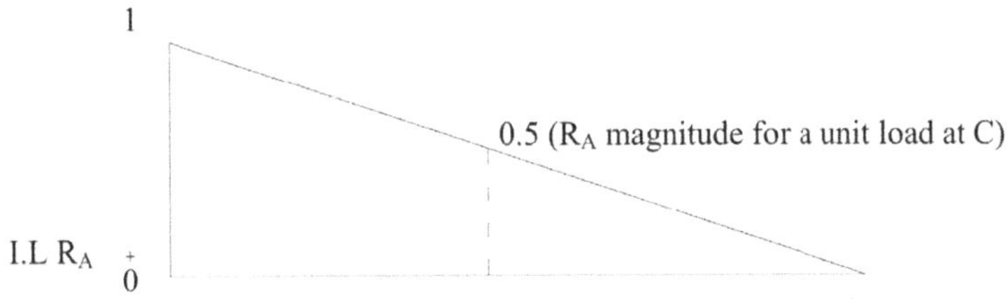

1

0.5 (R_A magnitude for a unit load at C)

I.L R_A $+\atop 0$

Fig. 6-3b. Influence line of the reaction at A (R_A)

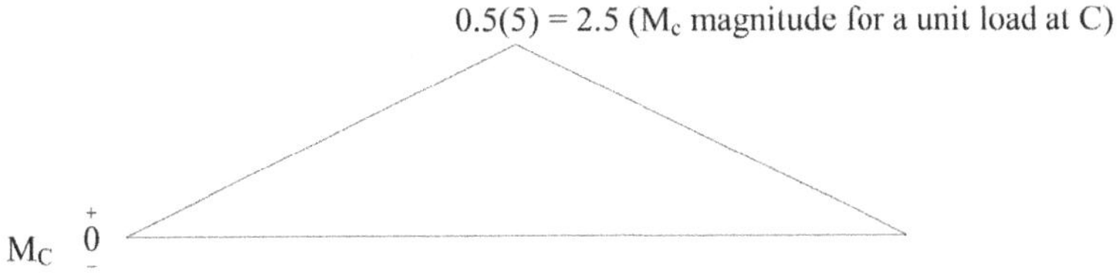

0.5(5) = 2.5 (M_c magnitude for a unit load at C)

M_C 0

Fig.6-3c. Influence lines of the moment at C (M_C)

$R_A = R_A$ magnitude for a unit load at C × actual load = 0.5(10) = **5 kN** ↑
$M_C = R_A$ magnitude for a unit load at C× distance from A to C × actual loads
$M_c = (0.5 \times 5)10 = $ **25 kN.m**

Example: 6-2

Construct the influence line diagram and determine the moment (M_{center}) for the beam shown in Fig. 6-4a.

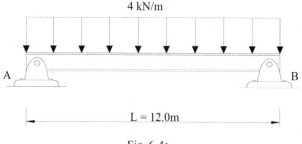

4 kN/m

A   B

L = 12.0m

Fig. 6-4a

SOLUTION

The influence line diagrams for the reaction at A (R_A) and the moment (M_C) are shown in Figs. 6-4b and 6-4c, respectively.

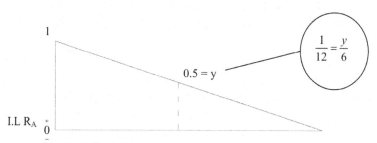

1

$$\frac{1}{12} = \frac{y}{6}$$

0.5 = y

I.L R_A 0

Fig.6-4b. Influence Line for Reaction at A (R_A)

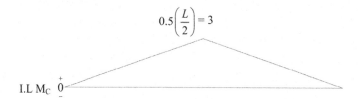

$$0.5\left(\frac{L}{2}\right) = 3$$

I.L M_C 0

Fig.6-4c. Influence Lines for Moment at C (M_C)

$R_A = W(A) = 4(1 \times 12 \times 0.5) = \textbf{24 kN} \uparrow$

Where W = uniform load intensity and A = area under the uniform load calculated from the influence line diagram of the reaction R_A.

$M_C = W(A) = 4 \left(0.5 \times \dfrac{L}{2} \times L \times 0.5\right) = \textbf{72 kN.m}$

Where W = uniform load intensity and A= area under the uniform load calculated from the influence line diagram of the moment M_c.

Examples

Example: 6-3

Construct the influence lines and determine $M_{C\,Max}$ for the beam shown in Fig 6-5a.

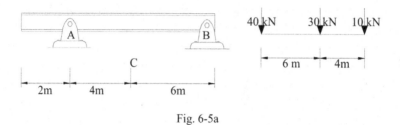

Fig. 6-5a

SOLUTION

The influence lines for the reaction at A (R_A) and the moment at C (M_C) for three load cases are shown in Figs. 6-5b, 6-5c, 6-5d, and 6-5e, respectively.

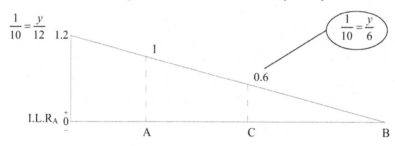

Fig.6-5b. Influence Line for Reaction at A (R_A)

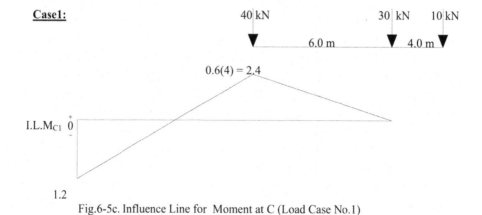

Fig.6-5c. Influence Line for Moment at C (Load Case No.1)

192

$M_{C1} = 40(2.4) =$ **96 kN.m**

Case 2:

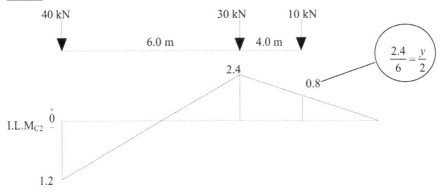

Fig.6-5d. Influence Line for Moment at C (Load Case No. 2)

$M_{C2} = -1.2(40) + 2.4(30) + 0.8(10) =$ **32 kN.m**

Case3:

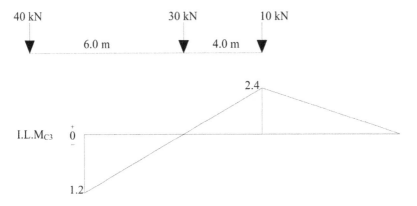

Fig. 6-5e. Influence Line for Moment at C (Load Case No. 3)

$M_{C3} = 10(2.4) =$ **24 kN.m**

The maximum moment at C is obtained in load case No. 1. Therefore,

$M_{C\,Max} =$ **96 kN.m**

Example: 6-4

Construct the influence lines and determine M_C, R_B, R_E for the beam shown in Fig 6-6a.

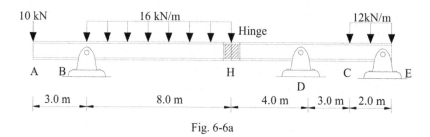

Fig. 6-6a

SOLUTION
The corresponding influence lines for the reaction at B (R_B), the reaction at E (R_E), and the moment at C (M_C) are shown in Figs. 6-6b, 6-6c, and 6-6d, respectively.

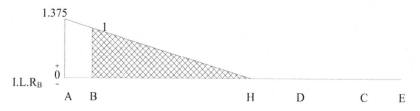

Fig.6-6b. Influence Line for Reaction at B (R_B)

Reaction at B:
$$R_B = P (1.375) + A_1(W_1) = 10(1.375) + (1 \times 8 \times 0.5)(16) = \mathbf{77.75\ kN} \uparrow$$

Where P = concentrated load at A, W_1 = uniform load of 16 kN/m along span BH, and A_1 = area under the uniform load of span BH

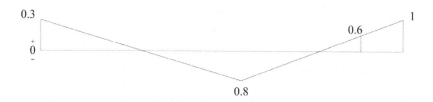

0.8

Fig.6-6c. Influence Line for Reaction at E (R_E)

Reaction at E:

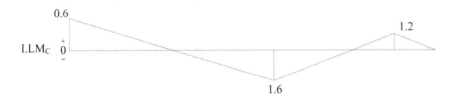

$$R_E = 10(0.3) - (16)(8 \times 0.8 \times 0.5) + 12\left[\left(\frac{0.6+1}{2}\right)2\right] = -29 = \textbf{29kN} \downarrow$$

Span BH — W_1 A_1

Span CE — W_2 A_2

Fig.6-6d. Influence Line for Moment at C (M_c)

Moment at C:

$$M_C = 10(0.6) - 16(1.6 \times 8 \times 0.5) + (1.2 \times 2 \times 0.5)(12) = -82\text{kN.m} = \textbf{82kN.m}$$

Example: 6-5

Construct the influence lines and determine M_D, R_B, R_E for the beam shown in Fig 6-7a.

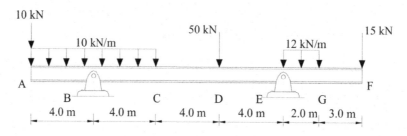

Fig. 6-7a

SOLUTION

The influence lines for the reaction at B (R_B), the reaction at E (R_E), and the moment at D (M_D) are shown in Fig. 6-7b, 6-7c, and 6-7d, respectively.

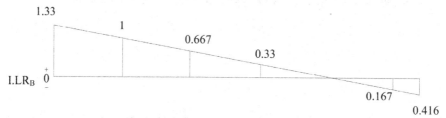

Fig.6-7b. Influence Line for Reaction at B (R_B)

Reaction at B:

$$R_B = 10(1.333) + 10\left[\left(\frac{1.333+0.667}{2}\right)8\right] + 50(0.333) - 12\left(0.167\times\frac{2}{2}\right) - 15(0.416)$$

$$R_B = \textbf{101.73kN} \uparrow$$

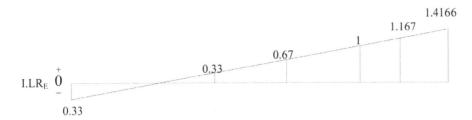

$$I.LR_E$$

Fig.6-7c. Influence Line for Reaction at E (R_E)

Reaction at E:

$$R_E = -10(0.333) - (10 \times \frac{0.333}{2}) + (10 \times \frac{0.333}{2}) + (50 \times 0.667) + \left(\frac{1.1667 + 1}{2}\right) \times 2 \times 12$$

$$+(15 \times 1.4166) = \textbf{77.27 kN} \uparrow$$

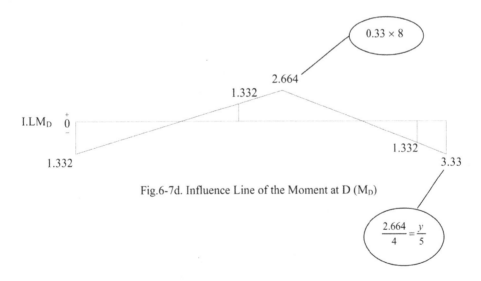

Fig.6-7d. Influence Line of the Moment at D (M_D)

Moment at D:

$$M_D = -10(1.332) + 2.664(50) - (12)(1.332 \times 2 \times 0.5) - (15 \times 3.33) = \textbf{54 kN.m}$$

Example: 6-6

Construct the influence lines and determine $M_{C\,Max}$ for the beam shown in Fig 6-8.a.

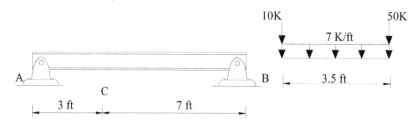

Fig. 6-8a

SOLUTION

The influence lines for the reaction at A (R_A) and the moment at C (M_C) for three load cases are shown in Fig. 6-8b, 6-8c, 6-8d, 6-8e, and 6-8f, respectively.

I.L.R_A

Fig. 6-8b. Influence Line for Reaction at A (R_A)

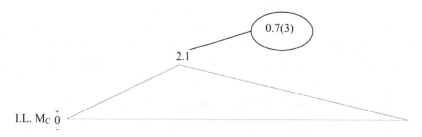

I.L. M_C

Fig. 6-8c. Influence Line for Moment at C (M_C)

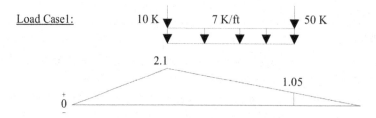

Load Case1:

Fig.6-8d. Influence Line for Moment at C (M_C) (Load Case No.1)

198

$$M_{C1} = 10(2.1) + 7\left[\frac{2.1+1.05}{2} \times 3.5\right] + (50 \times 1.05) = \mathbf{112.09 \ K.ft}$$

Load Case 2:

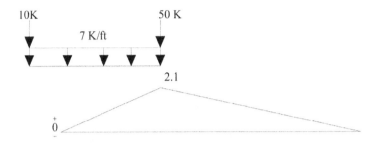

Fig. 6-8e. Influence Line for Moment at C (M_C) (Load Case No.2)

$$M_{C2} = (7 \times 2.1 \times 3 \times 0.5) + (50 \times 2.1) = \mathbf{127.05 \ K.ft}$$

Load Case3:

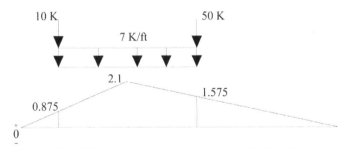

Fig. 6-8f.   Influence Line for Moment at C (M_C) (Load Case No.3)

$$M_{C3} = (10 \times 0.875) + 7\left(\frac{2.1+1.575}{2}\right)(1.75) + 7\left(\frac{2.1+0.875}{2}\right)(1.75) + (50 \times 1.575)$$

$$MC3 = \mathbf{128.23 \ K.ft}$$

The maximum moment at C is produced by Load Case no.3. Therefore,

$$\mathbf{M_{C \, Max} = 128.23 \ K.ft}$$

Summary

- Influence lines are used to establish where to position a moving load or a variable length of uniformly distributed live load on a structure to maximize the value of an internal force at a particular section of a beam, truss, or other type of structure.
- Influence lines are constructed for an internal force or a reaction at a particular point in a structure by evaluating the value of the force at the particular point as a unit load moves over the structure. The value of the internal force for each position of the unit load is plotted directly below the position of the unit load.
- Influence lines consist of a series of straight lines for determinate structures and curved lines for indeterminate structures.
- Quantitative influence lines can be most easily generated by a computer analysis in which a unit load is positioned at intervals of one-fifteenth to one-twentieth of the span of individual members. As an alternate to constructing influence lines, the designer can position the live load at successive positions along the span and use a computer analysis to establish the forces at critical sections. Influence lines for indeterminate structures are composed of curved lines.

Problems
....

1-   Determine M_C, R_A, and R_E using influence lines. Show all diagrams.

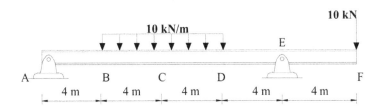

2-   Determine M_D, R_B, and R_E using influence lines. Show all diagrams.

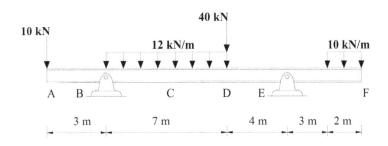

3-   Determine M_C, R_B, and R_D using influence lines. Show all diagrams.

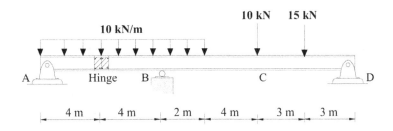

4-   Determine M_C and M_D using influence lines. Show all diagrams.

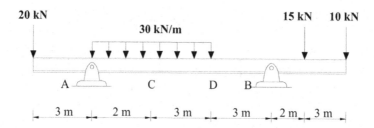

5-   Determine R_A and M_D using influence lines. Show all diagrams.

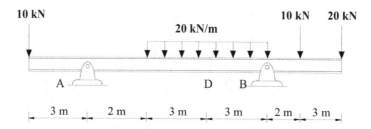

6-   Determine R_A, M_C, and R_B using influence lines. Show all diagrams.

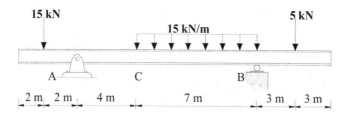

Chapter 6. Influence Lines

7-   Determine M_D using influence lines. Show all diagrams.

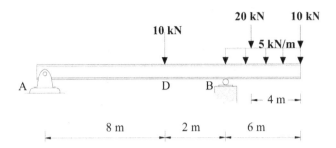

8-   Determine M_D, M_E, and M_B using influence lines. Show all diagrams.

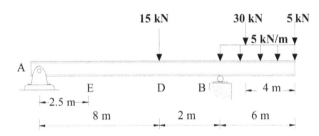

9-   Determine R_A, M_E, and M_D using influence lines. Show all diagrams.

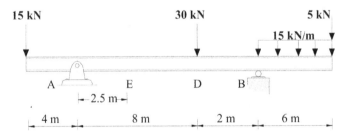

10-   Determine R_B, M_C, and R_D using influence lines. Show all diagrams.

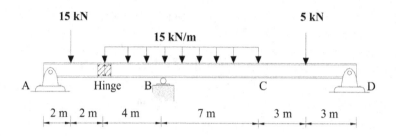

11-   Determine $M_{C\ MAX}$ using influence lines. Show all diagrams.

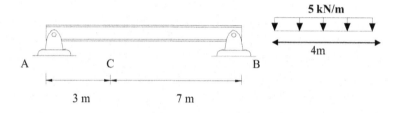

ARCHES AND CABLES

KSA – Madeena

Chapter 7.   Arches and Cables

7.1   ARCHES

Although arches are often used in scenic locations because of their attractive form, they also produce economical designs for long-span structures that (1) support large, uniformly distributed dead load and (2) provide a large unobstructed space under the arch (suitable for convention halls or sports arenas or in the case of a bridge providing clearance for tall ships), Fig. 7-1. Arched achieve their structural strength by developing axial compression stresses because of their curved shape.   That's why arches have been widely used in the past as structural elements with low tensile strengths and high compressive strengths, Fig. 7-2. Vertical loads including arch's own weight tend to push its supports outward. Therefore both supports must resist vertical and horizontal forces. Arches have a relatively small moment because of their horizontal reactions. Fig. 7-3 shows various types of arches which are as classified as:

1) Three-hinged arch with hinges at the supports and at the crown.
2) Two- hinged arch with hinges at the supports.
3) Fixed arches.

(a)

(b)

Fig. 7-1 Examples of Arches

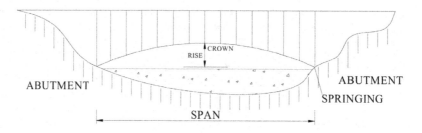

Figure 7-2a. Bridge Arch

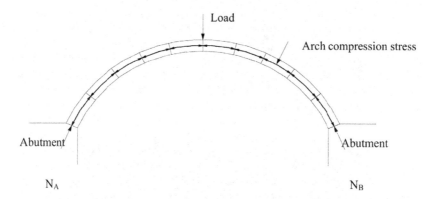

Figure 7-2b Stone Arch

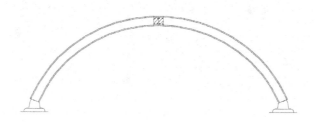

Figure 7-3a. Three-Hinged Arch

Figure 7-3b. Two-Hinged Arch

Figure 7-3c. Fixed Arch

Let us consider the section at the location D of the arch shown in Fig. 7-4a. The following terms are first defined:

1. V = shear at D.
2. H = horizontal thrust at D.
3. N = normal thrust at D.
4. S = radial shear at D.
5. M = bending moment at D.

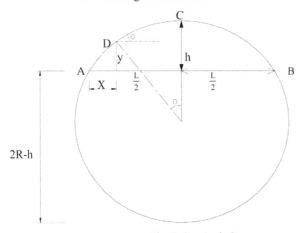

Fig. 7-4a. Arch Geometry

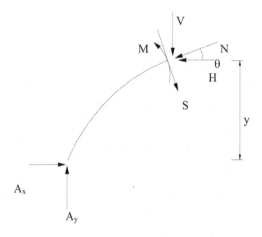

Figure 7-4b. Arch Internal Forces

The normal to the section makes an angle θ with the horizontal. Therefore, the normal thrust at any point along the arch is given by following equation:

$$N = V\,sin\theta + H\,cos\theta \qquad\qquad [7\text{-}1]$$

The radial shear at any point along the arch is given by the following equation:

$$S = V\,cos\theta - H\,sin\theta \qquad\qquad [7\text{-}2]$$

The moment at D can be determined by applying equilibrium equations to one part of the arch considering all forces including the reactions. The height at any point along a circular arch can be computed using the following equations:

$$R = \text{Radius} = \frac{L^2}{8h} + \frac{h}{2} \qquad\qquad [7\text{-}3]$$

$$y = h - R(1 - cos\theta) \qquad\qquad [7\text{-}4]$$

$$\theta = sin^{-1}\left[\frac{\dfrac{L}{2} - x}{R}\right] \qquad\qquad [7\text{-}5]$$

Example: 7-1

Determine the support reactions, the radial shear (S), the normal thrust (N), and the bending moment (M) at D for the circular arch shown Fig 7-5a.

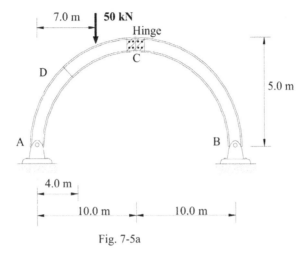

7.0 m | **50 kN**

5.0 m

4.0 m

10.0 m   10.0 m

Fig. 7-5a

SOLUTION

The equilibrium equations are used to determine the reactions of the arch as shown in Figs. 7-5b and 7-5c.

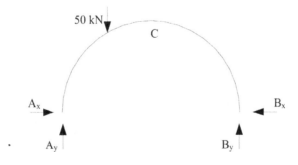

50 kN

A_x B_x

A_y B_y

Fig.7-5b. Arch ACB

Arch ACB

$$\sum M_B = 0 + \circlearrowleft \quad -50(13) + A_y(20) = 0 \Rightarrow A_y = 32.5 \text{ kN} \uparrow$$

$$\sum F_y = 0 \uparrow + \qquad -50 + 32.5 + B_y = 0 => B_y = 17.5 \text{ kN} \uparrow$$

Arch Segment AC

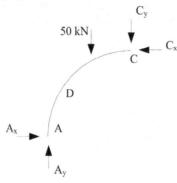

Fig.7-5c. Arch Segment AC

$$\sum M_C = 0 \; \circlearrowright + \qquad 32.5(10) - 50\,(3) - A_X\,(5) = 0 => A_x = 35 \text{ kN} \rightarrow$$

Arch ACB

$$\sum F_x = 0 \rightarrow + \qquad 35 - B_x = 0 \; => B_x = 35 \text{ kN} \leftarrow$$

The values of y_D and θ are first determined in order to compute the radial shear, the normal thrust, and the bending moment at D as shown in Figs. 7-5d and 7-5e.

Arch Segment AD

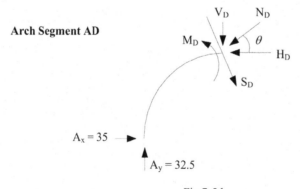

Fig.7-5d

$$\sum F_Y = 0 \uparrow + \; 32.5 - VD = 0 \qquad \Rightarrow V_D = 32.5 \text{ kN} \downarrow$$
$$\sum F_X = 0 \rightarrow + \; 35 - HD = 0 \qquad \Rightarrow H_D = 35 \text{ kN} \leftarrow$$

212

$$\text{R circle} = \frac{L^2}{8h} + \frac{h}{2} = \frac{20^2}{8(5)} + \frac{5}{2} = 12.5 \text{ m}$$

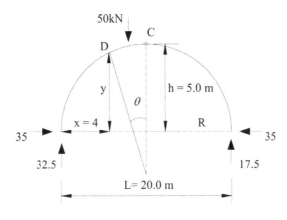

Fig.7-5e

$$\theta = \sin^{-1}\left[\frac{\frac{L}{2} - x}{R}\right] = \sin^{-1}\left[\frac{\frac{20}{2} - 4}{12.5}\right] = 28.685°$$

The normal thrust at D is computed using Equation 7-1 as follows.

$N_D = V_D \sin\theta + H \cos\theta = 32.5 (\sin28.685) + 35(\cos28.685) \Rightarrow \mathbf{N_D = 46.3 \ kN}$

The radial shear at D is computed using Equation 7-2 as follows.

$S_D = V_D \cos\theta - H \sin\theta = 32.5(\cos28.685) - 35(\sin28.685) \Rightarrow \mathbf{S_D = 11.71 \ kN}$

Since the arch is circular, y_D is determined using Equation 7-3

$y_D = h - R (1 - \cos\theta) = 5 - 12.5 (1 - \cos28.685) = 3.47 \text{ m}$

The moment at D is determined using the equilibrium equations as follows.

$\Sigma M_D = 0 \ \circlearrowleft + \qquad 32.5(4) - 35 (3.47) - M_D = 0 \Rightarrow \mathbf{M_D = 8.55 \ kN.m} \ \circlearrowleft$

Example: 7-2

Determine the support reactions, the radial shear (S), the normal thrust (N), and the bending moment (M) at D for the three-hinged parabolic arch shown in Fig 7-6a.

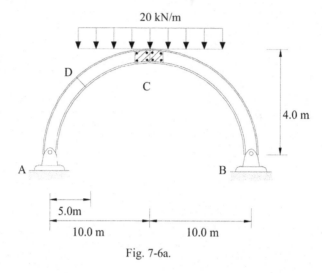

Fig. 7-6a.

SOLUTION

The equilibrium equations are used to determine the reactions as shown in Figs. 7-6b and 7-6c.

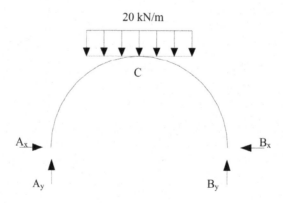

Fig.7-6b. Arch ACB

Arch ACB

$$\sum M_B = 0 + \circlearrowleft \qquad A_y = 200 \text{ kN} \uparrow$$

$$\sum F_y = 0 \uparrow + \qquad -400 + 200 + B_y = 0 \Rightarrow B_y = 200 \text{ kN} \uparrow$$

Arch Segment AC

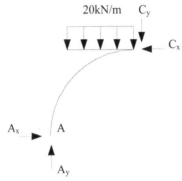

Fig.7-6c

$$\sum M_C = 0 \circlearrowleft + \qquad A_X = 250 \text{ kN} \rightarrow$$

$$\sum F_x = 0 \rightarrow + \qquad B_x = 250 \text{ kN} \leftarrow \qquad \textbf{Arch ACB}$$

The values of y_D and θ are first determined in order to compute the radial shear, the normal thrust, and the bending moment at D as shown in Figs. 7-6d and 7-6e.

Arch Segment AD

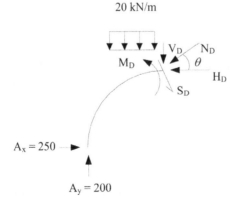

Fig.7-6d.

$\sum F_Y = 0 \uparrow +$ $V_D = 100$ kN $\downarrow$
$\sum F_X = 0 \rightarrow +$ $H_D = 250$ kN $\leftarrow$

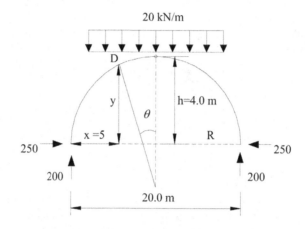

Fig.7-6e.

For a parabolic arch, y_D is given by the following equation:

$$y_D = \frac{4hx(L-x)}{L^2}$$ [7-6]

$$y_D = \frac{4(4)(5)(20-5)}{(20)^2} = 3.0 \text{ m}$$

For a parabolic arch, θ is given by the following equation:

$$\theta = \tan^{-1}\theta\left(\frac{dy}{dx}\right) = \tan^{-1}\theta\left[\frac{4h}{L^2}(L-2x)\right]$$ [7-7]

$$\theta = \tan^{-1}\theta\left[\frac{(4)(4)}{20^2}(20-2(5)\right]$$

$$\theta_D = 21.8^0$$

The normal thrust at D is computed using Equation 7-1 as follows.

$N_D = V_D \sin\theta + H_D \cos\theta = 100(\sin 21.8) + 250(\cos 21.8) \Rightarrow \mathbf{N_D = 269.258 \text{ kN}}$

The radial shear at D is computed using Equation 7-2 as follows.

$S_D = V_D \cos\theta - H_D \sin\theta = 100(\cos 21.8) - 250(\sin 21.8) \Rightarrow \mathbf{S_D = 0}$ (Why) ?

The moment at D is computed using the equilibrium equations as follows.

$\sum M_D = 0 \;\curvearrowright +$ $200(5) - 250(3) - 20(5)\left(\frac{5}{2}\right) - M_D = 0 \Rightarrow \mathbf{M_D = 0}$

Example: 7-3

Determine the support reactions, the radial shear (S), the normal thrust (N), and the bending moment (M) at D for the parabolic arch shown Fig 7-7a.

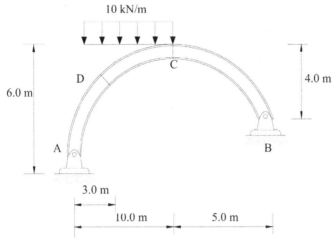

10 kN/m

C

D

4.0 m

6.0 m

A

B

3.0 m

10.0 m   5.0 m

Fig. 7-7a.

SOLUTION
The equations of equilibrium are used to determine the reactions as shown in Figs. 7-7b. and 7-7c.

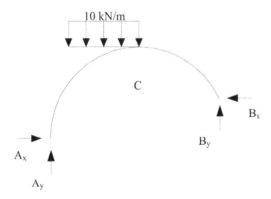

10 kN/m

C

B_x

B_y

A_x

A_y

Fig.7-7b

Arch ACB

$$\sum M_A = 0 \circlearrowleft \qquad -B_x(2) - B_y(15) + 100(5) = 0 \qquad [1]$$

Arch Segment AC

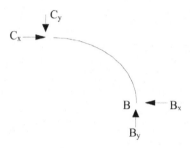

Fig.7-7c

$$\sum M_C = 0 \curvearrowright + \qquad B_X(4) - B_y(5) = 0 \qquad\qquad [2]$$

From [2] $\qquad\qquad B_x = \left(\dfrac{5}{4}\right) B_y \qquad\qquad\qquad [3]$

Solving Equations [1] and [2] simultaneously or substituting Equation [3] into [1], gives the following results:

$$B_x = 35.71 \text{ kN } \rightarrow$$
$$B_y = 28.57 \text{ kN } \uparrow$$

Arch ACB

$\sum F_Y = 0 \uparrow + \qquad A_y = 71.43 \text{ kN } \uparrow$
$\sum F_X = 0 \rightarrow + \qquad A_x = 35.71 \text{ kN } \leftarrow$

The values of y_D and θ are determined before computing the radial shear, the normal thrust, and the bending moment at D as shown in Figs. 7-7d. and 7-7e.
Since there is a difference in length, the arch segment AD can be treated as a parabolic of span $L = 2(10) = 20$ as shown in Figs. 7-7d and 7-7e.

10 kN/m

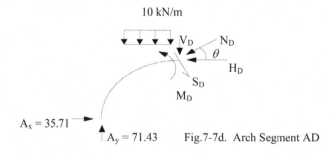

$A_x = 35.71$

$A_y = 71.43$ Fig.7-7d. Arch Segment AD

218

$F_Y = 0 \uparrow +$ $V_D = 41.43$ kN $\downarrow$
$F_X = 0 \rightarrow +$ $H_D = 35.71$ kN $\leftarrow$

For a parabolic arch $y_D = \dfrac{4h_1 x\,(L-x)}{L^2}$ [7-6]

$$y_D = \frac{(4)(6)(3)(20-3)}{20^2} = 3.06\text{m}$$

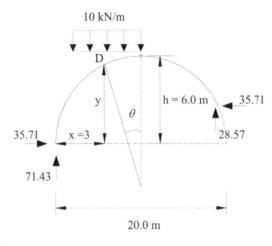

Fig.7-7e.

$$\theta = \tan^{-1}\theta\left(\frac{dy}{dx}\right) = \tan^{-1}\theta\left[\frac{4h}{L^2}(L-2x)\right] \qquad [7\text{-}7]$$

$$\theta = \tan^{-1}\theta\left[\frac{(4)(5)}{20^2}[20-2(3)]\right]$$

$$\theta_D = 40.030^0$$

The normal thrust at D is computed using Equation 7-1 as follows.

$N_D = V_D \sin\theta + H_D \cos\theta = 41.43\,(\sin 40.030) + 35.71(\cos 40.030) \Rightarrow \mathbf{N_D = 54\ kN}$

The radial shear at D is computed using Equation 7-2 as follows.

$S_D = V_D \cos\theta - H_D \sin\theta = 41.43\,(\cos 40.030) - 35.71\,(\sin 40.030) \Rightarrow \mathbf{S_D = 8.76\ kN}$

The moment at D is computed using equilibrium equations as follows.

$\sum M_D = 0 \;\circlearrowright\, + \quad 71.43(3) - 35.71\,(3.06) - 10(3)\left(\dfrac{3}{2}\right) - M_D = 0 \Rightarrow \mathbf{M_D = 60.02\ kN.m}\;\circlearrowleft$

7.2   CABLES

Cables,which are used in suspension bridges and transmission lines, come in the form of wires, ropes, chains, and many other forms. Because of their flexibility, cables develop only tension stresses and do not resist bending or compression stresses. The cable shape is adjusted to the applied vertical loads in such a manner that tension stresses are sufficient to withstand the loads.   Cables have funicular shapes (i.e., shapes consisting of linear segments) when subjected to concentrated loads (Fig. 7-8a). They have parabolic shapes when subjected to uniform loads along their horizontal span (Fig. 7-8b) and the catenary shapes when subjected to uniform loads along their length (7-8c).

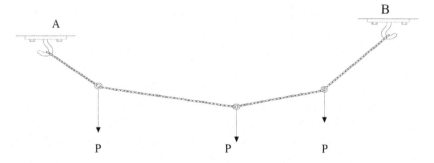

Fig. 7-8a. Cable with Concentrated Loads

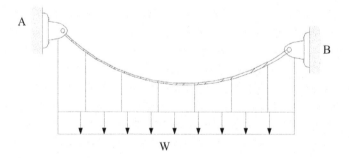

Fig. 7-8b. Cable with Uniform Load Along the Horizontal Span (Cable Ends at Different Levels)

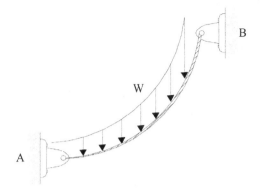

Fig.7-8c. Cable with Uniform Load Along its Length

Let us consider the cable segment shown in Fig. 7-8d. The following terms are first defined:

1. V = shear at D.
2. H = horizontal thrust at D.
3. T = cable tension at D.
4. M = moment at D (M= zerio)
5. θ = inclination angle at D.

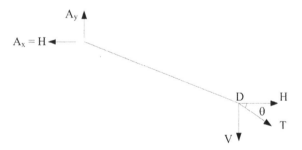

Figure 7-8d. Free Body Diagram of a Cable Segment

The horizontal thrust H is computed by applying the equations of equilibrium to the uniformly loaded cable along its horizontal span as follows.

$$H = \frac{wL^2}{8h} \quad \text{for cable ends at the same level} \qquad [7\text{-}8]$$

$$H = \frac{wL^2}{2(\sqrt{h_1} + \sqrt{h_2})^2} \quad \text{for cable ends at different levels} \qquad [7\text{-}9]$$

Chapter 7. Arches and Cables

The cable length L_c is computed based on the shape of the uniformly loaded cable along its horizontal span using the following equations:

For cable ends at the same level

$$L_c = L + \frac{8h^2}{3L}$$

[7-10]

For cable ends at different levels

$$L_c = L + \frac{2}{3}\frac{h_1^2}{l_1} + \frac{2}{3}\frac{h_2^2}{l_2}$$

[7-11]

The horizontal thrust of uniformly loaded cable along its horizontal span with both ends at the same level is affected by a temperature change as follows.

$$\Delta_h = \frac{3L^2\alpha\Delta t}{16h}$$

[7-12]

Where α = coefficient of thermal expansion, and Δ_T = change in temperature

The new horizontal thrust H_T is given by the following equation:

$$H_T = H + \Delta_H$$

[7-13]

Example: 7-4

Determine the reactions, the sag in the cable under the 90 kN concentrated load, and the maximum tension for the cable shown in Fig 7-9a.

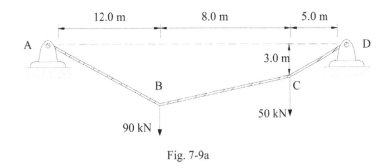

Fig. 7-9a

SOLUTION

The equilibrium equations are used to determine the reactions as shown in Fig. 7-9b.

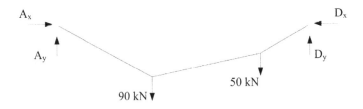

Fig.7-9b

$$\sum M_A = 0 + \circlearrowright \qquad 50(20) + 90(12) - D_y(25) = 0 \Rightarrow D_y = 83.2 \text{ kN} \uparrow$$

$$\sum F_Y = 0 \uparrow + \qquad 83.2 - 90 - 50 + A_y = 0 \Rightarrow A_y = 56.8 \text{ kN} \uparrow$$

The summation of the moments about C in segment CD is used to determine the horizontal reaction at D as shown in Fig.7-9c.

Segment CD

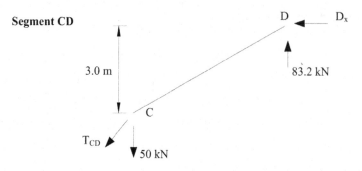

Fig.7-9c Segment CD

$$\Sigma\, M_C = 0 \;\circlearrowleft + \qquad -83.2(5) - D_X(3) = 0 \quad => \quad D_X = 138.67\ \text{kN} \rightarrow$$

The summation of forces in the cable horizontal direction is used to determine the horizontal reaction at A, as shown in Fig.7-9c.

$$\Sigma\, F_X = 0 \rightarrow + \quad A_X - 138.67 = 0 \Rightarrow A_X = 138.67\ \text{kN} \leftarrow$$

Segment AB

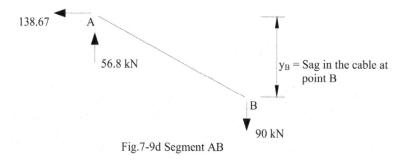

Fig.7-9d Segment AB

The summation of the moments about B in segment AB is used to determine the sag y_B under the 90kN concentrated load as shown in Fig.7-9d.

$$\Sigma\, M_B = 0 \;\circlearrowleft + \qquad 56.8(12) - 138.67(y_B) = 0 \quad => \quad y_B = 4.92\ \text{m}$$

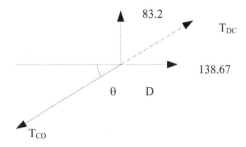

Fig.7-9e.

The maximum tension in the cable, which occurs at the supports where θ is maximum, is equal to the resultant of the reactions. Since $D_Y > A_y$, the maximum tension occurs at the support D as shown in Fig.7-9e.

$$T_{CD} = T_{MAX} = \sqrt{(83.2)^2 + (138.67)^2} = 161.7\,kN$$

The tension in the cable segment BC is computed by analyzing the cable segment DCB at point C as shown in Fig. 7-9f.

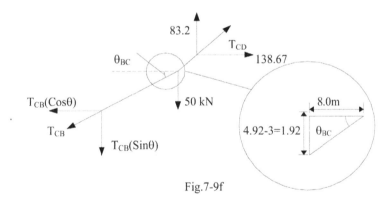

Fig.7-9f

$$\Sigma F_X = 0 \rightarrow + \quad T_{CB}\cos\theta - 138.67 = 0 \implies T_{CB} = 142.6\,kN$$

Note

The summation of the horizontal or vertical forces at point C in the segment DCB or at B in the segment ABC is used to determine the tension in cable BC. Likewise, the tension in CD can also be determined by analyzing point D.

Example: 7-5

Determine the reactions and the sag in the cable under at the 60 kN concentrated load as shown in Fig 7-10a.

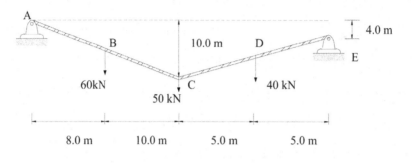

Fig. 7-10a

SOLUTION

The reactions of the unleveled cable structure are obtained by summing the moments about E and C as shown in Fig. 7-10b.

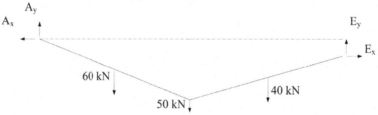

Fig.7-10b

$\Sigma\, M_E = 0 + \circlearrowright$ $-40(5)-50(10)-60(20)-A_x(4)+A_y(28) = 0$

$A_y(28) - A_x(4) - 1900 = 0$ [1]

$\Sigma\, M_C = 0 + \circlearrowleft$ $A_y(18) - A_x(10) - 60(10) = 0$ [2]

The values of A_y and A_x are obtained by solving equations [1] and [2] simultaneously as follows.

$A_y = 79.8$ kN   ↑

$A_x = 83.65$ kN   ←

The equations of equilibrium are used to determine the reactions of the unleveled cable structure as follows.

$\sum F_X = 0 \rightarrow +$ $E_X - 83.65 = 0$ $\Rightarrow$ $E_X = 83.65$ kN $\rightarrow$

$\sum F_Y = 0 \uparrow +$ $E_Y - 40 - 50 - 60 + 79.8 = 0 \Rightarrow E_Y = 70.2$ kN $\uparrow$

Summing the moments about B in the cable segment AB gives the sag at B (y_B) as shown in Fig.7-10c.

$\sum M_B = 0$ $\circlearrowleft +$ $-83.65(y_B + 1.143) + 79.8(8) = 0 \Rightarrow y_B = \mathbf{5.677m}$

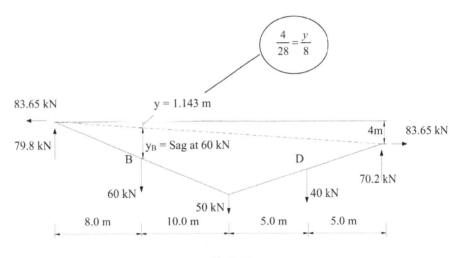

Fig.7-10c

Example: 7-6

Determine the reactions, the maximum tension and the length of the uniformly loaded cable shown in Fig.7-11a. What is the change in the maximum tension for a temperature rise (ΔT) of 20^0c?

α = Coefficient of thermal expansion = 12 x 10^{-6} ^0C.

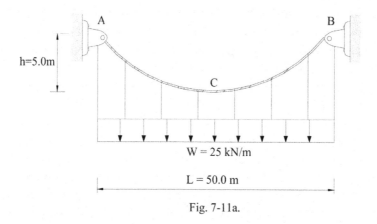

Fig. 7-11a.

SOLUTION

The equations of equilibrium as well as the symmetry are used as shown in Figs. 7-11b and 7-11c.

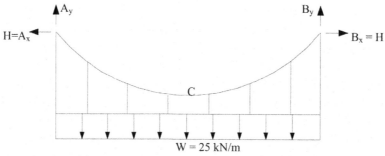

Fig.7-11b

$$A_Y = By = \frac{Wl}{2} = \frac{25(50)}{2} = 625 \text{ kN} \uparrow$$

Segment AC

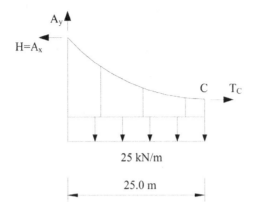

Fig. 7-11c

$$\sum M_C = 0 \quad \circlearrowright + \qquad H = \frac{Wl^2}{8h} = \frac{25(50)^2}{8(5)} = 1562.5 \, kN$$

$$T \, max = \sqrt{(625)^2 + (1562.5)^2} = 1682.9 \, kN$$

$$\text{Length of the cable} = L + \frac{8}{3}\frac{h^2}{l} = 50 + \frac{8}{3}\frac{(5^2)}{50}$$

$$L_C \quad = 51.3 \, m$$

The length increase of the cable, which is due to the temperature change, induces an increase in the vertical height (h) at the central point C.

$$\Delta_h = \frac{3L^2\alpha\Delta t}{16h} = \frac{3(25)^2(0.000012)(20)}{16(5)} = \mathbf{0.00563 \, m}$$

The change in the horizontal reaction (Δ_H) is given by

$$\Delta_H = \frac{3}{16}\left(\frac{l^2}{h^2}\right)\alpha\Delta T\,H = \frac{3}{16}(\frac{50^2}{5^2})(0.000012)(20)(1562.5) = 7 \, kN$$

The new horizontal reaction is given by:

$$H_T = H + \Delta_H = 1562.5 + 7 = 1569.5 \, kN$$

The new maximum tension is given by:

$$T^{'} = \sqrt{H_T^2 + Y_A^2} = \sqrt{1569.5^2 + 625^2} = \mathbf{1689.4 \, kN}$$

The change in the maximum cable tension is given by :

$$T^{''} = 1689.4 - 1682.9 = \mathbf{6.5 \, kN}$$

Example: 7-7

Determine the reactions, the maximum tension, the cable length, and the angle of inclination where the maximum tension occurs for the uniformly loaded cable shown in Fig.7-12a.

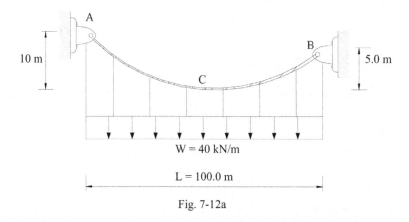

Fig. 7-12a

SOLUTION

At the lowest point C, the tensile force, which is horizontal, is equal to H. The equations of equilibrium are applied as shown in Fig. 7-12b.

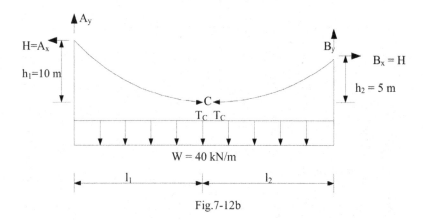

Fig.7-12b

Segment AC

$$\Sigma M_A = 0 \circlearrowleft + \qquad -H(10) + 40(\frac{l_1^2}{2}) = 0 \qquad\qquad [1]$$

Segment BC

$$\Sigma M_B = 0 \circlearrowleft + \qquad H(5) - 40(\frac{l_2^2}{2}) = 0 \qquad\qquad [2]$$

Since $l_1 + l_2 = 100$, Equation 2 can be re- written as

$$H(5) - 40\left(\frac{\left(100 - l_1^2\right)}{2}\right) = 0 \qquad\qquad [3]$$

Solving equations [1] and [3] simultaneously gives the following:

$H = 6862.9$ kN
$l_1 = 58.6$ m
$l_2 = 100 - l_1 = 41.4$m

Or using equation [7-9]

$$H = \frac{wl^2}{2\left(\sqrt{h_1} + \sqrt{h_2}\right)^2} \Rightarrow H = \frac{40(100)^2}{2\left(\sqrt{10} + \sqrt{5}\right)^2} = 6862.9 \text{ kN}$$

Segment AC

$$\Sigma F_Y = 0 \uparrow + \quad A_y - 40(58.6) = 0 \quad \Rightarrow A_y = 2344 \text{ kN} \uparrow$$

Segment BC

$$\Sigma F_Y = 0 \uparrow + \quad B_y - 40(41.4) = 0 \quad \Rightarrow B_y = 1656 \text{ kN} \uparrow$$

Since Ay > By $\Rightarrow T_{MAX} = \sqrt{A_y^2 + H^2} = \sqrt{(2344)^2 + (6862.9)^2} = 7252^{kN}$

Angle of inclination at A $\Rightarrow \theta_A = \cos^{-1} \frac{H}{T_{MAX}} = \cos^{-1} \frac{6862.9}{7252} = 18.85^0$

Length of the cable $= L + \frac{2}{3}\frac{h_1^2}{l_1} + \frac{2}{3}\frac{h_2^2}{l_2} = 100 + \frac{2}{3}\left[\frac{10^2}{58.6}\right] + \frac{2}{3}\left[\frac{5^2}{41.4}\right] = 101.54$ m

Example: 7-8

Determine the maximum tension and its cable slope as well as the bending moment at D (M_D) for the three hinged girder shown in Fig.7-13a.

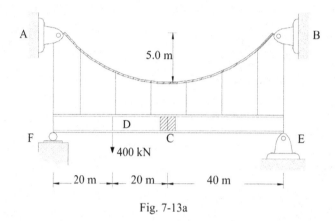

Fig. 7-13a

SOLUTION

The cable and the girder are connected by number of hinges. If they are closely spaced, the load on the cable and the girder could be taken as a uniform load (W_e), Fig. 7-13b.

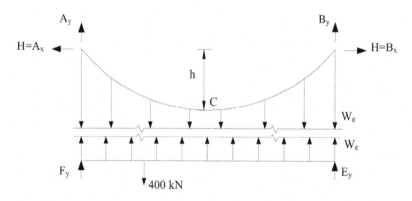

Fig.7-13b. Equivalent load W_e

The equations of equilibrium of the girder are used to determine the beam reactions as follows.

$\Sigma M_F = 0 +$ => $E_y = 100$ kN ↑

$\Sigma F_y = 0 ↑ +$ => $F_y = 300$ kN ↑

232

The summation of the moments about the hinge at C in the segment EC is used to determine the equivalent load.

$$\Sigma M_C = 0 + \circlearrowleft \quad \Rightarrow W_e(40)(\frac{40}{2}) - E_y(40) = 0$$

$$W_e(800) - 100(40) = 0 \Rightarrow W_e = 5 \text{ kN/m}$$

The moment at D (M_D) is equal to the summation of the moments due to actual loads and moment due to equivalent load (W_e) as shown in Figs.7-13c and 7-13d.

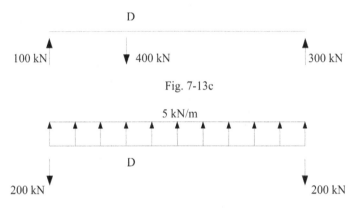

Fig. 7-13c

Fig. 7-13d

$$\Sigma M_D = 0 + \circlearrowleft \quad 100(20) - 200(20) + \frac{5(20)^2}{2} + M_D = 0 \quad \Rightarrow M_D = 1000 \text{ kN.m} \circlearrowleft$$

The equations of equilibrium in the cable are used to determine the remaining reactions.

$$A_y = \frac{W_e l}{2} = \frac{(5)(80)}{2} = 200 \text{ kN} \uparrow$$

$$\Sigma M_C = 0 + \circlearrowleft \quad H = \frac{W_e L^2}{8h}$$

$$H = \frac{5(80)^2}{8(5)} = 800 \text{ kN}$$

$$T_{MAX} = \sqrt{A_y^2 + H^2} = \sqrt{200^2 + 800^2} = 824.6 \text{ kN}$$

Slope of $T_{MAX} \Rightarrow \theta = \cos^{-1}\frac{H}{T_{MAX}} = \cos^{-1}\frac{800}{824.6} = 15.6^0$

Chapter 7. Arches and Cables

Summary
- Although short masonry arches are often used in scenic locations because of their attractive form, they also produce economical designs for long-span structures that (1) support large, uniformly distributed dead load and (2) provide a large unobstructed space under the arch (suitable for convention halls or sports arenas or in the case of a bridge providing clearance for tall ships).
- Arches are classified as : 1) three-hinged arch where the hinges are present at the supports and at the crown, 2) two-hinged arch where the hinges are present at the supports, and 3) fixed arches.
- The normal thrust and the radial shear at any point of an arch is given by Eqs. 7-1 and 7-2, respectively.
- The height of any point of circular and parabolic arches is given by Eqs. 7-3 and 7-4, respectively.
- For a given set of loads, the funicular shape of arch can be established using cable theory.
- Cables are used to construct long-span structures such as suspension and cable-stayed bridges, as well as roofs over large arenas (sports stadiums and exhibition halls) that require column-free space.
- Since they are flexible, cables can undergo large changes in geometry under moving loads; therefore, stabilizing elements must be provided to prevent excessive deformations. Also the supports at the of cables must be capable of anchoring large forces. If bedrock is not present for anchoring the ends of suspension bridge cables, massive abutments of reinforced concrete may be required.
- Because cables have no bending stiffness, the moment is zero at all sections along the cable. The general cable theorem establishes a simple equation to relate the horizontal thrust and the cable sag to the moment that develops in a fictitious, simply supported beam with the same span as the cable

Problems

■■

1-   Determine the reactions for the arch shown below.

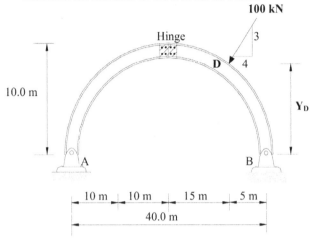

2-   Determine the bending moment at D (M_D).

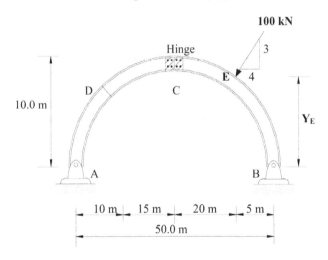

3-   Determine the reactions and draw the moment diagram.

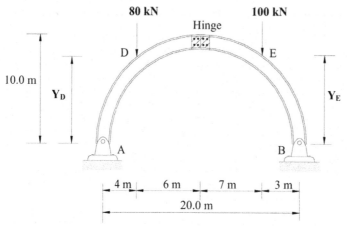

4-   Determine the reactions for the arch shown below.

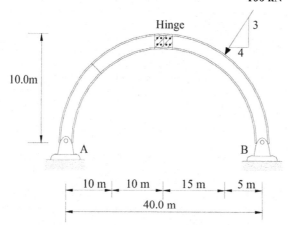

5-   Determine the maximum cable force.

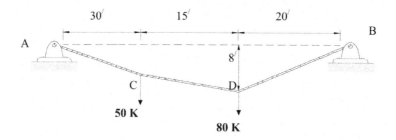

6-   Determine the reactions of the cable shown below and the sag at point B.

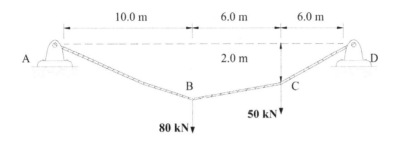

7-   Determine L_1, the diameter, and the length of the cable shown below.

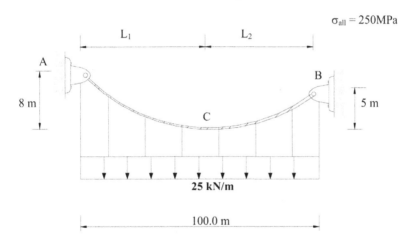

Chapter 7.   Arches and Cables

CHAPTER - 8
INDETERMINATE STRUCTURES

Baghdad -Iraq

Chapter 8. Indeterminate Structures

8.1   INTRODUCTION

Most structures are indeterminate. A structure is classified as indeterminate if the equations of equilibrium are insufficient to determine the reactions and the internal forces (shear, moment, axial, and torsion) of the structure. Continuous beams, trusses with double diagonals, and tall buildings are examples of indeterminate structures. The **advantages** of indeterminate structures are as follows.

1) Deflections and maximum stresses in indeterminate structures are smaller than those of its determinate counterparts.
2) Stresses in indeterminate structures are redistributed in the event of overloads (e.g., wind or earthquake loads).
3) Lateral forces and sideway effects can be better resisted by indeterminate structures.

Indeterminate structures have the following **disadvantages**:

1. The analysis and the design of indeterminate structures is more complex than the analysis and the design of determinate structures.
2. Differential settlements, temperature changes, and fabrication errors cause additional internal stresses in indeterminate structures.

The force and the displacement methods are commonly used for the analysis of indeterminate structures.

8.2   The Force Method

The force method is based on compatibility, which is based on the fact that deformed structural components must fit together in a way satisfying equilibrium, force – deformation relations, and boundary conditions. Compatibility conditions (deflection constraint conditions) are enforced throughout the structure by superposing a set of partial solutions that results in a set of equations with the forces as the unknowns and the flexibility quantities as the coefficients.

Redundant forces are those that can be removed from the structure without affecting the stability of the structure. Excess reaction components as well as excess internal force components are redundants.

The superposition principle, which state that " the effects of individual actions can be superimposed to determine the total effects on the structure" is used to determine the flexibility coefficients.   The force or flexibility method is referred to as the compatibility method **or the method of consistent deformation.**

8.3   The Displacement Method

The displacement method is based on joint equilibrium constraint conditions, which result in the solution of a set of equations where the joint deformations are the unknowns and the stiffness quantities as the coefficients. The displacement method is referred to as stiffness method or deformation method.

Summary

- Structures are mostly indeterminate structures, which have the following advantages:
    1. Smaller deflection and stresses than those of its determinate counterparts.
    2. Stress redistribution in the event of an over-load.
    3. Efficient lateral forces and side-sway effects resistance.
- Indeterminate structures also have the following disadvantages:
    1. Analysis and design complexity
    2. Additional internal stresses in an event of differential settlement, temperature change, and fabrication error.
- The analysis methods of indeterminate structures are as follows.
    1. The force method is based on deflection constraint conditions, which result in the solution of a set of equations where the forces (i.e., redundant reactions) are the unknowns and the flexibility quantities as the coefficients.
    2. The displacement method is based on joint equilibrium constraint conditions, which result in the solution of a set of equations where the joint deformations are the unknowns and the stiffness quantities as the coefficients.

CONSISTENT DEFORMATION METHOD

UAE - Abu Dhabi

Chapter 9.   Consistent Deformation Method

9.1   CONSISTENT DEFORMATION METHOD

The basic idea of solving statically indeterminate structures is to remove sufficient number of unknown forces such as reactions and internal members (**redundancies**) to create statically determinate structures (**primary structures**). At the location where the redundant is removed, the superposition principle is applied to determine the deflection of the primary structure due to the applied loads and to the redundant forces. The deflection constraint condition (compatibility condition) of the structure requires that the deflection at the desired location must be equal to the original displacement of the actual indeterminate structure at that location.   The redundant forces at that location are determined directly from the deflection equation.

The consistent deformation method consists of the following computational steps:

1) Determine the degree of indeterminacy of the structure as shown in Fig. 9-1a.

2) Remove enough reaction forces so that the remaining structure (primary structure) is statically determinate and geometrically stable as shown in Fig. 9-1b.

3) Compute the deformation caused by the actual loads on the primary structure. That is the vertical displacement at B $\left(\delta_B\right)$ as shown in Fig. 9-1c.

4) Apply the redundant force $\boldsymbol{B}_y$ on the primary structure after removing all the actual structure loads. The deformation at B (Δ_{BR}) due to the force $\boldsymbol{B}_y$ in the case of a linear elastic structure behavior is given by the following equation:

$$\Delta_{BR} = \delta_{BR}\boldsymbol{B}y \qquad\qquad [9\text{-}1]$$

Where δ_{BR} is the flexibility coefficient, Fig. 9-1d. The displacements at B $\left(\delta_B\right)$ and $\left(\delta_{BR}\right)$ can both be determined using the methods which were developed in Chapter 5.

5) The compatibility condition requires that the sum of the displacements calculated in steps 3 and 4 must be equal to the original displacement at point B $\left(\Delta_B\right)$ in the actual indeterminate structure. The deflection equation for the beam shown in Fig. 9-1a is given by the following equation:

$$\Delta_B = \boldsymbol{B}y\,\delta_{BR} - \delta_B = 0 \qquad\qquad [9\text{-}2]$$

The redundant force (**B**$_y$) is determined directly from the deflection equation (Eq. 9-2).

6) Use the equations of equilibrium to determine the remaining unknown forces. Another way for determining the remaining unknown forces is to apply the superposition principle

$$M_A = M_{A_B} + M_{A_{BR}} \qquad\qquad [9\text{-}3]$$

$$A_y = Ay_B + Ay_{BR} \qquad\qquad [9\text{-}4]$$

Where M_{A_B} and Ay_B are the moment and reaction in the primary structure due to actual loads, $M_{A_{BR}}$ and Ay_{BR} are the moment and reaction in the primary structure due to the force B_y only as shown in Figs. 9-1c and 9-1d.

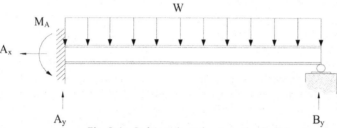

Fig. 9-1a. Indeterminate beam to the first degree

Fig. 9-1b. Primary structure

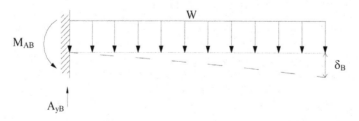

Fig. 9-1c. Primary structure deflected under actual loads

246

Fig. 9-1d. Primary structure deflected under redundant loads

The following examples demonstrate the approach of the method of consistent deformation using the moment area, conjugate beam, and virtual work method, respectively.

Example: 9-1

Draw the shear and moment diagrams for the indeterminate beam shown in Fig 9-2a.
Use both the consistent deformation method and the moment area method.

$E = 400.10^6 \text{ kN/m}^2, I = 200.10^6 \text{mm}^4$

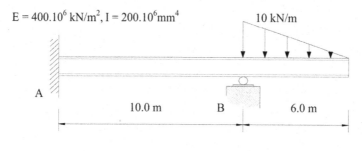

Fig. 9-2a

SOLUTION

The beam is statically indeterminate to the **first degree**. The super position principle
is applied to determine the flexibility coefficients by selecting the reaction (B_y) as
redundant (see Figs. 9-2b and 9-2c).

Deformation Caused by the Actual Loads

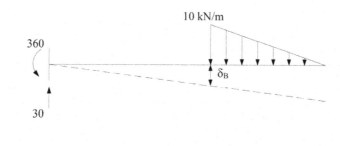

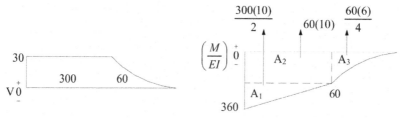

Fig. 9-2b

The Deformation caused by the actual loads on the primary structure is equal to:

$$\delta_B = \frac{1}{EI}\left[A_1\left(\frac{2}{3}\right)(10) + A_2(5) \right]$$

$$\delta_B = 13000 \frac{kN.m^3}{EI}$$

Deformation Caused by the Redundant Force B$_y$

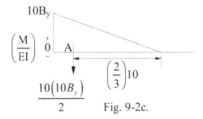

Fig. 9-2c.

The deformation, which is caused by the redundant force B$_y$, is equal to:

$$\Delta_{BR} = \delta_{BR}\, B_y = \frac{1}{EI}\left[A_1\left(\frac{2}{3}\right)(10) \right] B_y$$

$$\Delta_{BR} = 333.333 \;\frac{kN.m^3}{EI}\, B_y$$

The redundant force is computed by substituting the flexibility coefficient in the compatibility equation (Eq. 9-2).

$$\Delta_B = \delta_{BR} \ B_y \ - \delta_B$$

$$0 = B_y \ \frac{333.333}{EI} \ - \ \frac{13000}{EI}$$

$$B_y = \textbf{39 kN} \uparrow$$

The remaining reactions can be now determined using equations of equilibrium as shown in Fig. 9-2d.

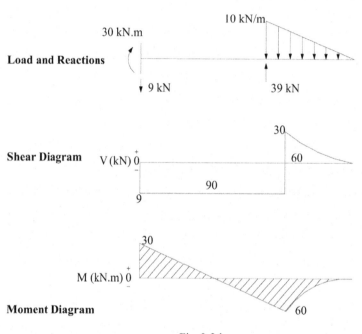

Fig. 9-2d

Example: 9-2

For the indeterminate beam shown in Fig. 9-3a, determine the following:
  a) The reactions. [$E = 200.10^6$ kN/m², $I = 100.10^6$ mm⁴]
  b) The reactions when the support B settles by 50 mm.
  c) The tension in a cable which replaces the support B.
     [$A_{cable} = 200$ mm², $L_{cable} = 10$ m, $E_{cable} = 200(10^6)$ kN/m²]
Use both the consistent deformation method and the moment area method.

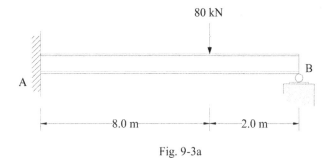

80 kN

A   B

—8.0 m—   —2.0 m—

Fig. 9-3a

SOLUTION

First Question:

The beam is statically indeterminate to the **first degree**. The super position principle
is applied to determine the flexibility coefficients. by selecting the reaction B_y as the
redundant, (see Figs. 9-3b and 9-3c).

Deformation Caused by the Actual Loads

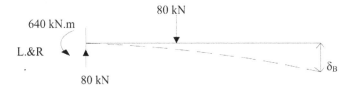

80 kN

640 kN.m

L.&R

80 kN

δ_B

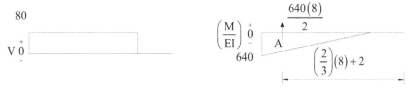

80

V 0

$\left(\dfrac{M}{EI}\right) \begin{matrix} + \\ 0 \\ - \end{matrix}$

640

$\dfrac{640(8)}{2}$

A

$\left(\dfrac{2}{3}\right)(8)+2$

Fig. 9-3b

251

The deformation, which is caused by the actual loads on the primary structure, is equal to:

$$\delta_B = \frac{1}{EI} \left[A \left(\frac{2}{3} \right) (8) + 2 \right]$$

$$\delta_B = 18773.333 \; \frac{kN.m^3}{EI}$$

Deformation Caused by the Redundant Force B_y

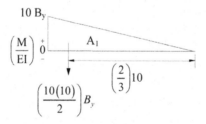

Fig. 9-3c

The deformation, which is caused by the redundant force B_y, is equal to:

$$\Delta_{BR} = \frac{1}{EI} \left[A_1 \left(\frac{2}{3} \right) (10) \right] B_y$$

$$\Delta_{BR} = \frac{333.333}{EI} \; B_y$$

The redundant force B_y is determined by substituting the flexibility coefficient in the compatibility equation (Eq. 9-2).

$$\Delta_B = \Delta_{BR} - \delta_B$$
$$0 = B_y \frac{333.333}{EI} - \frac{18773.333}{EI}$$
B_y = 56.32 kN $\uparrow$

The remaining reactions are determined by using the equilibrium equations as shown in Fig. 9-3d.

Fig. 9-3d. Applied Loads and Reactions

Second Question:

The support B settles by 50 mm. Therefore, Δ_B = 50 mm = 0.05 m.

The redundant force B_y is determined by using the deflection equation [Eq. 9-2]:

$$\Delta_B = \delta_B - \Delta_{BR}$$
$$0.05 = \frac{18773.333}{(EI = 20000)} - B_y \frac{333.333}{(EI = 20000)}$$
B_y = 53.32 kN $\uparrow$

The remaining reactions are determined by using the equilibrium equations as shown in Fig. 9-3e.

Fig. 9-3e

253

Third Question:

The support B is replaced by a cable as shown in Fig. 9-3f. The cable deflection is given by the following equation:

$$\Delta_{BC} = \frac{T\,L_c}{A_c\,E_c}$$

80 kN

Cable   10.0 m

Fig. 9-3f.

The redundant force T is determined using the compatibility equation (Eq. 9-2) as follows.

$$\Delta_{BC} + \Delta_{BR} = \delta_B$$

$$\frac{T\,(10)}{(200\times10^{-6})(200\times10^{6})} + \frac{333.33}{20000}T = \frac{18773.333}{20000}$$

T = 55.5 kN

Example: 9-3

Draw the shear and moment diagrams for the indeterminate beam shown in Fig. 9-4a.
Use both the consistent deformation and the conjugate beam methods.
[EI = Constant]

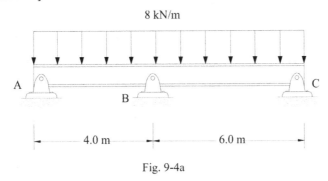

Fig. 9-4a

SOLUTION

B_y is selected as the redundant force as shown in Figs. 9-4b and 9-4c.

Deformation Caused by the Actual Loads

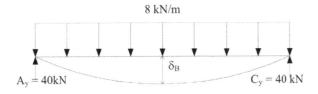

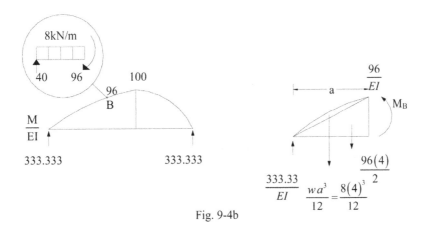

Fig. 9-4b

The Deformation caused by the actual load on the primary structure is equal to:

$$\delta_B = M_B = \frac{992}{EI}$$

The Deformation Caused by the Redundant Force B_y

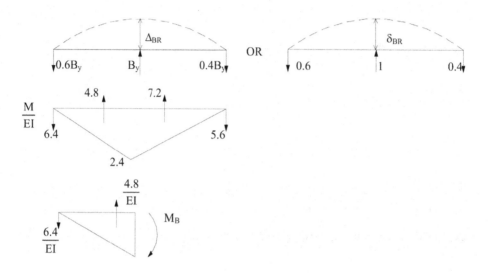

Fig. 9-4c

The deformation, which is caused by the redundant force B_y, is equal to:

$$\delta_{BR} = M_B = \frac{19.2}{EI}$$

$$\Delta_{BR} = \frac{19.2}{EI} B_y$$

The redundant B_y is determined using the compatibility equation (Eq. 9-2) as follows.

$$\Delta_B = \delta_B - B_y \; \delta_{BR}$$

$$0 = \frac{992}{EI} - B_y \; \frac{19.2}{EI}$$

$B_y = 51.67 \text{ kN} \uparrow$

The remaining equations are determined using the equilibrium equations as shown in Fig. 9-4d.

8 kN/m

Loads and Reactions

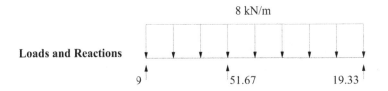

9   51.67   19.33

**Shear Diagram
kN**

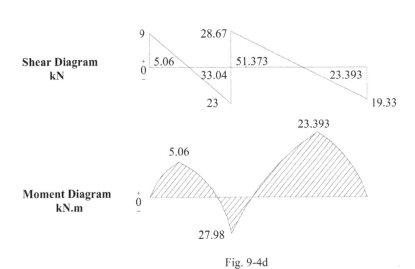

9   28.67

+ 5.06   51.373
0
–   33.04   23.393

23   19.33

23.393

5.06

**Moment Diagram
kN.m**

+
0
–

27.98

Fig. 9-4d

Example: 9-4

Draw the shear and moment diagrams for the indeterminate beam shown in Fig. 9-5a.
Use both the consistent deformation and the conjugate beam methods.
[EI = Constant]

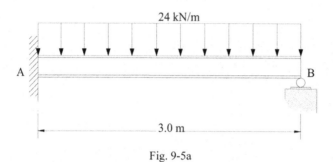

24 kN/m

A

B

3.0 m

Fig. 9-5a

SOLUTION

The beam is statically indeterminate to the first degree. The moment at A (M_A) is the
selected redundant as shown in Figs. 9-5b and 9-5c.

Rotation Caused by the Actual Loads

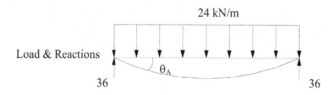

24 kN/m

Load & Reactions

θ_A

36   36

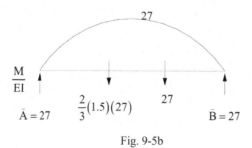

27

$\dfrac{M}{EI}$

$\bar{A} = 27$ $\dfrac{2}{3}(1.5)(27)$ 27 $\bar{B} = 27$

Fig. 9-5b

The rotation, which is caused by the actual loads on the primary structure, is equal to:

$$\theta_A = \bar{A_y} = \frac{27}{EI}$$

Rotation Caused by the Redundant Moment M$_A$

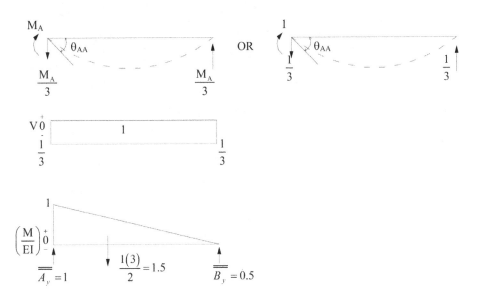

Fig. 9-5c

The rotation, which is caused by the redundant moment M$_A$, is equal to:

$$\theta_{AA} = \bar{\bar{A_y}} = \frac{1}{EI}$$

$$\theta_{AR} = M_A \theta_{AA} = \frac{M_A}{EI}$$

The redundant M$_A$ is determined using the compatibility equation [9-2a] as follows.

$$\theta_A = \bar{\theta}_A \pm M_A \theta_{AA}$$

Because it is positive (clockwise), the moment M$_A$ is given by the following equation:

$$\theta_A = \bar{\theta}_A + M_A \theta_{AA}$$

$$\theta_A = 0.0 = \frac{27}{EI} + \frac{M_A}{EI} \qquad\qquad (\theta_A = 0.0, \text{ Fixed-End Support})$$

M$_A$ = -27 kN.m = 27 kN.m

The remaining reactions are determined by using the equilibrium equations. The shear and moment diagrams are shown in Fig. 9-5d.

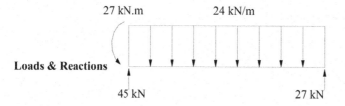

Loads & Reactions

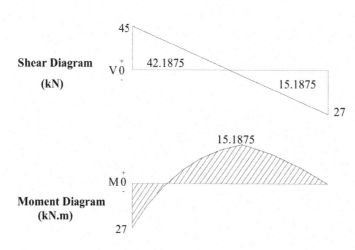

Shear Diagram

(kN)

Moment Diagram
(kN.m)

Fig. 9-5d

Example: 9-5

Determine the reactions and the spring settlement at B for the indeterminate beam shown in Fig 9-6a. Use both the consistent deformation and the conjugate beam methods.

$[E = 200.10^6 \text{ kN/m}^2, I = 200.10^6 \text{ mm}^4, K_{Spring} = 200 \text{ kN/m}]$

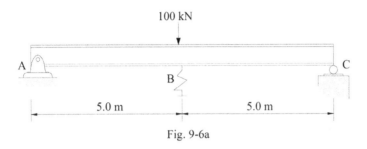

Fig. 9-6a

SOLUTION

The beam is statically indeterminate to the **first degree**. The spring force (B_y) is the selected redundant force as shown in Figs. 9-6b and 9-6c.

Deformation Caused by the Actual Loads

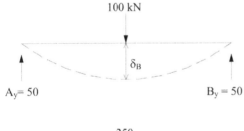

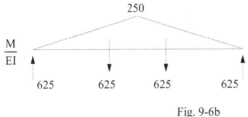

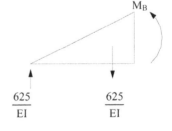

Fig. 9-6b

The deformation, which is caused by the actual loads on the primary structure, is

given by : $\delta_B = M_B = \dfrac{2083.333}{EI}$

Deformation Caused by the Redundant Force B$_y$

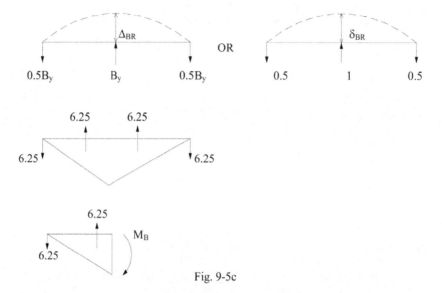

Fig. 9-5c

The deformation, which is caused by the redundant force B$_y$, is given by:

$$\delta_{BR} = M_B = \frac{20.83}{EI}$$

$$\Delta_{BR} = \frac{20.83}{EI} B_y$$

The total spring deflection at B under the force B$_y$ is equal to the sum of the beam deflection Δ_B and the spring displacement Δ_{SP}, which is in turn given by the following equation:

$$\Delta_{sp} = B_y \, \delta_{sp} = B_y \frac{1}{K_{sp}}$$

The redundant B$_y$ is determined using the compatibility equation (Eq. 9-2):

$$\Delta_B = \Delta_{BR} + \Delta_{sp} - \delta_B$$

$$0 = B_y \left[\delta_{BR} + \delta_{SP} \right] - \delta_B$$

$$0 = B_y \left[\frac{20.83}{EI} + \frac{1}{K_{sp}} \right] - \frac{2083.333}{EI} = B_y \left[\frac{20.83}{40000} + \frac{1}{200} \right] - \frac{2083.333}{EI}$$

B$_y$ = 9.434 kN ↑

The remaining reactions are determined using the equilibrium equations as shown in Fig. 9-6d

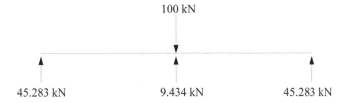

100 kN

45.283 kN   9.434 kN   45.283 kN

Fig. 9-6d. Applied Loads and Reactions

The settlement of the spring is equal to:

$$\Delta_{sp} = B_y \quad \delta_{sp} = B_y \quad \frac{1}{K_{sp}}$$

$$\Delta_{sp} = (9.434) \quad \frac{1}{200}$$

$\Delta_{SP} = \mathbf{0.0427\ m}$

The total displacement at B is given by the following equation

$$\Delta_B = B_y \left[\delta_{BR} + \delta_{sp}\right] = 9.434 \left[\frac{20.83}{40000} + \frac{1}{200}\right] = \mathbf{0.0521\ m} \downarrow$$

What do you think of this deflection? Is it large? And if it is , what to do about it?

Example: 9-6

Using the methods of consistent deformation and virtual work determine the reactions for the indeterminate frame shown in Fig. 9-7a.

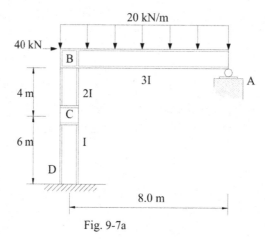

Fig. 9-7a

Hint: Because of the changes in the frame cross section, it is preferable to divide the frame into three parts.

SOLUTION

The frame is statically indeterminate to the **first degree**. The reaction A_y is the selected redundant force as shown in Figs. 9-7b, 9-7c, and 9-7d.

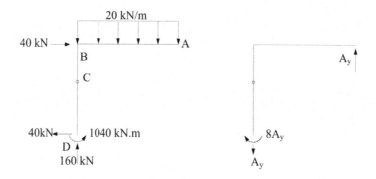

Fig. 9-7b. Applied Loads and Reactions   Fig. 9-7c. Redundant Force A_y

Chapter 9. Consistent Deformation Method

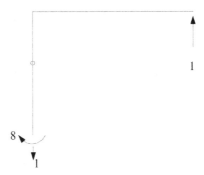

Fig. 9-7d. Unit Loads

Table 9-1 represents the energy table based on the redundant force A_y.

Table 9-1. **Energy Table**

member	EI	X = 0	M	M_R	$m_{\delta v}$	$\int_a^b$
DC	1	D	$40x - 1040$	$8\,A_y$	8	$\int_0^6$
CB	2	C	$40(6 + x) - 1040$	$8\,A_y$	8	$\int_0^4$
BA	3	A	$\dfrac{20\,x^2}{2} = 10x^2$	$A_y\,x$	x	$\int_0^8$

The displacement at A due to the actual loads is given by the following equation (See Fig. 9-7b):

$$\delta_A = \int \frac{M\,m_{\delta v}}{EI}\,dx = \left[\frac{1}{EI}\int_0^6 (40x - 1040)\,(8) + \frac{1}{2EI}\int_0^4 (40(6+x) - 1040)\,(8) + \frac{1}{3EI}\int_0^8 (10x^2)(-x)\right]$$

$$\delta_A = \frac{-59090}{EI}$$

The displacement at A due to the redundant forcee A_y is given by the following equation:

$$\Delta_{AR} = \int \frac{M_R\,m_{\delta V}}{EI} = \left[\frac{1}{EI}\int_0^6 (8\,A_y)(8) + \frac{1}{2EI}\int_0^4 (8\,A_y)\,(8) + \frac{1}{3EI}\int_0^8 (-A_y\,x)\,(-x)\right]$$

$$\Delta_{AR} = \frac{568.889}{EI}\,A_y$$

The redundant force A_y is determined using the compatibility equation (Eq. 9-2a) as follows.

$$\delta_A + \Delta_{AR} = 0$$

$$\frac{-59090}{EI} + \frac{568.889}{EI} A_y = 0$$

$A_y = 103.869$ kN ↑

The remaining reactions are computed using the equilibrium equations as shown in Fig. 9-7e.

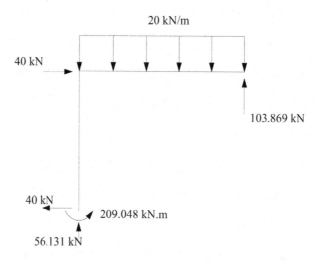

Fig. 9-7e. Applied Loads and Reactions

Example: 9-7

Using the methods of consistent deformation and virtual work, determine the bar forces and reactions for the truss shown in Fig. 9-8a. [AE=Constant]

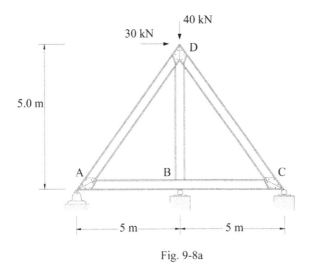

Fig. 9-8a

SOLUTION

The truss is externally indeterminate to the first degree. F_{BD} is the selected redundant force as shown in Figs. 9-8b and 9-8c.

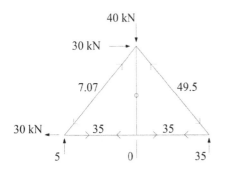

Fig. 9-8b. Bar forces F with actual loads and $F_{DB}=0$

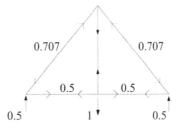

Fig. 9-8c. Bar forces f with F_{DB}=1kN

Table 9-2

member	F	f	L	FfL	f²L
AD	-7.07	-0.707	7.07	35.339	3.534
CD	-49.5	-0.707	7.07	247.425	3.534
AB	35	0.5	5	87.5	1.25
BC	35	0.5	5	87.5	1.25
BD	0	1	5	0	5
SUM				457.764	14.568

The redundant F_{BD} is determined using the compatibility equation as follows.

$$\delta_F + \delta_f F_{BD} = 0 \qquad [9\text{-}5]$$

$$\sum \frac{FfL}{AE} + \sum \frac{f^2 L}{AE} F_{BD} = 0 \qquad [9\text{-}6]$$

$$\frac{457.764}{AE} + \frac{14.568}{AE} F_{BD} = 0$$

$F_{BD} = -31.423$ kN $= 31.423$ kN $\underline{C}$

The remaining bar forces can be determined using either the joint method or the force equation [9-4a] (see Table 9-3).

$$F_i^* = F_i + R f_i \qquad [9\text{-}4a]$$

Where,

F_i^* = bar force;
F_i = initial bar force with redundant force = 0.0, Fig. 9-8b;
f_i = initial bar force with redundant force = 1.0, Fig. 9-8c;
R = redundant

Table 9-3

Member	F_i	$R = F_{BD}$	f_i	F_i^* (kN) + (Tension) - (Compression)
AD	-7.07	-31.423	-0.707	-7.07+(-31.423(-0.707)) = 15.146
CD	-49.5	-31.423	-0.707	-27.284
AB	35	-31.423	0.5	19.289
BC	35	-31.423	0.5	19.289
BD	0	-31.423	1	-31.423

Since the bar forces are known, then the reactions could easily be determined by applying principle of statics, Fig. 9-8d.

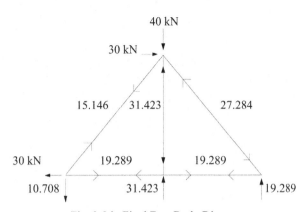

Fig. 9-8d. Final Free Body Diagram

Example: 9-8

Using the methods of consistent deformation and virtual work, determine the bar forces and reactions for the truss shown in Fig. 9-9a.

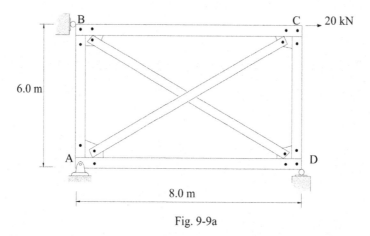

Fig. 9-9a

SOLUTION

The truss is indeterminate to the **first degree** internally and externally. This requires two compatibility equations with coupling terms to determine the redundants B_x and F_{AC} as shown in Figs. 9-9b, 9-9c, and 9-9d.

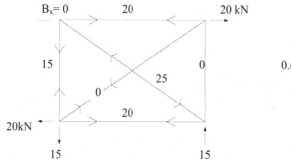

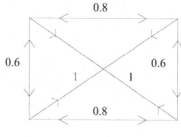

Fig. 9-9b. Bar forces F with actual loads   Fig. 9-9c. Bar forces f_1 with $F_{AC}=1$
and $F_{AC}=0$, and $B_x=0$ and $B_x = 0$

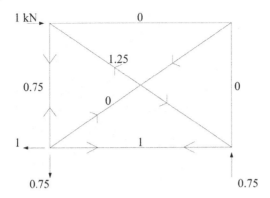

Fig.9-9d. Bar forces f_2 with $B_x = 1$ and $F_{AC}=0$

Table 9-4

member	F	f_1	L	$F f_1 L$	$f_1^2 L$	f_2	$F f_2 L$	$f_2^2 L$	$f_1 f_2 L^2$
AB	15	-0.6	6	-54	2.16	0.75	67.5	3.375	-2.7
BC	20	-0.8	8	-128	5.12	0	0	0	0
CD	0	-0.6	6	0	2.16	0	0	0	0
DA	20	-0.8	8	-128	5.12	1	160	8	-6.4
DB	-25	1	10	-250	10.0	-1.25	312.5	15.625	-12.5
AC	0	1	10	0	10.0	0	0	0	0
SUM				-560	34.56		540	27	-21.6

The compatibility equations are as follows.

$$\delta_{F_1} + \delta_{f_1}F_{AC} + \delta_{f_1 f_2}B_x = 0 \qquad\qquad [9\text{-}7]$$

$$\delta_{F_2} + \delta_{f_2 f_1}F_{AC} + \delta_{f_2}B_x = 0 \qquad\qquad [9\text{-}8]$$

The redundants are determined by solving equations [9-7] and [9-8] as follows.

$$-\frac{560}{AE} + \frac{34.56}{AE}F_{AC} + \frac{-21.6}{AE}B_x = 0 \qquad\qquad [9\text{-}7a]$$

$$\frac{540}{AE} + \frac{-21.6}{AE}F_{AC} + \frac{27}{AE}B_x = 0 \qquad\qquad [9\text{-}8a]$$

$F_{AC} = 7.407 \text{ kN} \qquad \underline{T}$

$B_x = -14.674 \text{ kN} = 14.674 \text{ kN} \leftarrow$

The remaining bar forces can be determined using either the principles of statics or the force equation [9-4b].

$$F_i^* = F_i + R_1 f_{1i} + R_2 f_{2i} \qquad\qquad [9\text{-}4b]$$

Table 9-5

Member	F (kN) Fig. 9-9b	f₁ (kN) Fig. 9-9c	$R_1 = F_{AC}$	f₂ (kN) Fig. 9-9d	$R_2 = B_x$	F* (kN) + Tension - Comp.
AB	15	-0.6	7.407	0.75	-14.074	0
BC	20	-0.8	7.407	0	-14.074	+14.074
CD	0	-0.6	7.407	0	-14.074	- 4.444
DA	20	-0.8	7.407	1	-14.074	0
DB	-25	1	7.407	-1.25	-14.074	0
AC	0	1	7.407	0	-14.074	+7.407

The reactions and bar forces are shown in Fig. 9-9e.

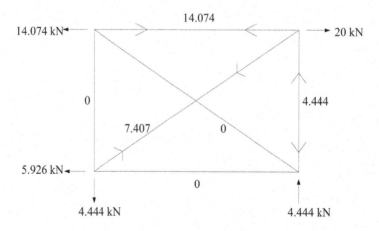

Fig. 9-8e. Reactions and Bar Forces

Example: 9-9

Using the methods of consistent deformation and conjugate beam, determine the
reactions for the beam shown in Fig 9-10a. [EI=Constant]

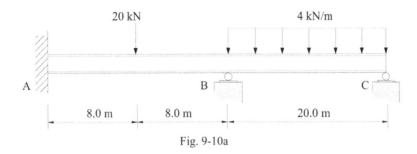

Fig. 9-10a

SOLUTION

The beam is statically indeterminate to the **2^{ed} degree**. This requires two
compatibility equations with coupling terms to determine the redundants B_y and C_y as
shown in Figs. 9-10b, 9-10c, 9-10d, and 9-10e.

Deformation Caused by the Concentrated Load

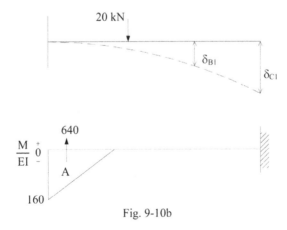

Fig. 9-10b

273

$$\delta_{B_1} = \frac{1}{EI} \left[A \left(\left(\frac{2}{3} \right) (8) + (8) \right) \right]$$

$$\delta_{c_1} = \frac{1}{EI} \left[A \left(\left(\frac{2}{3} \right) (8) + (28) \right) \right]$$

Deformation Caused by the Uniform Load

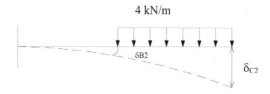

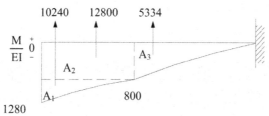

Fig. 9-10c

$$\delta_{B_2} = \frac{1}{EI} \left[A_1 \left(\frac{2}{3} \right) (16) + A_2 (8) \right]$$

$$\delta_{c_2} = \frac{1}{EI} \left[A_1 \left(\left(\frac{2}{3} \right) (16) + 20 \right) + A_2 (8+20) + A_3 \left(\frac{3}{4} (20) \right) \right]$$

Deformation caused by the redundant force B$_y$

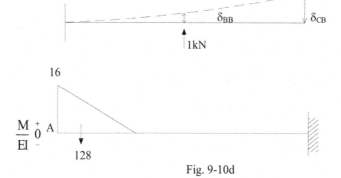

Fig. 9-10d

274

$$\delta_{BB} = \frac{1}{EI} \left[A \left(\frac{2}{3}\right) (16) \right]$$

$$\delta_{CB} = \frac{1}{EI} \left[A \left(\left(\frac{2}{3}\right) (16) + 20 \right) \right]$$

Deformation Caused by the Redundant C_y

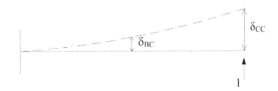

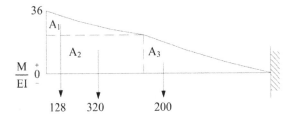

Fig. 9-10e

$$\delta_{CC} = \frac{1}{EI} \left[A_1 \left(\left(\frac{2}{3}\right)16 + 20 \right) + A_2 \ (8+20) + A_3 \left(\frac{2}{3}(20)\right) \right]$$

$$\delta_{BC} = \frac{1}{EI} \left[A_1 \left(\frac{2}{3}\right) (16) + A_2 \ (8) \right]$$

The compatibility equations are as follows.

$$\delta_{BB} \ B_y + \ \delta_{BC} \ C_y \ - \ \left(\delta_{B_1} + \ \delta_{B_2}\right) = 0 \qquad\qquad [9\text{-}6]$$

$$\delta_{CB} \ B_y + \ \delta_{CC} \ C_y \ - \ \left(\delta_{C_1} + \ \delta_{C_2}\right) = 0 \qquad\qquad [9\text{-}7]$$

The redundants are determined by simultaneously solving the two equations:

$$\frac{1363}{EI} \ B_y \ + \ \frac{3925}{EI}C_y \ - \ \frac{220160}{EI} = 0$$

$$\frac{3925}{EI} \ B_y \ + \ \frac{15552}{EI}C_y \ - \ \frac{773770}{EI} = 0$$

275

B$_y$ = 66.4 kN ↑
C$_y$ = 33 kN ↑

The remaining reactions are determined by using the equilibrium equations as shown in Fig. 9-10f.

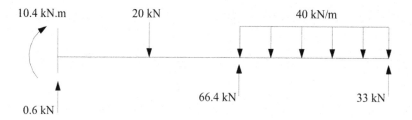

Fig. 9-10f. Actual Loads and Reactions

Example: 9-10

Using the methods of consistent deformation and virtual work, determine the bar force in member BD (F_{BD}) for the truss shown in Fig 9-11a when:-
   a) The temperature of member BD goes up by 20^0c ($\alpha = 12(10^{-6})$ / ^0c).
   b) The member BD is 2 mm (δ_L) longer than the required length due to fabrication error.

[A= 3000 mm^2, E = 200.10^6 kN/m^2, A = 0.003 m^2]

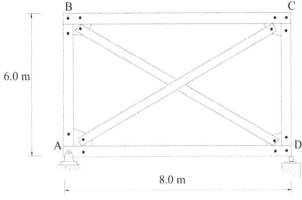

B

C

6.0 m

A

D

8.0 m

Fig. 9-11a

SOLUTION

The bar force F_{BD} is the selected redundant as shown in Fig. 9-11b.

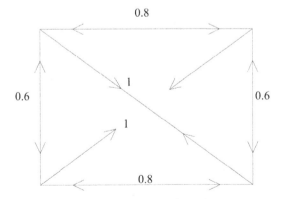

0.8

0.6

1

0.6

1

0.8

Fig. 9-11b. Bar Forces with F_{BD} =1 kN

The energy table, which is shown in Table 6, is set to determine the flexibility coefficients.

Table 9-6. Energy Table

member	f	L	f²L	(Answers to part a) *F$_i$ = (F$_1$ + R f$_i$) (kN)	(Answers to part b) (F$_i$ = 0) no external load
AB	-0.6	6	2.16	-25	-20.8
BC	-0.8	8	5.12	-33.36	-27.76
CD	-0.6	6	2.16	-25	-20.8
DA	-0.8	8	5.12	-33.36	-27.76
DB	1	10	10	41.7	34.7
AC	1	10	10	41.7	34.7
SUM			34.56		

(a) First Question:

The redundant F$_{BD}$ is determined for temperature effects from the compatibility equation::

$$\delta_{Temp} = \delta_f \ F_{BD}$$

$$L \ \alpha \ \Delta T \ - \ \sum \frac{f^2}{A} \frac{L}{E} \ F_{BS} = 0 \qquad\qquad [9\text{-}9]$$

$$(10)(12 \times 10^6)(20) \ - \ \frac{34.56}{600000} \ F_{BD} = 0$$

$$F_{BD} = 41.7 \text{ kN} \ \underline{T}$$

(b) Second Question:

Using the bar forces in Fig. 9-11b and setting the compatibility equation for fabrication error, the redundant force F$_{BD}$ is obtained as follows.

$$\delta_L = \delta_f \ F_{BD}$$

$$0.002 \ - \ \sum \frac{f^2}{A} \frac{L}{E} \ F_{BD} = 0 \qquad\qquad [9\text{-}10]$$

$$F_{BD} = 34.7 \text{ kN} \ \underline{T}$$

The bar forces, which are obtained for parts (a) and (b), are listed in the last two columns of Table 9-6. They are also shown in Fig. 9-11c and 9-11d, respectively.

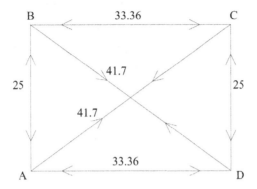

Fig. 9-11c. Bar Forces due to Temperature Effects

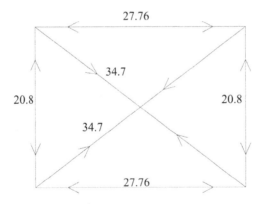

Fig. 9-11d. Bar forces due to Fabrication Error

9.2 THREE-MOMENT EQUATION

The three-moment equation method is generally used for continuous beams subjected to applied loads and / or support settlements. It is based on the compatibility condition that the slope at the middle support is the same for the two adjacent spans. This condition leads to an equation in term of the moment at three adjacent supports. For continuous beams with more than three spans, the equation is repeated until the number of equations is sufficient to determine the unknown moments at the supports.

The consistent deformation equation for the continuous beam shown in Fig. 9-12. is as follows.

$$\theta_{CA} + \theta_{CB} = 0$$

$$(\theta P_{CA} + \theta W_{CA} + \theta M_{CA}) + (\theta P_{CB} + \theta W_{CB} + \theta M_{CB}) = 0 \qquad [9\text{-}11]$$

The rotations at the center, which are obtained by applying the conjugate beam method and by using super position principle, are as follows.

$$\theta_{CA} = \frac{M_L L_L}{6 EI_L} + \frac{M_{Ce} L_L}{3 EI_L} + \frac{A_L \bar{x}_L}{EI_L L_L}$$

$$[9\text{-}12]$$

$$\theta_{CB} = \frac{M_{CR} L_R}{6 EI_R} - \frac{M_R L_R}{6 EI_R} - \frac{A_R \bar{x}_R}{EI_R L_R}$$

The general form of the moment equation is therefore equal to:

$$M_L \frac{L_L}{I_L} + 2 M_{Ce} \left(\frac{L_L}{I_L} + \frac{L_R}{I_R} \right) + M_R \left(\frac{L_R}{I_R} \right) = -\sum \frac{6 A_L \bar{x}_L}{I_L L_L} - \sum \frac{6 A_R \bar{x}_R}{I_R L_R} \qquad [9\text{-}13]$$

For the special case where $I_L = I_R$, Equation [9-13] becomes as follows.

$$M_L L_L + 2 M_{Ce} \left(L_L + L_R \right) + M_R L_R = -\sum \frac{6 A_L \bar{x}_L}{L_L} - \sum \frac{6 A_R \bar{x}_R}{L_R} \qquad [9\text{-}14]$$

For the special case where $I_L = I_R$ and $L_R = L_L$, Equation [9-13] becomes as follows.

$$M_L + 4 M_{Ce} + M_R = -\frac{6}{L^2} \sum A \bar{x} \qquad [9\text{-}15]$$

If the actual loads are included, the general form of the moment equation becomes:

$$M_L \frac{L_L}{I_L} + 2 M_{Ce} \left(\frac{L_L}{I_L} + \frac{L_R}{I_R} \right) + M_R \left(\frac{L_R}{I_R} \right) =$$

$$-\sum \frac{P_L L_L^2}{I_L} \left(z_L - z_L^3 \right) - \sum \frac{P_R L_R^2}{I_R} \left(z_R - z_R^3 \right) - \frac{W_L L_L^3}{4 I_L} - \frac{W_R L_R^3}{4 I_R} \qquad \text{[9-16]}$$

For the special case where, $I_L = I_R$, Equation [9-16] becomes as follows.

$$M_L L_L + 2 M_{Ce} \left(L_L + L_R \right) + M_R L_R =$$

$$-\sum P_L L_L^2 \left(z_L - z_L^3 \right) - \sum P_R L_R^2 \left(z_R - z_R^3 \right) - \frac{W_L L_L^3}{4} - \frac{W_R L_R^3}{4} \qquad \text{[9-17]}$$

If the rotations values tabulated in Table 9-7 are used, the moment equation becomes as follows.

$$M_L \frac{L_L}{I_L} + 2 M_{Ce} \left(\frac{L_L}{I_L} + \frac{L_R}{I_R} \right) + M_R \left(\frac{L_R}{I_R} \right) = -6E \left(\sum \theta_{Lj}'' + \sum \theta_{Rj}' \right) \qquad \text{[9-18]}$$

For the special case where $I_L = I_R$, Equation [9-18] becomes as follows.

$$M_L L_L + 2 M_{Ce} \left(L_L + L_R \right) + M_R L_R = -6E \left(\sum \theta_{Lj}'' + \sum \theta_{Rj}' \right) \qquad \text{[9-19]}$$

For the special case where $I_L = I_R$ and $L_R = L_L$

$$M_L + 4 M_{Ce} + M_R = -\frac{6EI}{L} \left(\sum \theta_{Lj}'' + \sum \theta_{Rj}' \right) \qquad \text{[9-20]}$$

If support settlements are considered, the three-moment equation becomes as follows.

$$M_L \frac{L_L}{I_L} + 2 M_{Ce} \left(\frac{L_L}{I_L} + \frac{L_R}{I_R} \right) + M_R \left(\frac{L_R}{I_R} \right) =$$

$$-\sum \frac{6 A_L \bar{x}_L}{I_L L_L} - \sum \frac{6 A_R \bar{x}_R}{I_R L_R} + 6E \left(\frac{\delta_L}{L_L} + \delta_{Ce} \left(\frac{1}{L_L} + \frac{1}{L_R} \right) + \frac{\delta_R}{L_R} \right) \qquad \text{[9-21]}$$

For the special case where $I_L = I_R$, Equation [9-21] becomes as follows.

$$M_L L_L + 2 M_{Ce} \left(L_L + L_R \right) + M_R L_R =$$

$$-\sum \frac{6 A_L \bar{x}_L}{L_L} - \sum \frac{6 A_R \bar{x}_R}{L_R} + 6E \left(\frac{\delta_L}{L_L} + \delta_{Ce} \left(\frac{1}{L_L} + \frac{1}{L_R} \right) + \frac{\delta_R}{L_R} \right) \qquad \text{[9-22]}$$

Where,

P_L, W_L, P_R, W_R	Applied loads of left and right spans
M_L, M_{Ce}, M_R	Internal Moments at left, center, and right supports
I_L, I_R	Moment of intertia of left and right spans
L_L, L_R	Length of left and right spans
z_L, z_R	Span length fraction that locate concntrated load
J	Support number
$\theta_{Lj}^{\prime\prime}, \theta_{Rj}^{\prime}$	Rotations at the left and right of support j
$\delta_L, \delta_{Ce}, \delta_R$	Movement of left, right, and center.

Note:

Substituting actual loads or tabulated values of rotation will change equations 9-21 and 9-22 to have the same first two terms of the right half as in equations 9-16 through 9-20. the following examples illustrate how to use these equations and their technique.

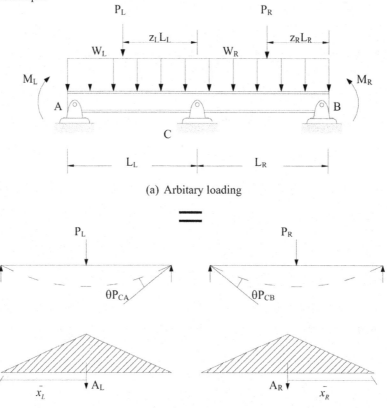

(a) Arbitary loading

(b) Beam with Applied P_{Loads} and Bending Moment Diagrams.

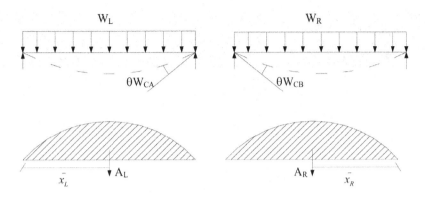

(c) Beam with applied W_{Loads} and Bending Moment Diagrams

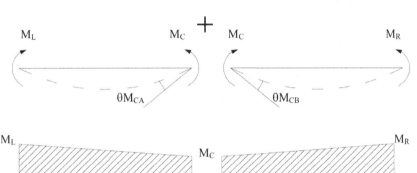

(d) Beam with Applied End Moment and Bending Moment Diagrams.

Fig. 9-12. Two-Span Continuous Beam

Table 9-7 Deflections and End Rotations

Case	θ'' (left support)	θ' (Right support)	δ_{max}
	$\dfrac{wL^3}{24EI}$	$-\dfrac{wL^3}{24EI}$	$\dfrac{5wL^4}{384EI}$
	$\dfrac{PL^2}{16EI}$	$-\dfrac{PL^2}{16EI}$	$\dfrac{PL^3}{48EI}$
	$\dfrac{Pa}{6EIL}(L-a)(2L-a)$	$-\dfrac{Pa}{6EIL}\left(L^2-a^2\right)$	$\dfrac{Pa^3b^3}{3EIL}$
	$\dfrac{7wL^4}{360EI}$	$-\dfrac{8wL^4}{360EI}$	$\dfrac{5wL^5}{768EI}$
	$\dfrac{5wL^3}{192EI}$	$-\dfrac{5wL^3}{192EI}$	$\dfrac{wL^4}{120EI}$

Example: 9-11

Determine the moment $M_{Central}$ for the continuous beam shown in Fig 9-13a using Equations 9-15, 9-17, and 9-20.
[EI = Constant]

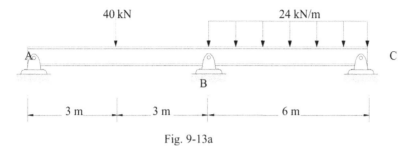

Fig. 9-13a

SOLUTION

1. Equation 9-15:
The required data is obtained from the moment diagrams as shown in Fig. 9-13b.

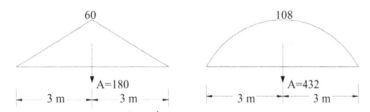

Fig. 9-13b. Simply Supported Spans

$M_L = M_R = 0$, $M_{Ce} = M_B$.
Substituting these moment values in Equation [9-15] gives the following:

$$0 + 4 M_{ce} + 0 \quad = \quad \frac{-6}{6^2}\left[(180)(3) + (432)(3)\right]$$

The Central Moment is therefore equal to:
$$M_B = -\, 76.5 \text{ kN.m} \quad = \quad 76.5 \text{ kN.m} \ \reflexbox{$\curvearrowright$}$$

2. Equation 9-17:

The required data is as follows.

Left hand side	Right hand side
$M_L = 0$	$M_R = M_B$
$L_L = 6.0$ m	$L_R = 6.0$ m
$W_L = 0$	$W_R = 24$ kN/m
$P_L = 40$ kN	$P_R = 0$
$Z_L = \dfrac{3}{6} = 0.5$	$Z_R = 0$

Substituting the above-mentioned values in Equation[9-17] gives the following:

$$0 + 2\ M_{Ce}\ (6 + 6) + 0\ = -40\ (6)^2\ (0.5 - 0.5^3) - 0 - \frac{24\left(6^3\right)}{4}$$

The Central Moment M_B is therefore equal to:

$M_B = -76.5$ kN.m $= 76.5$ kN.m

3. Equation 9-20:

The required rotation at the support B is obtained from Table 9-7.

$$\theta_{LJ}^{//} = \theta_{LB}^{//} = \frac{P}{16}\ \frac{l^2}{EI} = \frac{40(6)^2}{16EI}$$

$$\theta_{RJ}^{/} = \theta_{RB}^{/} = \frac{wl^3}{24EI} = \frac{24(6)^3}{24EI}$$

Substitute these values in equation 9-20 gives the following:

$$0\ +\ 4\ \ M_{CE}\ +\ 0 = \frac{-6EI}{6}\left(\frac{40(6)^2}{16EI} + \frac{24(6)^3}{24EI}\right)$$

The moment M_B is therefore equal to :

$M_B = -76.5$ kN.m $= 76.5$ kN.m

The reactions are determined using the equilibrium equations of statics as shown in Figs. 9-13c and 9-13d.

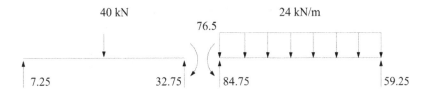

Fig. 9-13c. Free-Body Diagram

Fig. 9-13d. Applied Loads and Reactions

Example: 9-12

Using the three-moment equation method, determine the reactions for the beam shown in Fig. 9-14a.

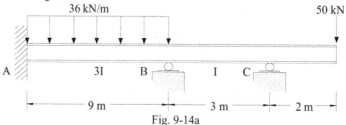

Fig. 9-14a

SOLUTION

Equation [9-16] will be used for the solution since (I) is not constant. To apply three-moment equation the structure has to be modified because of the fixed end as shown in Figs. 9-14b and 9-14c.

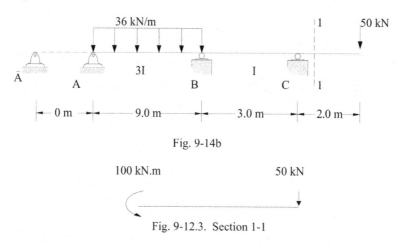

Fig. 9-14b

Fig. 9-12.3. Section 1-1

For Spans Ā A and AB

Left hand side	Right hand side	Central
$M_L = 0$	$M_R = M_B$	$M_{ce} = M_A$
$L_L = 0$	$L_R = 9.0$ m	
$W_L = 0$	$W_R = 36$	
$P_L = 0$	$P_R = 0$	
$Z_L = 0$	$Z_R = 0$	
$I_L = \infty$	$I_R = 3I$	

Substituting in equation [9-16].

$$0 + 2\,M_A\left(0 + \frac{9}{3I}\right) + M_B\left(\frac{9}{3I}\right) = 0 - 0 - 0 - \frac{36(9)^3}{4(3I)}$$

$$6\,M_A + 3\,M_B = -2187 \qquad\qquad\qquad [1]$$

For Spans AB and BC

Left hand side	Right hand side	Central
$M_L = M_A$	$M_R = M_C = -100$	$M_{ce} = M_B$
$L_L = 9.0\ m$	$L_R = 3.0\ m$	
$W_L = 36\ kN.m$	$W_R = 0$	
$P_L = 0$	$P_R = 0$	
$Z_L = 0$	$Z_R = 0$	
$I_L = 3I$	$I_R = I$	

Substituting in equation [9-16].

$$M_A\left(\frac{9}{3I}\right) + 2\,M_B\left(\frac{9}{3I} + \frac{3}{I}\right) + (-100)\left(\frac{3}{I}\right) = -0 - 0 - \frac{36(9)^3}{4(3I)}$$

$$3\,M_A + 12\,M_B = -1887 \qquad\qquad\qquad [2]$$

Solving equation (1) and (2) simultaneously gives the following moments:

$M_A = -326.7\ kN.m = 326.7\ kN.m$

$M_B = -75.57\ kN.m = 75.57\ kN.m$

The reactions, which are shown in Figs. 9-14d and 9-14e, are obtained using the principles of statics.

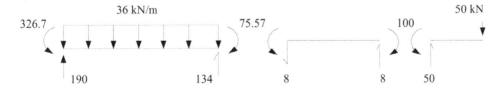

Fig. 9-14d. Member Free Body Diagram

289

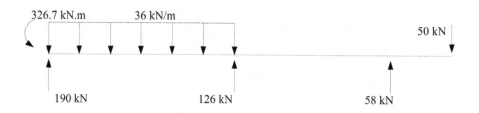

Fig. 9-14e. Applied Loads and Reactions

Example: 9-13

Determine the support reactions using the three moment equation method for the beam shown in Fig 9-15a if the support B settles by 50.0 mm. [EI=200.10⁹ N/m², I = 200.10⁻⁶ m⁴]

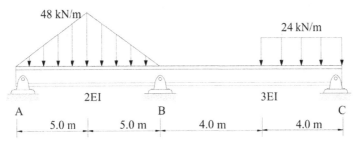

48 kN/m

24 kN/m

2EI   3EI

A   B   C

| 5.0 m | 5.0 m | 4.0 m | 4.0 m |

Fig. 9-15a

SOLUTION

Because of the presence of irregular types of loads, different inertia moments, and settlements, Equation 9-21 is modified using the information provided in Table 9-7 as follows .

$$M_L \frac{L_L}{I_L} + 2 \ M_{ce} \left[\frac{L_L}{I_L} + \frac{L_R}{I_R} \right] + M_R \ \frac{L_R}{I_R} = -6 \ E \ \left(\sum \theta'_{Lj} + \sum \theta_{Rj} \right) + 6 \ E \left[\frac{\delta_L}{L_L} - \delta_c \left(\frac{1}{L_L} + \frac{1}{L_R} \right) + \frac{\delta_R}{L_R} \right] \ [9-23]$$

The data, which is needed to solve Equation 9-23, is as follows.

Left hand side	Right hand side	Central
$M_L = 0$	$M_R = 0$	$M_{ce} = M_B$
$\delta_L = 0$	$\delta_R = 0$	$\delta_{ce} = \delta_B = 0.05m \downarrow$
$L_L = 10.0$ m	$L_R = 8.0$ m	
$\theta_{LB} = \dfrac{5 \ w \ L_L^3}{192 \ EI} = \dfrac{5(48)(10)^3}{192(2EI)}$	$\theta_{RB} = \dfrac{7 \ w \ L_R^3}{384(3 \ EI)}$	
$I_L = 2I$	$I_R = 3I$	

The data shown above is substituted into Equation 9-23 in order to determine the value of M_B as follows.

$$0 + 2 \ M_B \left(\frac{10}{2} + \frac{8}{3} \right) + 0 =$$

$$- 6 \ EI \left[\frac{5(48)(10)^3}{192(2 \ EI)} + \frac{7(24)(8)^3}{384(3 \ EI)} \right] + 6 \ EI \left[0 + 0.05 \left(\frac{1}{10} + \frac{1}{8} \right) + 0 \right]$$

$15.333 \ M_B = -6 \ [\ 625 + 74.667] + 27000$

M_B = 1487.4 kN.m $\circlearrowright$

The equilibrium equations are used to determine the reactions as shown in Figs. 9-15b and 9-15c.

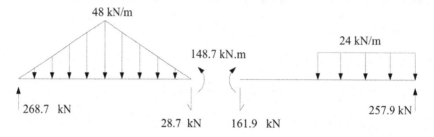

Fig. 9-15b. Member Free Body Diagram

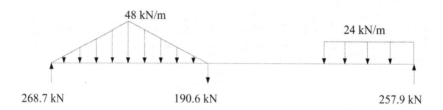

Fig. 9-15c. Applied Loads and Reactions

9.3   LEAST WORK METHOD

The method of least work, which is a special application of Castigliano's second theorem, is used to analyze statically indeterminate structures, Figs. 9-16 and 9-17. In this method, the unknown redundant forces are obtained through the minimization of the structure strain energy.

Least work theorem: The redundant forces in a linear elastic structure with a given system of loads, unyielding supports, and no induced displacements due to temperature, settlement, and fabrication error are obtained by minimizing the structure strain energy

For beams and frames

The least work theorem is expressed using the following equation:

$$\frac{\partial U}{\partial A_y} = 0 = \frac{1}{EI} \int_0^L M \left(\frac{\partial M}{\partial A_y} \right) dx \qquad\qquad [9\text{-}24]$$

where U = strain energy and A_y = selected redundant reaction

For trusses

The least work theorem is expressed using the following equation:

$$\frac{\partial U}{\partial F_{AB}} = 0 = \sum F_i \left(\frac{\partial F_i}{\partial F_{AB}} \right) \frac{L_i}{A_i E} \qquad\qquad [9\text{-}25]$$

where F_{AB} = selected bar force redundant, i = bar number, and L = bar length

The following examples illustrate how to use the least work method.

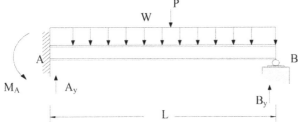

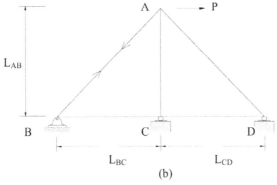

Fig. 9-16. Beam

(b)

Fig. 9-17. Truss

Example: 9-14

Determine the support reactions for the beam shown in Fig 9-18a.

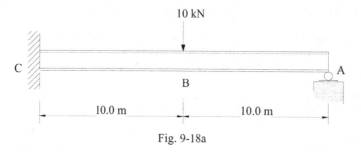

Fig. 9-18a

SOLUTION

The beam is statically indeterminate to the **first degree**. The least work theorem is applied to determine the redundant A_y as shown in Fig. 9-18b.

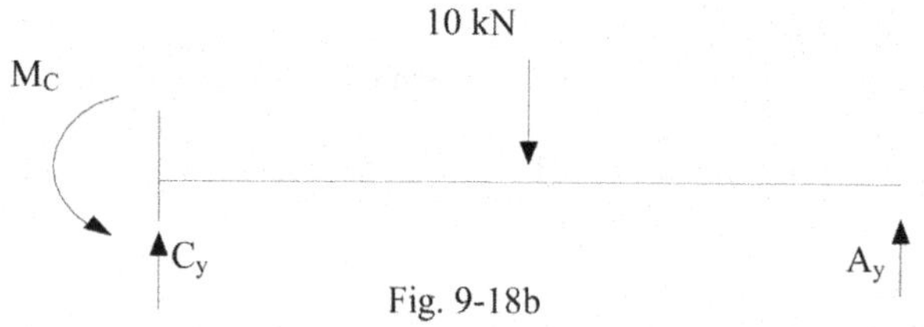

Fig. 9-18b

Table 9-8. Energy Table

Element	EI	X = 0	M	$\partial M / \partial A_y$	$\int_a^b$
AB	1	A	$A_y X$	X	$\int_0^{10}$
BC	1	B	$A_y(10 + X) - 10X$	$10 + X$	$\int_0^{10}$

The redundant reaction A_y is determined by using the least work theorem, Equation 9-24, and the data in Table 9-8 as follows.

$$\frac{\partial U}{\partial A_y} = 0 = \int_0^L \frac{M}{EI} \left(\frac{\partial M}{\partial A_y} \right) \partial x$$

$$0 = \frac{1}{EI} \int_0^{10} A_y x \,(x)\,dx + \frac{1}{EI} \int_0^{10} \left[A_y (10+x) - 10x \right](10+x)\,dx$$

$A_y = \textbf{3.125 kN} \uparrow$

The remaining reactions are then determined using the equilibrium equations as shown in Fig. 9-18c.

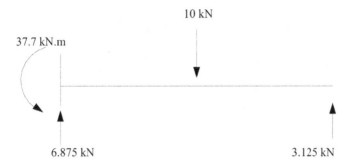

Fig. 9-18c. Applied Loads and Reactions

Example: 9-15

Use the least work method to determine the support reactions for the continuous beam shown in Fig 9-19a. [EI=Constant]

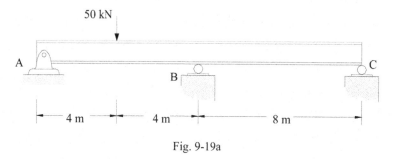

50 kN

A

B

C

4 m

4 m

8 m

Fig. 9-19a

SOLUTION

The beam is statically indeterminate to the first degree, and A_y is the selected redundant as shown in Fig. 9-19b.

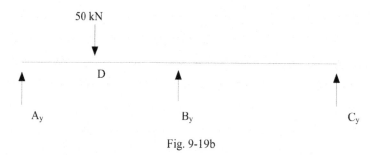

50 kN

D

A_y

B_y

C_y

Fig. 9-19b

The application of the least work method requires that the moment along the beam to be expressed as a function of A_y. Therefore, C_y is expressed in terms of A_y by summing the moments about B.

$$C_y = A_y - 25$$

[9-26]

Table 9-9. Energy Table

Element	EI	X = 0	M	$\partial M / \partial A_y$	$\int_a^b$
AD	1	A	$A_y X$	X	$\int_0^4$
DB	1	D	$A_y(4 + X) - 50X$	$4 + X$	$\int_0^4$
BC	1	C	$-(A_y - 25)X$	$-X$	$\int_0^8$

296

The redundant reaction A_y is determined by using the least work theorem, Equation 9-24, and the data in Table 9-9 as follows.

$$\frac{\partial U}{\partial A_y} = 0 = \int_0^L \frac{M}{EI} \left(\frac{\partial M}{\partial A_y} \right) \partial x$$

$$0 = \frac{1}{EI} \int_0^4 A_y x(x)dx + \frac{1}{EI} \int_0^4 \left[A_y(4+x) - 50x \right](4+X)dx + \frac{1}{EI} \int_0^8 -\left(A_y - 25 \right)x(-x)dx$$

$A_y = 20.313$ kN ↑

The remaining reactions are determined by using the equilibrium equations as shown in Fig. 9-19c.

50 kN

20.313 kN   34.375 kN   4.688 kN

Fig. 9-19c. Applied Loads and Reactions

Example: 9-16

Use the least work method to compute the support reactions for the beam shown in Fig 9-20a. [K_{SP}=400 kN/m, I= 200.10^6 mm^4, E= 200.10^6 kN/m^2]

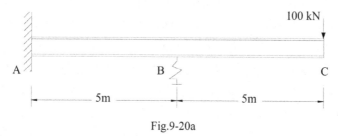

Fig.9-20a

SOLUTION

The beam is statically indeterminate to the first degree. The spring reaction B_y is the selected redundant as shown in Fig. 9-20b. Taking into consideration the spring settlement Δ_{sp} the least work method is applied to determine the redundant.

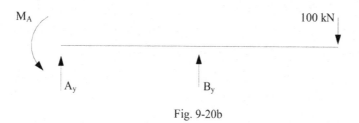

Fig. 9-20b

Table 9-10. Energy Table

Element	EI	X = 0	M	$\partial M \ / \ \partial B_y$	$\int_a^b$
CB	1	C	100 X	0	$\int_0^5$
BA	1	B	100(5 + X) −B_y X	- X	$\int_0^5$

The redundant reaction A_y is determined by using the least work theorem, Equation 9-24, and the data in Table 9-10 as follows.

$$\frac{\partial U}{\partial B_y} = 0 = \int_0^L \frac{M}{EI} \left(\frac{\partial M}{\partial B_y} \right) \partial x + \Delta_{SP}$$

$$0 = \frac{1}{EI} \int_0^5 (100X)(0)dx + \frac{1}{EI} \int_0^5 [100(5+x) - B_Y x](-x)dx + \left(\Delta_{SP} = \frac{B_Y}{K_{SP}} \right)$$

$$-\frac{B_y}{400} = 0 + \frac{1}{40000} \int_0^5 [100(5+x) - B_y x](-x)dx$$

$B_y = 73.5 \text{ kN} \uparrow$

The remaining reactions are determined by using the equilibrium equations as shown in Fig. 9-20c.

632.5 kN.m

100 kN

26.5 kN

73.5 kN

Fig. 9-20c. Applied Loads and Reactions

Example: 9-17

Determine the support reactions for the circular arch shown in Fig 9-21a.
[EI = Constant]

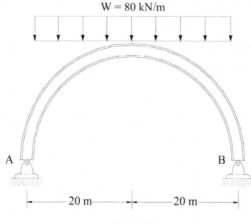

Fig. 9-21a

SOLUTION

The arch is indeterminate to the first degree. The least work method is applied to determine the redundant H shown in Fig. 9-21b. Taking the moment about an arbitrary point C along the arch, we have the following:

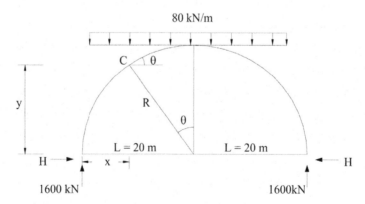

Fig. 9-21b

$$M_c = 1600x - Hy - 80\,\frac{x^2}{2}$$

Since $x = R - R\sin\theta = 20 - 20\sin\theta$

$y = R \cos \theta = 20 \cos \theta$

Thus $M_c = 1600 \ (\ 20 - 20 \sin \theta \ ) - H \ (\ 20 \cos \theta \ ) - \dfrac{80}{2} \ (\ 20 - 20 \sin \theta \ )^2$

$\dfrac{\partial M_c}{\partial H} = - 20 \cos \theta$

Note that $(\sin \theta)^{-1} = \dfrac{L}{R} = \dfrac{20}{20} = 1$

$\theta = 90^0 = \dfrac{\pi}{2}$ radians at H.

Applying least work theorem, Equation [9-(24)] becomes as follows.

$$\dfrac{\partial U}{\partial H} = 0 = \int_0^{\frac{\pi}{2}} \dfrac{M_C}{EI} \left(\dfrac{\partial M_C}{\partial H} \right) ds \qquad\qquad \text{[9-(24-a)]}$$

Where $d_s = R \ d\theta$ as shown in Fig. 9-21c.

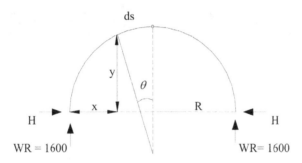

Fig.9-21c

Substituting in M_c in Equation [9-(24-a)], and integrating we have the following:

$$\mathbf{H} = \dfrac{4}{3} \left(\dfrac{WR}{\pi} \right) = \dfrac{4}{3} \left(\dfrac{1600}{\pi} \right) = \mathbf{679 \ kN} \ \rightarrow$$

The reactions are shown in Fig. 9-21d

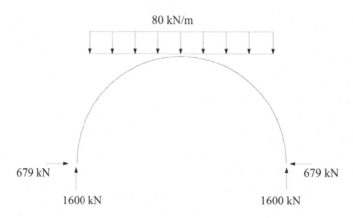

Fig. 9-21d. Applied Loads and Reactions

<u>Note:</u>

For the special cases of arch loading, the values of the reactions H are summarized in Table 9-11.

Case of loading	Figure	H
Concentrated	P ⬇ H → ⟵ L ─ L → ⟵ H	$\dfrac{P}{\pi}$
Concentrated	P ⬇ θ H → ⟵ L ─ L → ⟵ H	$\dfrac{P}{\pi}\cos^2\theta$
Uniform	w ↓↓↓↓↓↓↓ H → ⟵ L ─ L → ⟵ H	$\dfrac{2}{3}\left(\dfrac{w}{\pi}\right)$

Example: 9-18

Apply the least work method to determine the support reactions for the frame shown in Fig 9-22a.

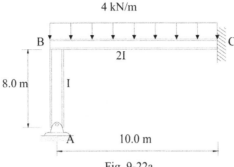

Fig. 9-22a

SOLUTION

The frame is statically indeterminate to the second degree. The reactions H and A_y are the selected redundants as shown in Fig. 9-22b.

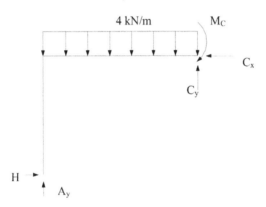

Fig. 9-22b.

Table 9-12. Energy Table

Element	EI	X = 0	M	$\partial M / \partial A_y$	$\partial M / \partial H$	$\int_a^b$
AB	1	A	- H X	0	- X	$\int_0^8$
BC	2	B	$A_y(X) - 8\,H - \dfrac{4X^2}{2}$	X	- 8	$\int_0^{10}$

The first redundant reaction A_y is determined by using the least work theorem, Equation 9-24, and the data in Table 9-12 as follows.

$$\frac{\partial U}{\partial A_y} = 0 = \int_0^L \frac{M}{EI} \left(\frac{\partial M}{\partial A_y} \right) \partial x$$

$$0 = \frac{1}{EI} \int_0^8 (-Hx)(0)dx + \frac{1}{2EI} \int_0^{10} \left[A_y(x) - 8H - \frac{4x^2}{2} \right](x)dx$$

$$\frac{500}{3} A_y - 200H - 2500 = 0 \qquad [1]$$

The second redundant reaction H is determined by using the least work theorem, Equation 9-24, and the data in Table 9-12 as follows.

$$\frac{\partial U}{\partial H} = 0 = \int_0^L \frac{M}{EI} \left(\frac{\partial M}{\partial H} \right) \partial x$$

$$0 = \frac{1}{EI} \int_0^8 (-Hx)(-x) \, dx + \frac{1}{2EI} \int_0^{10} \left[A_Y(x) - 8H - \frac{4x^2}{2} \right] (-8) \, dx$$

$$-200 A_y + \frac{1472}{3} H - \frac{8000}{3} = 0 \qquad [2]$$

The redundants Ay and H are determined by solving equations [1] and [2] simultaneously as follows.

$$\textbf{Ay = 16.6 kN} \uparrow \quad , \quad \textbf{H = 1.33 kN} \rightarrow$$

The remaining reactions are obtained using the equilibrium equations as shown in Fig. 9-22c.

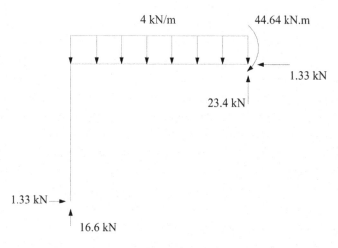

Fig. 9-22c. Applied Loads and Reactions

Example: 9-19

Determine the bar forces and support reactions for the truss shown in Fig 9-23a.
[EA=Constant]

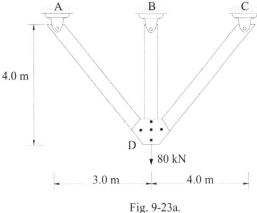

Fig. 9-23a.

SOLUTION

The truss is statically indeterminate to the first degree. The force in bar CD is the selected redundant. Applying least work method requires that the other bar forces F_{BD} and F_{AD} to be expressed in terms of the redundant F_{CD}, Fig. 9-23a. The horizontal equilibrium of joint A gives the following:

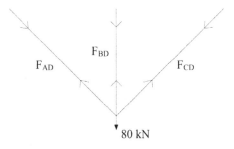

Fig. 9-23a.

$$\sum F_x = 0 \xrightarrow{+} - F_{AD}(0.6) + 0.707\, F_{CD} = 0$$
$$F_{AD} = 1.18\, F_{CD} \qquad\qquad [1]$$

The vertical equilibrium of joint A and equation [1] give the following equation:

$$\sum F_y = 0\uparrow + \quad -80 + (0.8)\,1.18\,F_{CD} + 0.707\,F_{CD} + F_{BD} = 0$$

$$F_{BD} = 80 - 1.651 F_{CD} \tag{2}$$

Once the bar forces F_{BD} and F_{AD} are expressed in terms of the redundant F_{CD}, the energy table is constructed as shown in Table 9-13.

Table 9-13. Energy Table.

Element	L	F	L / AE	$\partial F / \partial F_{CD}$	$F(\partial F / \partial F_{CD})\dfrac{L}{AE}$
AD	5	$1.18\,F_{CD}$	5 / const.	1.18	$6.962\,F_{CD}$
BD	4	$80 - 1.651\,F_{CD}$	4 / const.	-1.65	$-528.32 + 10.9\,F_{CD}$
CD	5.66	F_{CD}	5.66 / const.	1	$5.66\,F_{CD}$
					$\sum 23.522\,F_{CD} - 528.32$

The redundant bar force F_{CD} is determined by using the least work theorem, Equation 9-22, and the data in Table 9-13 as follows.

$$\frac{\partial U}{\partial F_{CD}} = 0 = \sum F_i \left(\frac{\partial F_i}{\partial F_{CD}} \right) \frac{L_i}{A_i E}$$

$$0 = 23.522\,F_{CD} - 528.32$$

$$F_{CD} = 22.46 \text{ kN} \quad \underline{T}$$

The bar forces FAD and FBD are determined by solving equations [1] and [2] simultaneously as follows.

$$F_{AD} = 26.5 \text{ kN} \quad \underline{T} \quad , \quad F_{BD} = 42.9 \text{ kN} \quad \underline{T}$$

The reactions are determined by applying the equation of equilibrium to joints A, B, and C as shown in Fig. 9-19.3

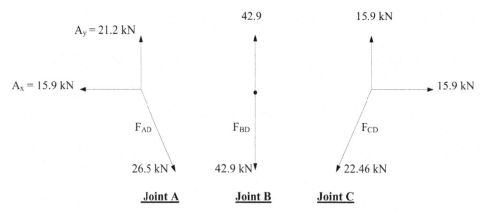

Fig. 9-23c. Reactions

Chapter 9. Consistent Deformation Method

Summary

- The flexibility method of analysis, also called the method of consistent deformations, is one of the oldest classical methods of analyzing indeterminate structures.
- Before the development of general-purpose computer programs for structural analysis, the flexibility method was the only method available for analyzing indeterminate trusses. The flexibility method is based on removing restraints until a stable determinate released structure is established. Since there are alternate choices with respect to which restraints to remove, this analysis aspect does not lend itself to the development of a general-purpose computer program.
- The flexibility method is still used to analyze certain types of structures in which the general configuration and components of the structure are standardized but the dimensions vary.

Problems
▄▄▄

1-   Use the consistent deformation method (Moment Area Method) to determine the following:

a) The reactions. [$E = 400.10^6$ kN/m^2, $I=300.10^6$ mm^4)
b) The reactions if support B settles 50 mm.
c) The tension in the cable if support B is replaced by a
     cable ($A_c = 200$ mm^2, $L_c = 10$ m, $E = 400.10^6$ kN/m^2)
d) The spring force if support B is replaced by a spring ($K_{sp} = 300$ kN/m).

12 kN/m

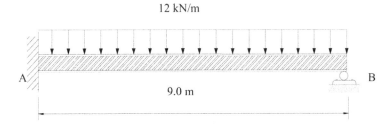

A   B

9.0 m

2-   Draw the shear and moment diagrams using the consistent deformation
      method (Moment Area Method). [EI = Constant]

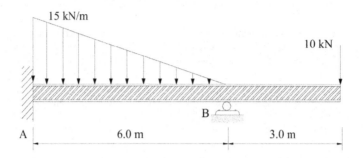

3-   Draw the shear and moment diagrams using the force method (Moment Area
      Method) for a 50 mm settlement of support C.
      $[E = 200.10^6 \text{ kN/m}^2, I = 400.10^6 \text{ mm}^4]$

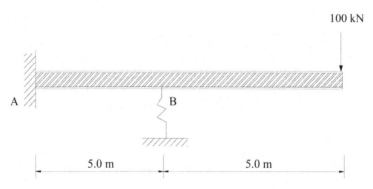

4-   Determine the reactions using the force method (Moment Area Method).
      $[E = 200.10^6 \text{ kN/m}, K_{sp} = 400 \text{ kN/m}, I= 200.10^6 \text{ mm}^4]$

5-   Draw the shear and moment diagrams using consistent deformation method
      (Moment Area). [EI = Constant]

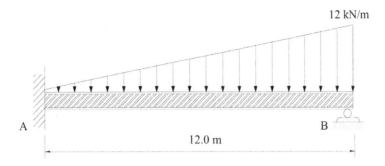

12 kN/m

A

12.0 m

B

6-   Determine the tension in the cable using the consistent deformation method
      (Moment Area). [E = 200.10^6 kN/m^2, I = 200.10^6 mm^4, A_c = 400mm^2]

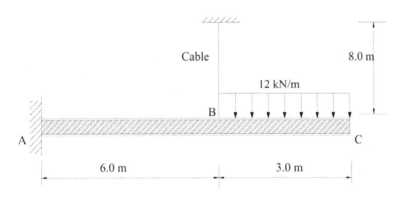

Cable

8.0 m

12 kN/m

A   B   C

6.0 m   3.0 m

7-   Draw the shear and moment diagrams using the consistent method (Moment
      Area) EI = constant.

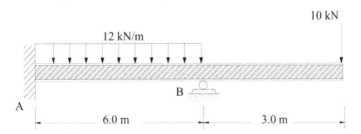

10 kN

12 kN/m

A   B

6.0 m   3.0 m

8- Determine the reactions using the force method (Moment area)
    [$K_{sp} = 400$ kN/m, I= 600.10^6 mm^4, E = 200.10^6 kN/m^2]

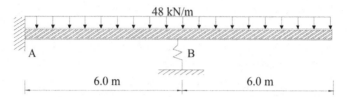

48 kN/m

A   B

6.0 m   6.0 m

9- Use the force method to determine the reactions. [E=200.10^6 kN/m^2,
    I = 100.10^6 mm^4, A$_{Cable}$ = 200 mm^2].

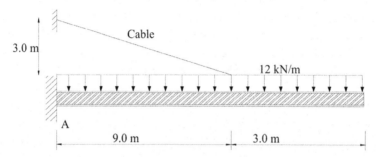

3.0 m

Cable

12 kN/m

A

9.0 m   3.0 m

10- Use the force method to determine the reactions. [E=200.10^6 kN/m^2,
    I = 200.10^6 mm^4, K$_{sp}$= 200 kN/m].

10 kN/m

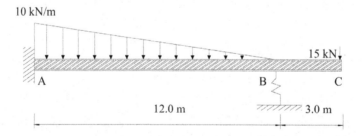

15 kN

A   B   C

12.0 m   3.0 m

11- Use the force method to determine the reactions (Moment Area Method).
    [E= 200.10^6 kN/m^2, I = 100.10^6 mm^4, A$_{cable}$ = 200 mm^2]

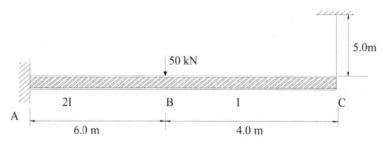

5.0m

50 kN

2I   B   I   C

A

6.0 m   4.0 m

Chapter 9. Consistent Deformation Method

12- Use the force method (Moment Area Method) to determine the reactions.
[E=200.10^6 kN/m^2, I = 400.10^6 mm^4, K$_{sp}$ = 400 kN/m].

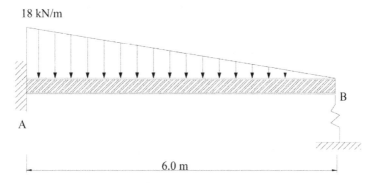

18 kN/m

A

B

6.0 m

13- Determine the reactions using the force method (Conjugate)
[K$_{sp}$ = 400 kN/m, I= 200.10^6 mm^4, E = 200.10^6 kN/m^2]

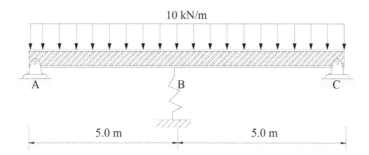

10 kN/m

A

B

C

5.0 m

5.0 m

14- Use the consistent method (conjugate beam) to draw the shear and moment
diagrams for a 50 mm settlement of the support C

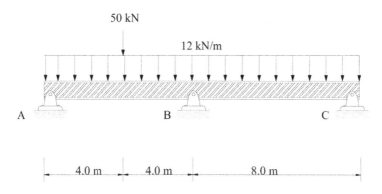

50 kN

12 kN/m

A

B

C

4.0 m

4.0 m

8.0 m

15- Draw the shear and moment diagrams if support B settles 40mm, Using consistent method (Conjugate beam). [$E=200.10^6$ kN/m^2, I = 200.10^6 mm^4]

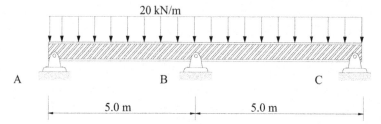

20 kN/m

A   B   C

5.0 m   5.0 m

16- Determine the bar force in member CD using the consistent deformation method (Virtual Work). [EA = Constant]

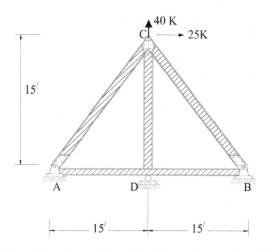

40 K

C ——→ 25K

15$'$

A   D   B

15$'$ 15$'$

17- Determine the force in Bar BD, using consistent deformation method (Virtual work). [AE=Constant]

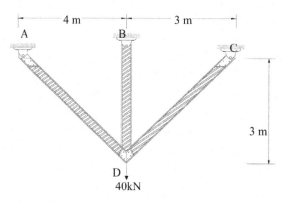

4 m   3 m

A   B   C

3 m

3 m

D
40kN

18-   Determine the bar forces and reactions using the virtual work method.
       [AE=Const]

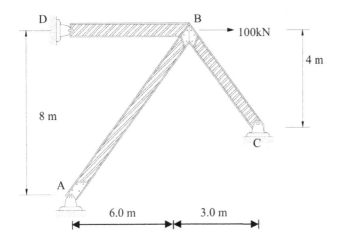

19-   Determine the reactions using the least work method. [EI=Constant]

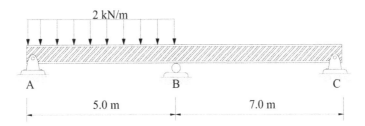

20-   Determine the reactions using the least work method. [EI=Constant]

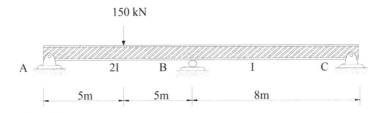

21- Determine the reactions using the least work method. [EI=Constant]

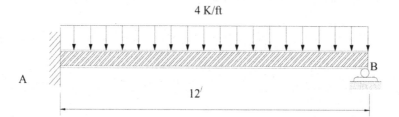

4 K/ft

A

B

12′

22- Determine the bar forces and reactions using the force method (Energy)

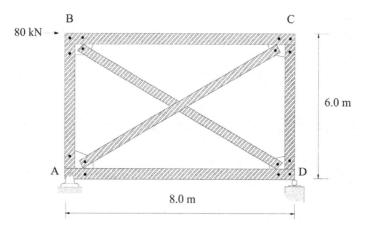

B

C

80 kN →

6.0 m

A

D

8.0 m

23- Draw the shear and moment diagrams using the least work method.

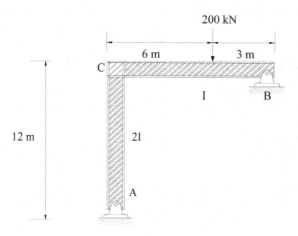

200 kN

6 m

3 m

C

I

B

12 m

2I

A

CHAPTER - 10
SLOPE DEFLECTION METHOD

Water Tower, Kuwait

Chapter 10.   Slope Deflection Method

Introduction

The displacement method requires the satisfaction of the equilibrium equations at all connectivity points of connectivity, which result in a set of equations having the displacements as the unknowns and the stiffness quantities as the coefficients. The computerized version (in matrix form) of the slope deflection method, which is also called the stiffness matrix method, is a very powerful method for the analysis of large structures using computers. The slope deflection method, which is an analytical method for statically indeterminate structures, expresses the end moments of structural members in terms of the end displacements (δ) and rotations (θ).

The analysis using the slope deflection method starts by defining the independent node displacements as degrees of freedom (the unknowns of the solution). The frame shown in Fig. 10-1a, has four degrees of freedom $\delta_R, \theta_R, \theta_C$, and θ_D.

In the slope deflection method, the following assumptions are made:

1. All joints are rigid, meaning that the angle between any two members in a joint stays the same before and after deformation.
2. The effects of axial deformation are neglected. Thus, For the frame shown in Fig. 10-1a, $\delta_B = \delta_C = \delta$.
3. The effects of shear deformation are neglected.

Sign Convention.

Clockwise end moments and rotations are positive and counter-clockwise end moments and rotations are negative. The left span of a settled support will have a negative rotation of the span cord (ψ) while the right span of a settled support will have a positive rotation of the span cord (ψ) as shown in Figs. 10-1c, 10-1d, and 10-1e.

Slope-Deflection Equations.

The slope deflection equations yield the final end moments of each member. The final end moments of the member are equal to the sum of the end moments corresponding to the member deformed shapes due to the actual loading, the member-end rotations, and the member-end translations as shown in Fig. 10-2a, 10-2b, and 10-2c.

The slope equations are obtained by expressing the end moments of each member as function of the deformations of that member using the moment area method as shown below.

$$M_{AB} = \frac{2EI}{L}\left(2\theta_A + \theta_B + \frac{3\delta}{L}\right) + FEM_{AB} \qquad [10\text{-}1]$$

$$M_{BA} = \frac{2EI}{L}\left(2\theta_B + \theta_A + \frac{3\delta}{L}\right) + FEM_{BA} \qquad \text{[10-2]}$$

Where M_{AB}, M_{BA} = final end moments; θ_A, θ_B = end rotations; I = Moment of inertia; and E = modulus of elasticity

$\dfrac{2EI}{L}(2\theta_A)$ = Moment due to the rotation contribution of the end A with end B held fixed.

$\dfrac{2EI}{L}(2\theta_B)$ = Moment due to the rotation contribution of the end B with the end A held fixed.

$\dfrac{2EI}{L}\left(\dfrac{3\delta}{L}\right)$ = end moment due to settlement contribution of the end B

FEM_{AB}, FEM_{BA} = end moments due to the actual loading of the beam as shown in Table 10-1.

the slope deflection equations can be re-written are as follows

$$M_{AB} = 2Ek\left(2\theta_A + \theta_B + \psi\right) + FEM_{AB} \qquad \text{[10-3]}$$

$$M_{BA} = 2Ek\left(2\theta_B + \theta_A + \psi\right) + FEM_{BA} \qquad \text{[10-4]}$$

Where $k = \dfrac{I}{L}$ (stiffness factor) and $\psi = \dfrac{\delta}{L}$ (span cord rotation)

The slope deflection method steps for determining the final end moments of continuous beams, frames without sides way Figs.10-3a and 10-3b, and frames with sides way, Fig. 10-3c are as follows.

1) Determine the fixed end moments for each structural member using Table 10-1.
2) Determine the end moments of each structural member using the slope deflection equations. The unknowns in these equations are the rotations and displacements of all degree of freedoms.
3) Determine the joint equilibrium equations for each unknown degree of freedom.
4) Determine the unknown degrees of freedom by substituting the slope deflection equations into the equilibrium equations.
5) Determined the final end moments by substituting the unknown values in the slope deflection equations.
6) Consider each structural member as simply supported beam subjected to given loads and end moments (final moments from previous step). Apply the principles of statics to determine the end reactions of each member and draw the shear and moment diagram.

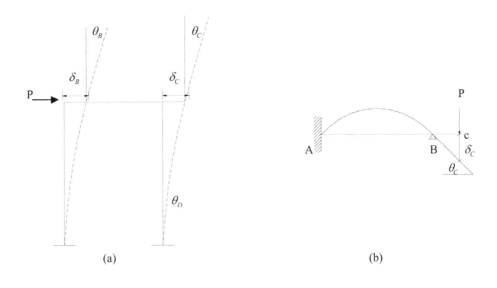

(a)

(b)

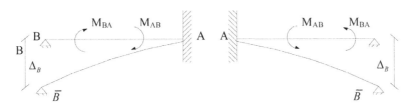

(c) Positive end moments

(d) Negative end moment

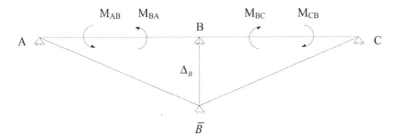

(e) Negative End Moments in Span AB

(f) Positive End Moment in Span BC

Fig. 10-1. Displacements, Rotations and End moments of structures.

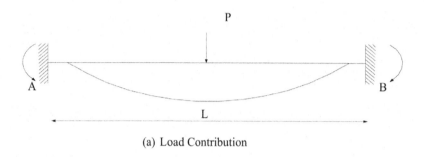

(a) Load Contribution

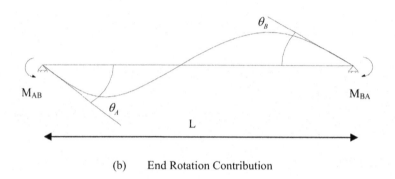

(b)   End Rotation Contribution

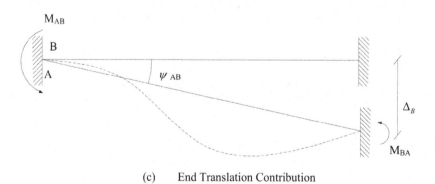

(c)   End Translation Contribution

Fig. 10-2. Member Deformation Shapes

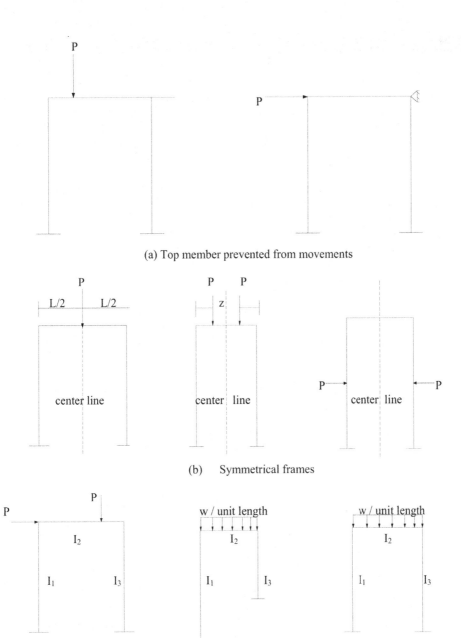

(a) Top member prevented from movements

(b)   Symmetrical frames

(c)   Frames with sway

Fig. 10-3. Frames

Table 10-1. Fixed-End Moments for beams with constant EI.

No.	Left Support	Loading	Right Support
1	$-\dfrac{a\,b^2}{L^2}P$		$\dfrac{a^2\,b}{L^2}P$
2	$-\dfrac{L}{8}P$		$\dfrac{L}{8}P$
3	$-\dfrac{a}{L}\left(L-a\right)P$		$\dfrac{a}{L}\left(L-a\right)P$
4	$-\dfrac{2}{9}L\,P$		$\dfrac{2}{9}L\,P$
5	$-\dfrac{W\,L^2}{12}$		$\dfrac{W\,L^2}{12}$
6	$-\dfrac{a^2}{12}\left[4\dfrac{b}{L}+2\dfrac{b^2}{L^2}+\dfrac{a^2}{L^2}\right]W$		$\dfrac{a^2}{12}\left[3-2\dfrac{b}{L}-\dfrac{b^2}{L^2}-2\dfrac{a^2}{L^2}\right]W$
7	$-\dfrac{11}{192}W\,L^2$		$\dfrac{5}{192}W\,L^2$
8	$-\dfrac{2}{3}C\left[2b\left(1-\dfrac{b^2}{L^2}-\dfrac{C^2}{L^2}\right)-a\left(1-\dfrac{a^2}{L^2}-\dfrac{C^2}{L^2}\right)\right]W$		$\dfrac{2}{3}C\left[2a\left(1-\dfrac{a^2}{L^2}-\dfrac{C^2}{L^2}\right)-b\left(1-\dfrac{b^2}{L^2}-\dfrac{C^2}{L^2}\right)\right]W$

Cont. Table 10-1 Fixed-End Moments for beams with constant EI.

No.	Left Support	Loading	Right Support
9	$-\dfrac{CL}{12}\left(3-4\dfrac{C^2}{L^2}\right)W$		$\dfrac{CL}{12}\left(3-4\dfrac{C^2}{L^2}\right)W$
10	$-\dfrac{a^2}{6}\left(3-4\dfrac{a}{L}\right)W$		$\dfrac{a^2}{6}\left(3-4\dfrac{a}{L}\right)W$
11	$-\dfrac{1}{30}WL^2$		$\dfrac{1}{20}WL^2$
12	$-\dfrac{5}{96}L^2W$		$\dfrac{5}{96}L^2W$
13	$-\dfrac{1}{32}WL^2$		$\dfrac{1}{32}WL^2$
14	$-\dfrac{3+2\alpha}{60}WL^2$		$\dfrac{2+3\alpha}{60}WL^2$
15	$-\dfrac{a^2}{30}\left[10-\dfrac{15a}{L}-\dfrac{6a^2}{L^2}\right]W$		$\dfrac{a^2}{180}\left[\dfrac{45a}{L}-\dfrac{36a^2}{L^2}\right]W$
16	$-\dfrac{a^2}{60}\left[2+\dfrac{6b}{L}+\dfrac{2b^2}{L^2}+\dfrac{a^2}{L^2}\right]W$		$\dfrac{a^2}{60}\left[4-\dfrac{3b}{L}-\dfrac{b^2}{L^2}-\dfrac{6a^2}{L^2}\right]W$

Cont. Table 10-1 Fixed-End Moments for beams with constant EI

No.	Left Support	Loading	Right Support
17	$-\dfrac{CL}{24}\left(3-2\dfrac{C^2}{L^2}\right)W$	W, C, C, L/2, L/2	$\dfrac{CL}{24}\left(3-2\dfrac{C^2}{L^2}\right)W$
18	$-\dfrac{17}{384}WL^2$	W, W, L/2, L/2	$\dfrac{17}{384}WL^2$
19	$-\dfrac{L}{96}\left[L+2C\right]$ $\left[5-\dfrac{4C^2}{L^2}\right]W$	C W C, L/2, L/2	$\dfrac{L}{96}\left[L+2C\right]$ $\left[5-\dfrac{4C^2}{L^2}\right]W$
20	$-\dfrac{1}{15}L^2W$	W (2nd Par), L/2, L/2	$\dfrac{1}{15}L^2W$
21	$-\dfrac{1}{60}WL^2$	W, 2nd Par, W, L/2, L/2	$\dfrac{1}{60}WL^2$
22	$-\dfrac{2}{\pi^3}WL^2$	W, Sine Curve, L/2, L/2	$\dfrac{2}{\pi^3}WL^2$
23	$\dfrac{b\left(2a-b\right)}{L^2}M$	M, a, b	$\dfrac{a\left(2b-a\right)}{L^2}M$
24	$\dfrac{M}{4}$	M, L/2, L/2	$\dfrac{M}{4}$

Cont. Table 10-1 End Reactions for beams with constant EI

No.	Left Support	Loading	Right Support
25	$M_A = -\dfrac{3PL}{16}$ $R_A = \dfrac{11P}{16}$		$R_B = \dfrac{5P}{16}$
26	$M_A = -\dfrac{P}{L^2}\left(b^2 a + \dfrac{a^2 b}{2}\right)$ $R_A = \dfrac{Pa^2}{2L^2}(b + 2L)$		$R_B = P - R_A$
27	$M_A = -\dfrac{PL}{3}$ $R_A = \dfrac{4P}{3}$		$R_B = \dfrac{2P}{3}$
28	$M_A = -\dfrac{wL^2}{8}$ $R_A = \dfrac{5w}{8}$		$R_B = \dfrac{5w}{8}$
29	$M_A = -\dfrac{9}{128}wL^2$ $R_A = \dfrac{37wL}{128}$		$R_B = \dfrac{7wL}{128}$
30	$M_A = -\dfrac{wL}{15}$ $R_A = \dfrac{2w\,L}{5}$		$R_B = \dfrac{wL}{10}$
31	$M_A = -\dfrac{5}{64}W\,L^2$ $R_A = \dfrac{21\,W\,L}{2(32)}$		$R_B = \dfrac{11\,WL}{2(32)}$

W = kN/m = unit weigth / length

Chapter 10.   Slope Deflection Method

Example: 10-1

Draw the shear and moment diagrams for the beam shown in Fig 10-4a.

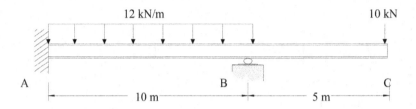

12 kN/m   10 kN

A   B   C
|————————— 10 m —————————|—— 5 m ——|

Fig. 10-4a

SOLUTION

Fixed End Moments.   The span AB is considered only because the span BC is an over-hang and the moment M_{BC} can be computed from statics as shown in Fig. 10-4b and in Table 10-1.

$$FEM_{AB} = -\frac{(12)(10)^2}{12} = -100\,kN.m$$

$$FEM_{BA} = \frac{(12)(10)^2}{12} = 100\,kN.m$$

The slope deflection Equations [10-1 and 10-2] can be written as follows.

$$M_{AB} = \frac{2EI}{10}\left[2\theta_A + \theta_B + \frac{3\delta}{10}\right] - 100$$

$\theta_A = 0$ **Fixed end at A,** $\delta = 0$ (**No settlement**)

$M_{AB} = 0.2\,EI\ \theta_B - 100$ [1]

$$M_{BA} = \frac{2EI}{10}\left[2\theta_B + \theta_A + \frac{3\delta}{10}\right] + 100$$

$M_{BA} = 0.4\,EI\ \theta_B + 100$ [2]

Equilibrium Equations. The over-hang span BC produces a moment M_{BC} as shown in Fig. 10-4b.

326

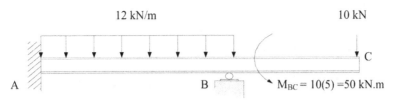

Fig. 10-4b

$$\Sigma M_B = 0 \; + \circlearrowright \quad M_{BA} + M_{BC} = 0 \qquad\qquad [3]$$

Substituting equation [2] into equation [3] yields the value of θ_B as follows.

$$\Sigma M_B = 0 \quad = 0.4 \, EI \, \theta_B + 100 - 50 = 0$$

$$EI \, \theta_B = -125 \quad \text{thus} \quad \theta_B = -\frac{125}{EI}$$

Final moments. Substituting θ_B in Equations [1] and [2] yields the values of the final moments M_{AB} and M_{BA} as follows.

$M_{AB} = -125 \text{ kN.m} = 125 \text{ kN.m} \; \circlearrowleft$	
$M_{BA} = 50 \text{ kN.m} \; \circlearrowright$	

The beam free-body diagram is shown in Fig. 10-4c while the reactions are shown in Fig. 10-4d. The shear and moment diagrams are shown in Figs. 10-4e and 10-4f, respectively.

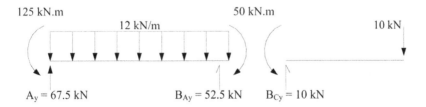

Fig. 10-4c. Member Free Body Diagram

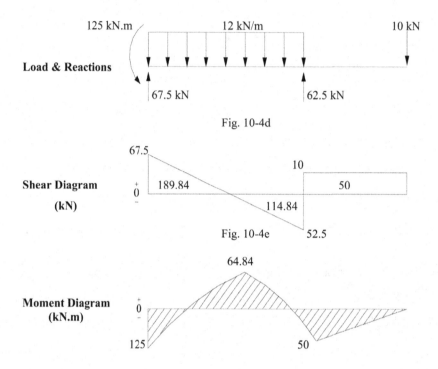

Load & Reactions

125 kN.m

12 kN/m

10 kN

67.5 kN

62.5 kN

Fig. 10-4d

Shear Diagram (kN)

67.5

189.84

0

10

50

114.84

52.5

Fig. 10-4e

Moment Diagram (kN.m)

64.84

0

125

50

Fig. 10-4f

Example: 10-2

Determine the reactions for the beam shown in Fig 10-5a.

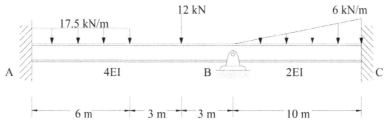

Fig. 10-5a

SOLUTION

Fixed End Moments.
The span AB is subjected to two loads. The fixed-end moments for the span AB, which are computed using Table 10-1, are as follows.

$FEM_{AB} = -144.375 - 6.75 = -151.125$ kN.m
$FEM_{BA} = 65.625 + 20.25 = 85.875$ kN.m

The fixed-end moments for the span BC, which are also computed using Table 10-1, are as follows.
$FEM_{BC} = -20$ kN.m
$FEM_{CB} = 30$ kN.m

Slope deflection Equations (Equations 10-1 and 10-2)

SPAN AB

$$M_{AB} = \frac{2E(4I)}{12}\left[2\theta_A + \theta_B + \frac{3\delta}{12}\right] - 151.125$$

The beam is fixed-ended at point A. Therefore, $\theta_A = 0$ and $\delta = 0$ (**No settlement**)

$$M_{AB} = \frac{8EI\theta_B}{12} - 151.125 \qquad [1]$$

and

$$M_{BA} = \frac{16EI\theta_B}{12} + 85.875 \qquad [2]$$

329

SPAN BC

$$M_{BC} = \frac{2(2EI)}{10}\left(2\theta_B + \theta_C + \frac{3\delta}{10}\right) - 20$$

The beam is fixed-ended at point C. Therefore, $\theta_C = 0$ and $\delta = 0$ (No settlement)

$$M_{BC} = \frac{8EI}{10}\theta_B - 20 \qquad\qquad\qquad [3]$$

$$M_{CB} = \frac{4EI}{10}\theta_B + 30 \qquad\qquad\qquad [4]$$

Equilibrium Equations.
 The moment equilibrium requirement can be stated as follows.

$$\Sigma\, M_B = 0 + \qquad\qquad M_{BA} + M_{BC} = 0 \qquad\qquad\qquad [5]$$

Substituting Equations [2] and [3] into Equation [5] yields the value of θ_B.

$$\theta_B = \frac{-30.88}{EI}$$

The final moments shown below are obtained using the Equations [1], [2], [3], and [4].

M_{AB} = -171.7 kN.m   = 171.7 kN.m
M_{BA} = 44.7 kN.m
M_{BC} = - 44.7 kN.m = 44.7 kN.m
M_{CB} = 17.65 kN.m

The final moment values are used to determine the reactions, which are shown in Figs. 10-5b and 10-5c.

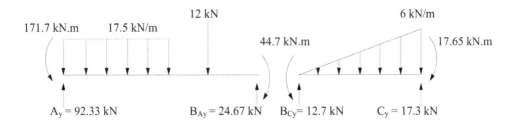

Fig. 10-5b, Member Free Body Diagram

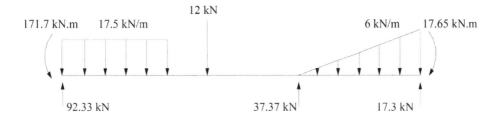

Fig. 10-5c. Applied Loads and Reactions

Example: 10-3

Determine the reactions for the beam shown in Fig 10-6a. [EI=Constant]

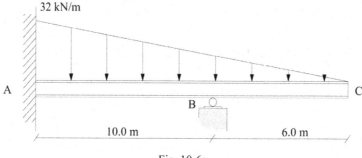

Fig. 10-6a

SOLUTION

Fixed End Moments.
Because the span BC is an over-hang, only the span AB is considered.   The moment M_{BC} can be calculated easily from statics as shown in Fig. 10-6b.

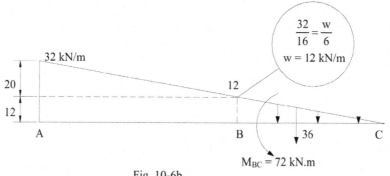

Fig. 10-6b

$$FEM_{AB} = -\left[\frac{12(10)^2}{12} + \frac{20(10)^2}{20}\right] = -200 \text{ kN.m}$$

$$FEM_{BA} = \left[\frac{12(10)^2}{12} + \frac{20(10)^2}{30}\right] = 166.67 \text{ kN.m}$$

$M_{BC} = -72$ kN.m

Slope deflection Equations (Equations 10-1 and 10-2)

The support A is fixed. Therefore, $\theta_A = 0$ and $\delta = 0$ (No settlement)

$$M_{AB} = 0.2 \text{ EI } \theta_B - 200 \tag{1}$$

$$M_{BA} = \frac{2 EI}{10}\left(2\theta_B + \theta_A + \frac{3\delta}{10}\right) + 166.67$$

$$M_{BA} = 0.4 \text{ EI } \theta_B + 166.67 \tag{2}$$

Equilibrium Equations.

The moment equilibrium at the support B yields the following equation:

$$\Sigma\, M_B = 0 \qquad M_{BA} + M_{BC} = 0 \tag{3}$$

The substitution of Equations [1] and [2] into Equation [3] yields the value of θ_B as follows.

$$0.4 \text{ EI } \theta_B + 166.67 - 72 = 0$$

$$\theta_B = \frac{-236.675}{EI}$$

Final Moments.

The substitution of the value of θ_B in Equations [1] and [2] yields the following values for the moments M_{AB} and M_{BA} :

$M_{AB} = -247.335$ kN.m $= 247.35$ kN.m

$M_{BA} = 72$ kN.m   (must be equal and opposite to the over-hang moment)

The reactions were computed using the final moments, Fig. 10-6c

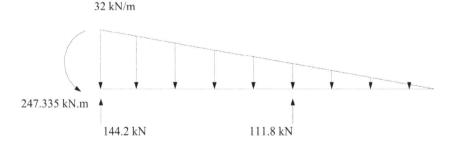

32 kN/m

247.335 kN.m

144.2 kN   111.8 kN

Fig. 10-6c. Applied Loads and Reactions

Example: 10-4

Draw the shear and moment diagrams for the beam shown in Fig. 10-7a.
[EI=Constant]

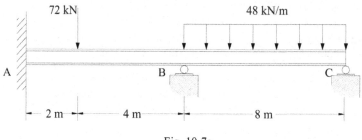

Fig. 10-7a

SOLUTION

Fixed End Moments.
Table 10-1 is used to determine the following fixed-end moments for the spans AB
and BC:

$$FEM_{AB} = -64 \text{ kN.m}$$
$$FEM_{BA} = 32 \text{ kN.m}$$
$$FEM_{BC} = -256 \text{ kN.m}$$
$$FEM_{CB} = 256 \text{ kN.m}$$

Slope deflection Equations (Equations [10-1] and [10-2])

SPAN AB

$$M_{AB} = \frac{2EI}{6}\left(2\theta_A + \theta_B + \frac{3\delta}{L}\right) - 64$$

The support A is fixed. Therefore, $\theta_A = 0$ and $\delta = 0$

$$M_{AB} = \frac{2EI}{6}\theta_B - 64 \qquad\qquad [1]$$

and

$$M_{BA} = \frac{4EI}{6}\theta_B + 32 \qquad\qquad [2]$$

Span BC

$$M_{BC} = \frac{2EI}{8}\left(2\theta_B + \theta_C + \frac{3\delta}{L}\right) - 256$$

There is no support settlement at B and C ($\delta = 0$). Therefore, the moments M_{BC} and M_{CB} can be re-written as follows.

$$M_{BC} = \frac{4EI}{8}\theta_B + \frac{2EI}{8}\theta_C - 256 \qquad\qquad [3]$$

and

$$M_{CB} = \frac{2EI}{8}\theta_B + \frac{4EI}{8}\theta_C + 256 \qquad\qquad [4]$$

Equilibrium Equations.
The equilibrium equations (moment) at the supports B and C are used to determine the values of the unknowns θ_B and θ_C as follows.

$$\Sigma\, M_B = 0 + \qquad M_{BA} + M_{BC} = 0 \qquad\qquad [5]$$

Substituting Equations [2] and [3] into Equation [5] yields the following equation:

$$1.167 \ EI \ \theta_B + 0.25 \ EI \ \theta_C = 224 \qquad\qquad [6]$$

Because it is pinned, the summation of the moments at the support C is equal to zero. The following equation can therefore be written:

$$\Sigma\, M_B = 0 \qquad M_{CB} = 0 \qquad\qquad [7]$$

Substituting Equation [4] in Equation [7] yields the following equation:

$$0.25 \ EI\theta_B + 0.5 \ EI\theta_C = -256 \qquad\qquad [8]$$

The following values of θ_B and θ_C are obtained by solving Equations [6] and [8] simultaneously:

$$\theta_B = \frac{337.909}{EI} \qquad , \qquad \theta_C = \frac{-680.955}{EI}$$

The following final moments are computed using Equations [1], [2], [3], and [4]:

$M_{AB} = 48.6$ kN.m
$M_{BA} = 257.3$ kN.m
$M_{BC} = -257.3 = 257.3$ kN.m
$M_{CB} = 0$ (pinned support)

The reactions, shear, and moment diagrams are shown in Figs. 10-7b, 10-7c, and 10-7d, respectively.

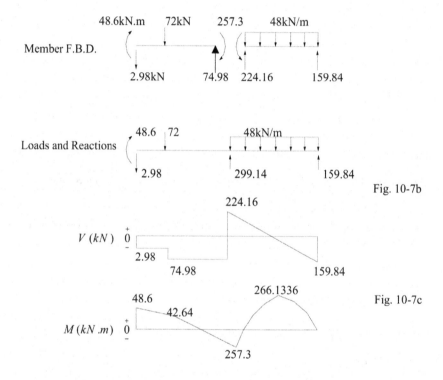

Member F.B.D.

48.6kN.m   72kN   257.3   48kN/m

2.98kN   74.98   224.16   159.84

Loads and Reactions

48.6   72   48kN/m

2.98   299.14   159.84

Fig. 10-7b

$V (kN)$ 0

224.16

2.98

74.98   159.84

Fig. 10-7c

$M (kN .m)$ 0

48.6

42.64

266.1336

257.3

Fig. 10-7d

Example: 10-5

Determine the reactions of the beam shown in Fig. 10-8a for a 50-mm settlement of the support B. [E= 200.10^9 N/m^2, I = 2000.10^{-6} m^4]

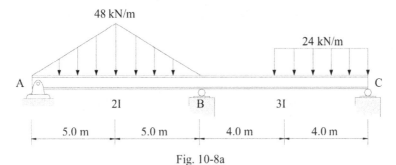

48 kN/m

24 kN/m

A

C

2I

B

3I

| 5.0 m | 5.0 m | 4.0 m | 4.0 m |

Fig. 10-8a

SOLUTION

Fixed End Moments.
Table 10-1 is used to determine the following fixed-end moments for the spans AB and BC:

$\text{FEM}_{AB} = -250 \text{ kN.m}$
$\text{FEM}_{BA} = 250 \text{ kN.m}$
$\text{FEM}_{BC} = -40 \text{ kN.m}$
$\text{FEM}_{CB} = 88 \text{ kN.m}$

Slope deflection Equations (Equations [10-1] and [10-2]).
The supports A and C are pinned (i.e., zero moments) . The equation [10-2] can therefore be simplified as follows.

$$M_{BA} = \frac{3EI}{L}\left[\theta_B + \frac{\delta}{L}\right] + FEM_{BA} - \frac{1}{2}FEM_{AB} \qquad [10-5]$$

SPAN AB

The slope deflection equation (Equation [10-5]) yields the following moment M_{BA}:

$$M_{BA} = \frac{3(2EI)}{10}\left(\theta_B - \frac{0.05}{10}\right) + 250 - \frac{1}{2}(-250)$$

$$M_{BA} = 240000 \; \theta_B - 825 \qquad [1]$$

SPAN BC

The slope deflection equation (Equation [10-5]) yields the following moment M_{BC}:

$$M_{BC} = \frac{3(3EI)}{8}\left[\theta_B + \frac{0.05}{8}\right] - 40 - \frac{1}{2}(88)$$

337

$$M_{BC} = 450000 \; \theta_B + 2728.5 \qquad\qquad [2]$$

Equilibrium Equations.
The moment equilibrium at the support B yields the following equation:
$$\Sigma \, M_B = 0 \; + \qquad M_{BA} + M_{BC} = 0 \qquad\qquad [3]$$

Substituting equations [1] and [2] into equation [3] yields the following value of θ_B :
$\theta_B = - 2758(10^{-6})$ rad.

Final Moments.
The following moment values are obtained by substituting the value of θ_B into
equations [1] and [2]:

$M_{AB} = \; 0$ kN.m (pinned support)
$M_{BA} = -1487.087$ kN.m $= 1487.087$ kN.m $\curvearrowleft$
$M_{BC} = 1487.087$ kN.m $\curvearrowright$
$M_{CB} = 0$ (pinned support)

The reactions, shear, and moment diagrams, which are determined using the final
moments values, are shown in Figs. 10-8b and 10-8c.

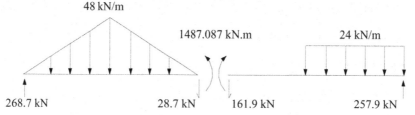

Fig. 10-8b. Member Free Body Diagrams

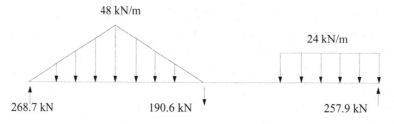

Fig. 10-8c. Applied Loads and Reactions

Example: 10-6

Draw the shear and moment diagrams for the frame shown in Fig 10-9a.
[EI=Constant]

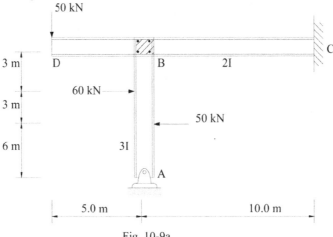

Fig. 10-9a

SOLUTION

Frame lateral movements are called sidesways or sway. The frame is restrained to
sway at the beam level by the support at C. Therefore, this frame is called a braced
frame or frames without sway (Fig. 10-3a).

Fixed End Moments.
Table 10-1 is used to determine the following fixed-end moments for the spans AB
and BC and for the over-hang BD.

$$FEM_{AB} = -33.75 + 75 \qquad = 41.25 \text{ kN.m}$$
$$FEM_{BA} = 101.25 - 75 \qquad = 26.25 \text{ kN.m}$$
$$FEM_{BC} = FEM_{CB} \qquad = 0 \qquad \text{kN.m}$$
$$M_{BD} \quad = 5(50) \qquad = 250 \quad \text{kN.m}$$

Slope deflection Equations (Equations [10-2], and [10-5]).
Because the support A is pinned, the moment M_{BA} is given by the following
equation (Equation [10-5]):

$$M_{BA} = \frac{3E(3I)}{12}\left[\theta_B + \frac{\delta}{12}\right] + 26.25 - \frac{41.25}{2}$$

Because the frame is braced (i.e., no sidesway and $\delta = 0$), the expression of the
moment M_{BA} can be further simplified as follows.

$$M_{BA} = \frac{3E(3I)}{12}(\theta_B + 0) + 5.625 \qquad [1]$$

The moment M_{BC} is given by the following equation:

339

$$M_{BC} = \frac{2E(2I)}{10}\left(2\theta_B + \theta_C + \frac{3\delta}{10}\right) + 0$$

Because the support C is fixed ($\theta_C = 0$) and the frame is braced ($\delta = 0$), the moments M_{BC} and M_{CB} can be expressed as follows.

$$M_{BC} = \frac{2E(2I)}{10}(2\theta_B) \qquad\qquad [2]$$

and

$$M_{CB} = \frac{2E(2I)}{10}(\theta_B) \qquad\qquad [3]$$

Equilibrium Equations.
The moment equilibrium at the support B can be expressed as follows.

$$\Sigma M_B = 0 \quad\quad M_{BA} + M_{BC} + M_{BD} = 0 \qquad\qquad [4]$$

The substitution of Equations [1], [2] and M_{BD} into Equation[4] yields the following value of θ_B:

$$\theta_B = \frac{-164.919}{EI}$$

Final Moments.
The following final moments are obtained by substituting the value of θ_B into Equations [1], [2] and [3]:

M_{AB} = 0 kN.m (Pinned End)
M_{BA} = -118.06 = 118.1 kN.m ⟲
M_{BC} = -131.9 = 131.9 kN.m ⟲
M_{CB} = -66   = 66 kN.m ⟲

The reactions, which are obtained using the final moments values, are shown in Fig. 10-8b. The shear and moment diagrams are shown in Fig. 10-9c and 10-9d, respectively.

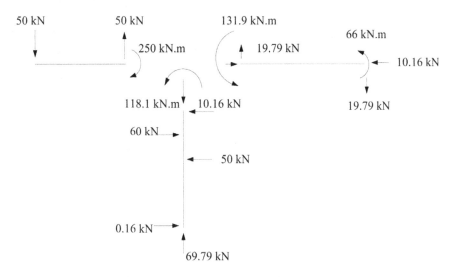

Fig. 10-9b. Member Free Body Diagram

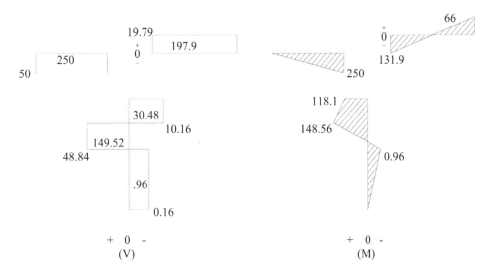

Fig. 10-9c. Shear Diagram, kN   Fig. 10-9d. Moment Diagram, kN.m

Example: 10-7

Determine the reactions for the frame shown in Fig 10-10a.

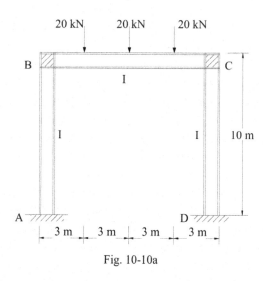

Fig. 10-10a

SOLUTION
Because its geometry and its loading are symmetrical about the vertical axis, the frame will not sway even though it is unrestrained, Fig. 10-3b.

Fixed End Moments.
Table 10-1 is used to determine the moment values in the spans AB, BC, and CD as follows.

$$\text{FEM}_{AB} = \text{FEM}_{BA} \quad = 0 \quad \text{kN.m}$$
$$\text{FEM}_{BC} = -75 \quad = -75 \quad \text{kN.m}$$
$$\text{FEM}_{CB} = 75 \quad = 75 \quad \text{kN.m}$$
$$\text{FEM}_{CD} = \text{FEM}_{DC} \quad = 0 \quad \text{kN.m}$$

Slope deflection Equations [10-1] and [10-2].

Note that:

$\theta_A = \theta_D = 0$	Fixed End
$\psi_{AB} = \psi_{BC} = \psi_{CD} = 0$	No Sway
$\theta_B = -\theta_C$	Symmetry
$M_{AB} = -M_{DC}$	Symmetry
$M_{BA} = -M_{CD}$	Symmetry
$M_{BC} = -M_{CB}$	Symmetry

342

$$M_{AB} = \frac{2EI}{10}[0 + \theta_B + 0] + 0 = \frac{2EI\,\theta_B}{10} \qquad [1]$$

$$M_{BA} = \frac{2EI}{10}[2\theta_B + 0 + 0] + 0 = \frac{4EI\,\theta_B}{10} \qquad [2]$$

$$M_{BC} = \frac{2EI}{12}[2\theta_B + \theta_C + 0] - 75$$

Making use of the symmetry conditions, the moment M$_{BC}$ can be re-written as follows.

$$M_{BC} = \frac{2EI}{12}[2\theta_B - \theta_B + 0] - 75 = \frac{2EI\,\theta_B}{12} - 75 \qquad [3]$$

Equilibrium Equations.
The moment equilibrium equation at the support B is expressed as follows.

$$\sum M_B = 0 \qquad M_{BA} + M_{BC} = 0 \qquad [4]$$

The value of unknown θ_B, which is obtained by substituting Equations [2] and [3] into Equation[4], is equal to the following:

$$\theta_B = \frac{132.35}{EI}$$

The final moments, which are obtained by substituting the value of the unknown θ_B into Equations [1], [2], and [3], are equal to the following:

M$_{AB}$ = 26.5 kN.m
M$_{BA}$ = 53.0 kN.m
M$_{BC}$ = -53.0 = 53.0 kN.m
M$_{CB}$ = 53.0 kN.m
M$_{CD}$ = 53 kN.m
M$_{DC}$ = 26.5 kN.m

The reactions, which are computed using the final moments, are shown in Fig. 10-10b.

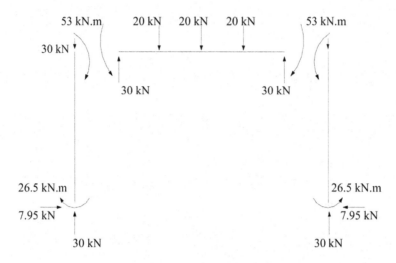

Fig. 10-10b. Member Free Body Diagram

Example: 10-8

Determine the reactions for the frame shown in Fig. 10-11a.

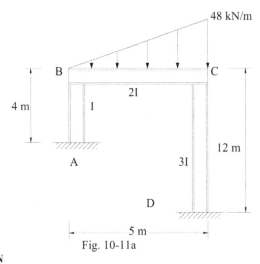

Fig. 10-11a

SOLUTION

Because its geometry and its loading are not symmetrical and because it is not restrained, the frame is not braced and will sway, Fig. 10-3c.

The axial deformation is neglected in member BC. Therefore, both columns will have the same lateral displacement as shown in Fig. 10-11b.

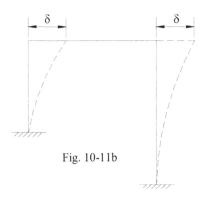

Fig. 10-11b

Fixed End Moments. The following moment values in the spans AB, BC, and CD are determined using Table 10-1:

FEM$_{AB}$ = FEM$_{BA}$	= 0	kN.m
FEM$_{BC}$ = – 40	= -40	kN.m
FEM$_{CB}$ = 60	= 60	kN.m
FEM$_{CD}$ = FEM$_{DC}$	= 0	kN.m

Slope deflection Equations [10-1] and [10-2].
The slope deflection equations [10-1] and [10-2] are used to determine the following moment expressions:

$$M_{AB} = \frac{2EI}{4}\left[2\theta_A + \theta_B - \frac{3\delta}{4}\right] + 0 = \frac{2EI\theta_B}{4} - \frac{6EI}{16}\delta \tag{1}$$

$$M_{BA} = \frac{2EI}{4}\left[2\theta_B + \theta_A - \frac{3\delta}{4}\right] + 0 = \frac{4EI\theta_B}{4} - \frac{6EI}{16}\delta \tag{2}$$

$$M_{BC} = \frac{2E(2I)}{5}\left[2\theta_B + \theta_C - 3(0)\right] - 40 \tag{3}$$

$$M_{CB} = \frac{2E(2I)}{5}\left[2\theta_C + \theta_B - 3(0)\right] + 60 \tag{4}$$

$$M_{CD} = \frac{2E(3I)}{12}\left[2\theta_C + \theta_D - \frac{3\delta}{12}\right] + 0 \tag{5}$$

$$M_{DC} = \frac{2E(3I)}{12}\left[2\theta_D + \theta_C - \frac{3\delta}{12}\right] + 0 \tag{6}$$

Because the supports A and D are fixed , the values of θ_A and θ_D are equal to zero ($\theta_A = \theta_D = 0$). Thus, the unknown values are θ_B, θ_C, and δ

Equilibrium Equations.
Three equations are needed for the solution because there are three unknowns,. Two equations are given by the moment equilibrium equations at the supports B and C.

$$\Sigma M_B = 0 + \circlearrowright \quad M_{BA} + M_{BC} = 0 \tag{7}$$

$$\Sigma M_C = 0 + \circlearrowright \quad M_{CB} + M_{CD} = 0 \tag{8}$$

The third equation will be derived from the summation of the frame forces in the horizontal direction as shown in Fig. 10-11c.

$$A_x + D_x = 0 \tag{9}$$

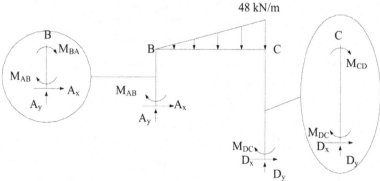

Fig. 10-11c

Taking the summation of the moments about the top joint of each column, Fig. 10-11c, yields the following equations:

$$\Sigma M_B = 0 \; \circlearrowright \qquad A_x = \frac{M_{AB} + M_{BA}}{4} \qquad\qquad [10]$$

$$\Sigma M_C = 0 \; \circlearrowright \qquad D_x = \frac{M_{CD} + M_{DC}}{12} \qquad\qquad [11]$$

Substituting Equations [10] and [11] into Equation [9] yield the following equation:

$$\frac{M_{AB} + M_{BA}}{4} + \frac{M_{CD} + M_{DC}}{12} = 0 \qquad\qquad [12]$$

To solve for the unknowns, Equations [2] and [3] are substituted into Equation[7], Equations [4] and [5] are substituted into Equation[8], and Equations [1], [2], [5], and [6] are substituted into Equation[12] as follows.

$$\frac{13}{5}\theta_B + \frac{4}{5}\theta_C - \frac{3}{8}\delta = \frac{40}{EI}$$

$$\frac{4}{5}\theta_B + \frac{13}{5}\theta_C - \frac{1}{8}\delta = \frac{-60}{EI}$$

$$\frac{3}{8}\theta_B + \frac{1}{8}\theta_C - \frac{5}{24}\delta = 0$$

The unknowns, which are obtained by solving simultaneously the three equations, are as follows.

$$EI\theta_B = 29.912, \qquad EI\theta_C = -30.574, \qquad EI\delta = 35.497$$

Final Moments.

The final moments, which are determined by substituting θ_B, θ_C, δ into Equations [1] to [6], are as follows.

$M_{AB} = 1.645$ kN.m $\circlearrowright$
$M_{BA} = 16.6$ kN.m $\circlearrowright$
$M_{BC} = -16.6 = 16.6$ kN.m $\circlearrowleft$
$M_{CB} = 35.0$ kN.m $\circlearrowright$
$M_{CD} = -35.0 = 35.0$ kN.m $\circlearrowleft$
$M_{DC} = -19.724 = 19.724$ kN.m $\circlearrowleft$

The reactions, which are computed using the final moments values, are shown in Fig. 10-11d.

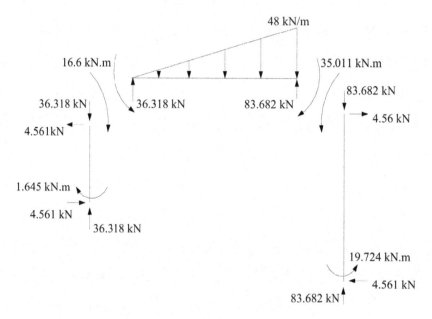

Fig. 10-11d. Member Free Body Diagram

Example: 10-9

Determine the reactions for the frame shown in Fig. 10-12a.

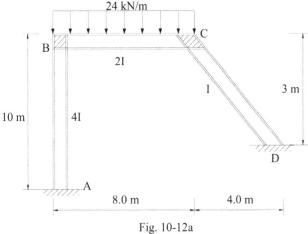

Fig. 10-12a

SOLUTION

The frame will sway because it is not restrained at the beam level and its geometry is not symmetric, Fig. 10-3c. The sloping member CD causes the other members to have different displacements that relate to each other through the angle of inclination, Fig. 10-12b.

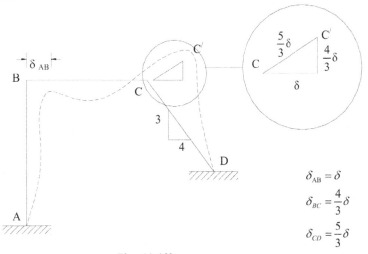

$$\delta_{AB} = \delta$$

$$\delta_{BC} = \frac{4}{3}\delta$$

$$\delta_{CD} = \frac{5}{3}\delta$$

Fig. 10-12b

Fixed End Moments.
The fixed-end moments for the spans AB, BC, and CD, which are determined using
Table 10-1, are listed below.

$$\text{FEM}_{AB} = \text{FEM}_{BA} \quad = 0 \qquad\qquad \text{kN.m}$$
$$\text{FEM}_{BC} = -128 \qquad = -128 \qquad \text{kN.m}$$
$$\text{FEM}_{CB} = 128 \qquad\quad = 128 \qquad\; \text{kN.m}$$
$$\text{FEM}_{CD} = \text{FEM}_{DC} \quad = 0 \qquad\qquad \text{kN.m}$$

Slope deflection Equations [10-1] and [10-2].

The slope deflection equations [10-1] and [10-2] are used to determine the following
moment expressions:

$$M_{AB} = \frac{2E\left(4I\right)}{10}\left[2\theta_A + \theta_B - \frac{3\delta}{10}\right] + 0 \qquad\qquad [1]$$

$$M_{BA} = \frac{2E\left(4I\right)}{10}\left[2\theta_B + \theta_A - \frac{3\delta}{10}\right] + 0 \qquad\qquad [2]$$

$$M_{BC} = \frac{2E\left(2I\right)}{8}\left[2\theta_B + \theta_C + \frac{3\delta\left(\frac{4}{3}\right)}{8}\right] - 128 \qquad\qquad [3]$$

$$M_{CB} = \frac{2E\left(2I\right)}{8}\left[2\theta_C + \theta_B + \frac{3\delta\left(\frac{4}{3}\right)}{8}\right] + 128 \qquad\qquad [4]$$

$$M_{CD} = \frac{2EI}{5}\left[2\theta_C + \theta_D - \frac{3\delta\left(\frac{5}{3}\right)}{5}\right] + 0 \qquad\qquad [5]$$

$$M_{DC} = \frac{2EI}{5}\left[2\theta_D + \theta_C - \frac{3\delta\left(\frac{5}{3}\right)}{5}\right] + 0 \qquad\qquad [6]$$

Because the supports A and D are fixed , the values of θ_A and θ_D are equal to zero
($\theta_A = \theta_D = 0$). Thus, the unknown values are θ_B, θ_C, and δ

Equilibrium Equations.

Since there are three unknowns, therefore three equations are needed. Two equations are given by the moment equilibrium equations for joints B and C.

$$\Sigma M_B = 0 \quad \quad M_{BA} + M_{BC} = 0 \quad\quad\quad\quad [7]$$

$$\Sigma M_C = 0 \quad \quad M_{CB} + M_{CD} = 0 \quad\quad\quad\quad [8]$$

The third equation will be derived from the summation of the forces in the horizontal direction (x-direction) for the frame, Fig. 10-12c.

$$A_x + D_x = 0 \quad\quad\quad\quad [9]$$

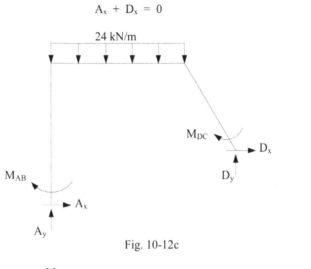

Fig. 10-12c

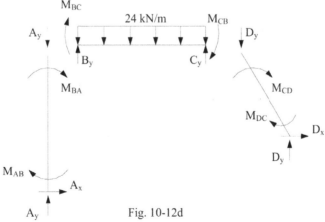

Fig. 10-12d

Taking moments about the top joint of each column, the following expressions for A_x and D_x can be written, Fig. 10-12d.

$$\sum M_B = 0 \quad \circlearrowright \qquad A_x = \frac{M_{AB} + M_{BA}}{10} \qquad\qquad [10]$$

$$\sum M_C = 0 + \circlearrowright \qquad D_x = \frac{M_{CD} + M_{DC} - D_y(4)}{3} \qquad [11]$$

The unknown reaction D_y is obtained by considering the equilibrium of the vertical forces at the joint C and the equilibrium of the moments at the joint B, as follows.

$$\Sigma\, F_y = 0 \uparrow + \qquad\qquad\qquad C_y - D_y = 0$$
$$\therefore\; C_y = D_y \qquad\qquad [12]$$

and

$$\Sigma\, M_B = 0 \quad \circlearrowright \qquad C_y = D_y = \frac{M_{BC} + M_{CB} + \left[24(8)\left(\dfrac{8}{2}\right) \right]}{8} \qquad [13]$$

Substituting Equation [13] into [11] yields the following expression for the reaction D_x:

$$D_x = \frac{M_{DC} + M_{CD}}{3} - \frac{4}{3}\left[\frac{M_{BC} + M_{CB} + 768}{8} \right] \qquad [14]$$

To determine the unknowns, equations [2] and [3] are substituted into Equation [7], Equations [4] and [5] are substituted into Equation[8], and Equations [1] to [6] are substituted into Equation[15], as follows.

$$\frac{13}{5}\theta_B + \frac{1}{2}\theta_C + \frac{1}{100}\delta = \frac{128}{EI}$$

$$\frac{1}{2}\theta_B + \frac{9}{5}\theta_C - \frac{3}{20}\delta = \frac{-128}{EI}$$

$$-\frac{1}{100}\theta_B + \frac{3}{20}\theta_C - \frac{199}{500}\delta = \frac{128}{EI}$$

Solving simultaneously the system of equations yields the values of the unknowns θ_B, θ_C, and δ, which are listed below.

$$EI\theta_B = 74.216,$$
$$EI\theta_C = -122.531,$$
$$EI\delta = -369.653$$

Final Moments.

The final moments, which are obtained by substituting θ_B, θ_C, δ into equations [1] to [6], are listed below.

$M_{AB} = 148.09$ kN.m

$M_{BA} = 207.463$ kN.m

$M_{BC} = -207.463 \quad = 207.463$ kN.m

$M_{CB} = -49.836 \quad = 49.836$ kN.m

$M_{CD} = 49.836$ kN.m

$M_{DC} = 98.849$ kN.m

The reactions, which are obtained using the final moments values, are shown in Fig. 10-12e.

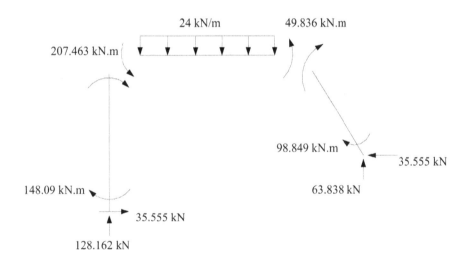

Fig. 10-12e. Member Free Body Diagram

Example: 10-10

Determine the reactions for the gable frame shown in Fig 10-13a.

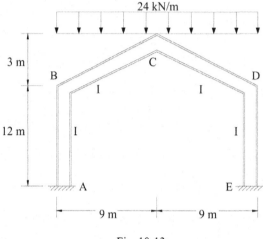

Fig. 10-13a

SOLUTION

The gable frame is symmetrical with respect to its load and its geometry. Because of symmetry, the joints B and D will move away from each other in the horizontal direction with an amount δ, and the joint C will only have a vertical displacement as shown in Fig. 10-13b.

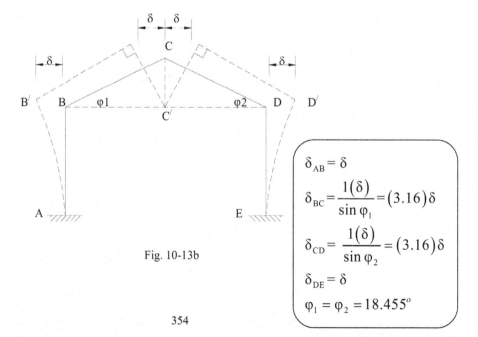

Fig. 10-13b

$$\delta_{AB} = \delta$$

$$\delta_{BC} = \frac{1(\delta)}{\sin \varphi_1} = (3.16)\delta$$

$$\delta_{CD} = \frac{1(\delta)}{\sin \varphi_2} = (3.16)\delta$$

$$\delta_{DE} = \delta$$

$$\varphi_1 = \varphi_2 = 18.455^o$$

Fixed End Moments.

The fixed-end moments for the spans AB, BC, CD, and DE, which are computed using Table 10-1, are listed below.

$$FEM_{AB} = FEM_{BA} = 0$$

$$FEM_{BC} = \frac{wl^2}{12} = \frac{24(9)^2}{12} = \text{-}162 \text{ kN.m}$$

$$FEM_{CB} = 162 \text{ kN.m}$$

$$FEM_{CD} = \text{-}162 \text{ kN.m}$$

$$FEM_{DC} = 162 \text{ kN.m}$$

$$FEM_{DE} = FEM_{ED} = 0$$

Slope deflection Equations [10-1] and [10-2].

The slope deflection equations [10-1] and [10-2] are used to determine the following moment expressions:

$$M_{AB} = \frac{2EI}{12}\left[2\theta_A + \theta_B + \frac{3\delta}{12}\right] + 0 \tag{1}$$

$$M_{BA} = \frac{2EI}{12}\left[2\theta_B + \theta_A + \frac{3\delta}{12}\right] + 0 \tag{2}$$

$$M_{BC} = \frac{2EI}{9.5}\left[2\theta_B + \theta_C - \frac{3\delta(3.16)}{9.5}\right] - 162 \tag{3}$$

$$M_{CB} = \frac{2EI}{9.5}\left[2\theta_C + \theta_B - \frac{3\delta(3.16)}{9.5}\right] + 162 \tag{4}$$

$$M_{CD} = \frac{2EI}{9.5}\left[2\theta_C + \theta_D + \frac{3\delta(3.16)}{9.5}\right] - 162 \tag{5}$$

$$M_{DC} = \frac{2EI}{9.5}\left[2\theta_D + \theta_C + \frac{3\delta(3.16)}{9.5}\right] + 162 \tag{6}$$

$$M_{DE} = \frac{2EI}{12}\left[2\theta_D + \theta_E - \frac{3\delta}{12}\right] + 0 \tag{7}$$

$$M_{ED} = \frac{2EI}{12}\left[2\theta_E + \theta_D - \frac{3\delta}{12}\right] + 0 \tag{8}$$

Because the supports A and E are fixed , the values of θ_A and θ_E are equal to zero ($\theta_A = \theta_E = 0$). Because of symmetry, the value of θ_C is equal to zero ($\theta_C = 0$) and the value of θ_B and θ_D are equal and opposite ($\theta_D = \text{-}\theta_B$).

Equilibrium Equations.

Because of symmetry there are only two unknowns. Therefore, two equations are needed. The moment equilibrium equation at joint B is as follows (Fig. 10-13c).

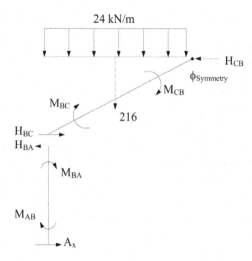

Fig. 10-13c

$$\sum M_B = 0 \quad \quad M_{BA} + M_{BC} = 0 \quad\quad\quad\quad\quad [9]$$

The equilibrium of the horizontal forces at the joint B yields the following equations:

$$\Sigma\, F_x = 0 \rightarrow+ \quad\quad H_{BA} = H_{BC} \quad\quad\quad\quad\quad [10]$$
$$\text{and} \quad\quad H_{BC} = H_{CB}$$

Taking moments about A in column AB yields the following expression for the internal force H_{BA}:

$$H_{BA} = \frac{M_{BA} + M_{AB}}{12} \quad\quad\quad\quad\quad [11]$$

Taking moments about B in inclined member BC yields the following expressions for the internal forces H_{CB} and H_{BC}:

$$H_{CB} = H_{BC} = \frac{M_{BC} + M_{CB} + 972}{3} \quad\quad\quad\quad\quad [12]$$

Substituting Equations [2] and [3] into Equation [9] and Equations [1] to [4] into Equation [10] yields the following equations:

$$0.754\theta_B - 0.1684\delta = \frac{162}{EI}$$

$$-0.169\theta_B + 0.147\delta = \frac{324}{EI}$$

Solving simultaneously these two equations yields the following values for the unknowns θ_B and δ: $EI\theta_B = 951.41$ and $EI\delta = 3298$.

Final Moments

The final moments, which are computed by substituting the values of θ_B and δ in Equations [1] to [8], are listed below.

$M_{AB} = 296$ kN.m
$M_{BA} = 454$ kN.m
$M_{BC} = -454 = 454$ kN.m
$M_{CB} = -331 = 331$ kN.m
$M_{CD} = 331$ kN.m
$M_{DC} = 454$ kN.m
$M_{DE} = -454 = 454$ kN.m
$M_{ED} = -296 = 296$ kN.m

The reactions, which are obtained using the final moments values, are shown in Fig. 10-13d .

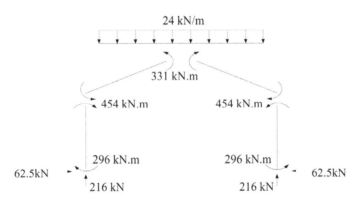

Fig. 10-13d

Example: 10-11

Determine the reactions for the frame shown in Fig. 10-14a if the support A is
displaced 50 mm down and 40 mm to the right, and is rotated 0.005 rad clock wise.
[$E = 200.10^6$ kN/m^2, $I = 400.10^{-6}$ m^4]

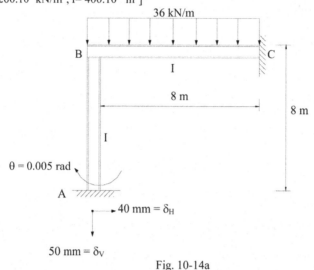

Fig. 10-14a

SOLUTION

Fixed End Moments.

The fixed-end moments for spans AB and BC, which are computed using Table 10-1,
are listed below.

SPAN AB (Column)

$$\text{FEM}_{AB} = \frac{6\,EI\,\delta_H}{L^2} + \frac{4\,EI\theta}{L} = 300 + 200 = 500 \text{ kN.m}$$

$$\text{FEM}_{BA} = \frac{6\,EI\,\delta_H}{L^2} + \frac{2\,EI\theta}{L} = 300 + 100 = 400 \text{ kN.m}$$

SPAN BC (Beam)

$$\text{FEM}_{BC} = -\frac{wl^2}{12} + \frac{6\,EI\delta_v}{L^2} = -192 + 375 = 183 \text{ kN.m}$$

$$\text{FEM}_{CB} = \frac{wl^2}{12} + \frac{6\,EI\delta_v}{L^2} = 192 + 375 = 567 \text{ kN.m}$$

Note:
The displacement direction affects the sign of the moment. In this example, the
displacements produce positive moments in both spans.

Slope deflection Equations [10-1,2].

The slope deflection equations [10-1] and [10-2] are used to determine the following moment expressions:

$$M_{AB} = \frac{2EI}{8}\left[2\theta_A + \theta_B + \frac{3\delta}{8}\right] + 500 \tag{1}$$

$$M_{BA} = \frac{2EI}{8}\left[2\theta_B + \theta_A + \frac{3\delta}{8}\right] + 400 \tag{2}$$

$$M_{BC} = \frac{2EI}{8}\left[2\theta_B + \theta_C + \frac{3\delta}{8}\right] + 183 \tag{3}$$

$$M_{CB} = \frac{2EI}{8}\left[2\theta_C + \theta_B + \frac{3\delta}{8}\right] + 567 \tag{4}$$

Because the supports A and C are fixed , the values of θ_A and θ_C are equal to zero ($\theta_A = \theta_C = 0$). Because the frame is braced (no sway allowed), the values of ψ_{AB} and ψ_{BC} are equal to zero ($\psi_{AB} = \psi_{BC} = 0$)

Equilibrium Equations.
The moment equilibrium equation for joint B is expressed as follows.

$$\sum M_B = 0 \; \circlearrowright \qquad M_{BA} + M_{BC} = 0 \tag{5}$$

Substituting Equations [2] and [3] into Equation [5] yields the following value for the unknown θ_B:

$$\theta_B = \frac{-583}{EI}$$

Final Moments.
The final moments, which are obtained by substituting the value of θ_B in Equations [1] to [4], are listed below.

$M_{AB} = 354.25$ kN.m $\circlearrowright$

$M_{BA} = 108.5$ kN.m $\circlearrowright$

$M_{BC} = -108.5 \; = 108.5$ kN.m $\circlearrowleft$

$M_{CB} = 421.25$ kN.m $\circlearrowright$

The reactions, which are obtained using the final moments values, are shown in Fig. 10-14b.

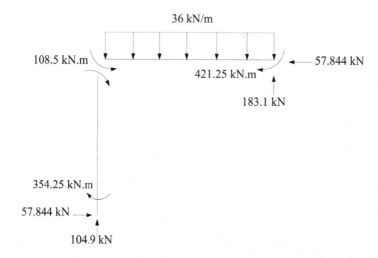

Fig. 10-14b. Member Free Body Diagram

Example: 10-12

Determine the reactions for the two-story frame shown in Fig 10-15a.

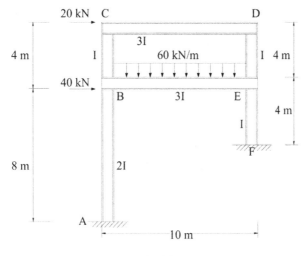

Fig. 10-15a

SOLUTION

The frame will have lateral displacements (sways) at both stories as shown in Fig. 10-15b.

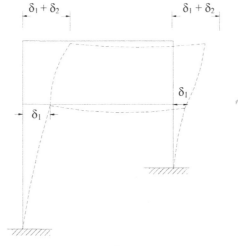

Fig. 10-15b

Fixed End Moments.

The fixed-end moments for the spans AB, BC, CD, DE, EB, and BF, which are computed using Table 10-1, are listed below.

$$\text{FEM}_{AB} = \text{FEM}_{BA} = \text{FEM}_{BC} = \text{FEM}_{CB} = \text{FEM}_{CD} = \text{FEM}_{DC} = 0$$
$$\text{FEM}_{EF} = \text{FEM}_{FE} = 0$$
$$\text{FEM}_{BE} = -500 \text{ kN.m}$$
$$\text{FEM}_{EB} = 500 \text{ kN.m}$$

Slope deflection Equations

The slope deflection equations [10-1] and [10-2] are used to determine the following moment expressions:

$$M_{AB} = \frac{2E(2I)}{8}\left[2\theta_A + \theta_B - \frac{3\delta_1}{8}\right] + 0 \tag{1}$$

$$M_{BA} = \frac{2E(2I)}{8}\left[2\theta_B + \theta_A - \frac{3\delta_1}{8}\right] + 0 \tag{2}$$

$$M_{BC} = \frac{2EI}{4}\left[2\theta_B + \theta_C - \frac{3\delta_2}{4}\right] + 0 \tag{3}$$

$$M_{CB} = \frac{2EI}{4}\left[2\theta_C + \theta_B - \frac{3\delta_2}{4}\right] + 0 \tag{4}$$

$$M_{CD} = \frac{2E(3I)}{10}\left[2\theta_C + \theta_D - 3(0)\right] + 0 \tag{5}$$

$$M_{DC} = \frac{2E(3I)}{10}\left[2\theta_D + \theta_C - 3(0)\right] + 0 \tag{6}$$

$$M_{DE} = \frac{2EI}{4}\left[2\theta_D + \theta_E - \frac{3\delta_2}{4}\right] + 0 \tag{7}$$

$$M_{ED} = \frac{2EI}{4}\left[2\theta_E + \theta_D - \frac{3\delta_2}{4}\right] + 0 \tag{8}$$

$$M_{EF} = \frac{2EI}{4}\left[2\theta_E + \theta_F - \frac{3\delta_1}{4}\right] + 0 \tag{9}$$

$$M_{FE} = \frac{2EI}{4}\left[2\theta_F + \theta_E - \frac{3\delta_1}{4}\right] + 0 \tag{10}$$

$$M_{BE} = \frac{2E(3I)}{10}\left[2\theta_B + \theta_E - 3(0)\right] - 500 \tag{11}$$

$$M_{EB} = \frac{2E(3I)}{10}\left[2\theta_E + \theta_B - 3(0)\right] + 500 \tag{12}$$

Because the supports A and F are fixed , the values of θ_A and θ_F are equal to zero $(\theta_A = \theta_F = 0)$. Therefore, the unknows are the values of $\theta_B, \theta_C, \theta_D, \theta_E, \delta_1, \delta_2$.

Equilibrium Equations.

The moment equilibrium equations for joints B, C, D, and E can be expressed as follows.

362

$$\sum M_B = 0 \quad \circlearrowright \qquad M_{BA} + M_{BC} + M_{BE} = 0 \qquad [13]$$

$$\sum M_C = 0 \quad \circlearrowright \qquad M_{CB} + M_{CD} \qquad\quad = 0 \qquad [14]$$

$$\sum M_D = 0 \quad \circlearrowright \qquad M_{DC} + M_{DE} \qquad\quad = 0 \qquad [15]$$

$$\sum M_E = 0 \quad \circlearrowright \qquad M_{EF} + M_{ED} + M_{EB} = 0 \qquad [16]$$

The other two equations will be derived from the summation of forces in the horizontal direction (x-direction) as shown in Fig. 10-15c and Fig. 10-15d.

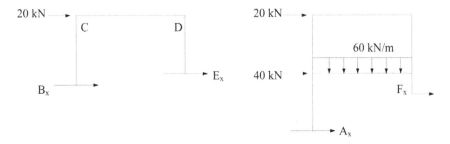

Fig. 10-15c. Upper Story   Fig. 10-15d. Lower Story

The summation of the horizontal forces on the upper story yields the following equation (Fig. 10-15c):

$\Sigma F_x = 0 \rightarrow +$

$$20 + B_x + E_x = 0$$

$$20 + \frac{M_{BC} + M_{CB}}{4} + \frac{M_{DE} + M_{ED}}{4} = 0 \qquad [17]$$

The summation of the horizontal forces on the lower level yields the following equation (Fig. 10-15d):

Lower story;.

$\Sigma F_x = 0 \rightarrow +$

$$60 + A_x + F_x = 0$$

$$60 + \frac{M_{AB} + M_{BA}}{8} + \frac{M_{EF} + M_{FE}}{4} = 0 \qquad [18]$$

Solving for the unknowns requires substituting Equations [1] to [12] into Equations [13] to [18] as follows.

$$\frac{16}{5}\theta_B + \frac{1}{2}\theta_C + 0\theta_D + \frac{3}{5}\theta_E - \frac{3}{16}\delta_1 - \frac{3}{8}\delta_2 = \frac{500}{EI}$$

$$\frac{1}{2}\theta_B + \frac{11}{5}\theta_C + \frac{3}{5}\theta_D + 0\theta_E + 0\delta_1 - \frac{3}{8}\delta_2 = 0$$

$$0\theta_B + \frac{3}{5}\theta_C + \frac{11}{5}\theta_D + \frac{1}{2}\theta_E + 0\delta_1 - \frac{3}{8}\delta_2 = 0$$

$$\frac{3}{5}\theta_B + 0\theta_C + \frac{1}{2}\theta_D + \frac{16}{5}\theta_E - \frac{3}{8}\delta_1 - \frac{3}{8}\delta_2 = -\frac{500}{EI}$$

$$\frac{3}{8}\theta_B + \frac{3}{8}\theta_C + \frac{3}{8}\theta_D + \frac{3}{8}\theta_E + 0\delta_1 - \frac{3}{8}\delta_2 = \frac{-20}{EI}$$

$$\frac{3}{16}\theta_B + 0\theta_C + 0\theta_D + \frac{3}{8}\theta_E - \frac{15}{64}\delta_1 + 0\delta_2 = \frac{-60}{EI}$$

Solving simultaneously these equations yields the following values for the unknows:
$EI\theta_B = 220.13$, $EI\theta_C = -49.83$, $EI\theta_D = 74.51$, $EI\theta_E = -177.75$, $EI\delta_1 = 147.7$, and
$EI\delta_2 = 120.39$

Final Moments.
The final moments , which are obtained by substituting the values of the six unknowns in Equations [1] to [12], are listed below.

$M_{AB} = 82.4$ kN.m
$M_{BA} = 192.4$ kN.m
$M_{BC} = 150$ kN.m
$M_{CB} = 15$ kN.m
$M_{CD} = -15 = 15$ kN.m
$M_{DC} = 59.5$ kN.m
$M_{DE} = -59.5 = 59.5$ kN.m
$M_{ED} = -185.64 = 185.64$ kN.m
$M_{EF} = -233.14 = 233.14$ kN.m
$M_{FE} = -144.26 = 144.26$ kN.m
$M_{BE} = -342.5 = 342.5$ kN.m

$M_{EB} = 418.78 \ kN.m$ ↻

The reactions, which are computed using the final moments values, are shown in Fig. 10-15e.

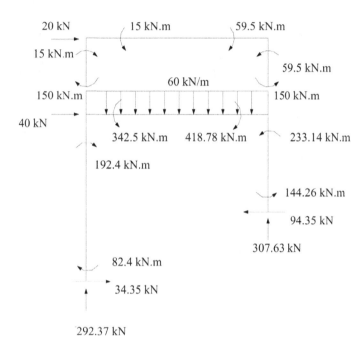

Fig. 10-15e. Member Free Body Diagram

Summary

- The slope-deflection procedure is an early classical method analyzing indeterminate beams and rigid frames, in which joint displacements are the unknowns.
- For highly indeterminate structures with a large number of joints, the slope-deflection solution requires the solution of a series of simultaneous equations equal in number to the unknown displacements —a time-consuming operation. While the use of the slope-deflection method to analyze structures is impractical given the availability of the computer programs, the familiarity with the method provides valuable insight into the behavior of structures.
- A variation of the slope-deflection procedure, the general stiffness method, used to prepare general-purpose computer programs, is presented in Chapter 16. This method utilizes stiffness coefficients – forces produced by unit joint displacements.

Problems

For the problems 1 to 3, determine the reactions using the slope deflection method.
EI = Const

1-

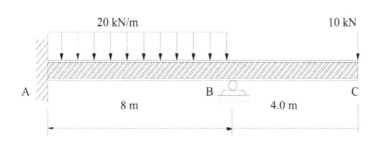

2-

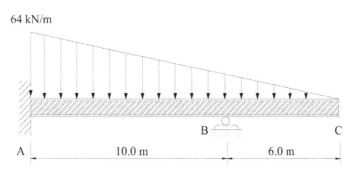

3-

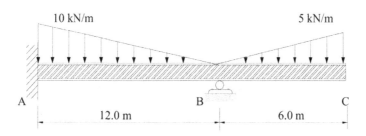

Chapter 10.   Slope Deflection Method

4-   Use the slope deflection method to draw the shear and moment diagrams for a
      0.05-meter settlement of the support B.
      [E = 200.10^6 kN/m^2, I =100.10^6 mm^4]

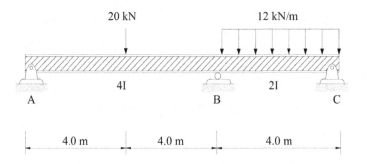

5-   Use the slope deflection method to determine the reactions for a 50-mm
      settlement of the support C.   [E = 200.10^9 N/m^2, I = 2000.10^-6 m^4]

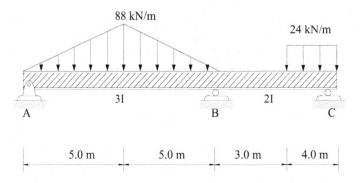

6-   Use the slope deflection method to determine the reactions. [E I = const.].

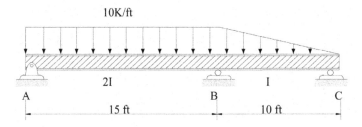

7-   Use the slope deflection method to draw the shear and moment diagrams.

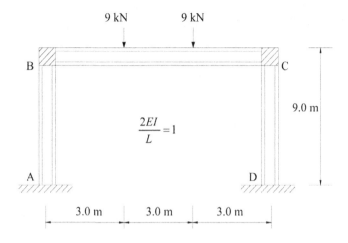

8-   Use the slope deflection method to draw the shear and moment diagrams.

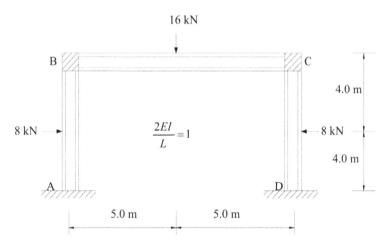

Chapter 10.   Slope Deflection Method

9-   Determine the reactions.

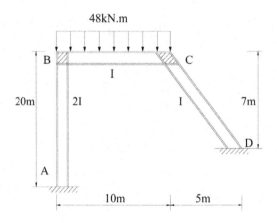

48kN.m

B

I

C

20m   2I   I   7m

A

10m   5m

D

10-   Determine the reactions.

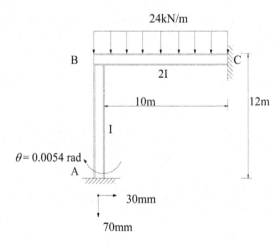

24kN/m

B   C

2I

10m   12m

I

θ = 0.0054 rad
A

30mm

70mm

CHAPTER-11
MOMENT DISTRIBUTION METHOD

Dubai-UAE

Introduction

The moment distribution method, which is an approximate procedure for analyzing indeterminate beams and frames, eliminates the need to write and solve the simultaneous equations required by the slope-deflection method.

The moment distribution method starts with the assumption that all free joints are restrained by clamps, producing fixed-end conditions. When loads are applied, fixed-end moments are induced. The solution is completed by unlocking and relocking joints in succession and distributing moments to both ends of all the members framing every free joint until all joints are in equilibrium. The moment distribution is achieved by using a series of iterations using stiffness factors, distribution factors, carry over factors, and carry over moments.

The time required to complete the analysis increases significantly if the frames are free to side-sway. The method can be extended to non-prismatic member by using available standard tables of fixed-end moments.

Although it provides students with an insight into the behavior of continuous structures, the moment distribution method has currently limited use in practice because computer solutions are faster and more accurate.

Member Stiffness Factors

The member stiffness factor is the moment required to produce a unit rotation at one end A (Fig. 11-1).
The stiffness factor for a beam AB is given by the following equation:

$$k = \frac{M}{\theta} \qquad\qquad [11 - 1]$$

Applying conjugate beam method, the stiffness factor is given by the following equation (Fig. 11-2):

$$k = \frac{4EI}{L} \qquad \text{Fixed far end} \qquad [11 - 2]$$

For a continuous beam, the relative stiffness factors are given by the following equations:

$$k = \frac{I}{L} \qquad \text{Fixed far end} \quad M_F \neq 0 \qquad [11$$
$$- 3]$$

$$k = \frac{3}{4}\frac{I}{L} \qquad \text{Pinned far end} \quad M_F = 0 \qquad [11$$
$$- 4]$$

Distribution Factors

The distribution factor, which is the ratio of the stiffness factor of the selected beam to the sum of the stiffness of all the members connected at that specific joint, is given by the following equation:

$$DF = \frac{k}{\sum k} \qquad\qquad [11 - 5]$$

Note that the summation of all distribution factors at a joint must equal 1.

Carry Over Moments

The carry over moment is the moment $(\overline{M})$ developed at a fixed far end due to the applied moment (M) at near end as shown in Fig. 11–1.

Carry Over Factors

The ratio of the carry over moment at the far end due to the applied moment is given by the following equation (Fig. 11-2):

$$\text{Carry over factor} = \frac{\overline{M}}{M} = \frac{1}{2} \qquad \text{for prismatic member}$$

The determination of the final end moments of continuous beams, frames without sideway, and frame with sideway using the moment distribution method consists of the following steps:

1. Determine the fixed end moments for each member using Table10-1.
2. Calculate the stiffness factors for all members.
3. Calculate the distribution factors for all members.
4. Distribute the negative value of the unbalanced moment at the joint to various members meeting at that joint proportional to their distribution factors. Repeat the process for all joints.
5. Carry over moments to the far ends of the members causing unbalanced joints.
6. Repeat the steps 4 and 5 until the distributed moments converge to the desired accuracy.
7. Add all moments at each member end to get final joint moments.
8. Steps 3 through 7 have to be repeated for sway frames loaded with end moments caused by an arbitrary Δ (Frame moments). Once the frame moments with the desired accuracy are determined the final moments are calculated using the Correction Factor (CF).
9. Consider each member of the structure as simply supported beam subjected to given loads and end moments (final moments from the previous step). Apply the principles of statics to determine the end reactions of each member and draw the shear and moment diagrams.

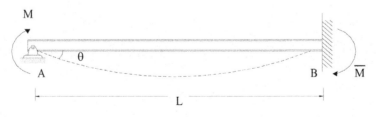

Fig. 11-1

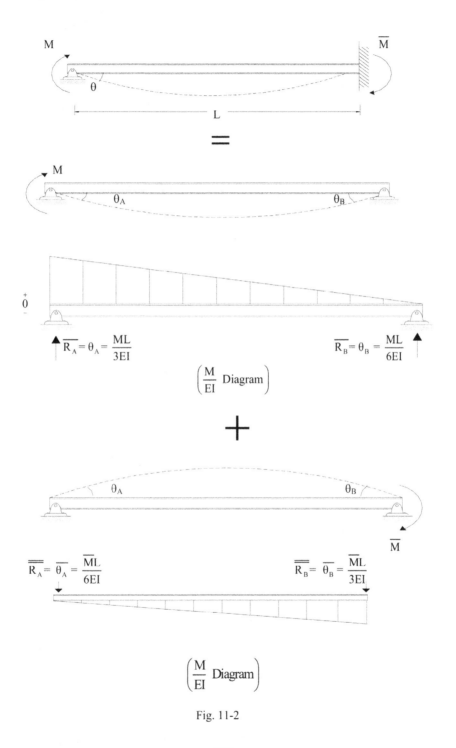

$$=$$

$$\left(\frac{M}{EI} \; \text{Diagram} \right)$$

$$+$$

$$\left(\frac{M}{EI} \; \text{Diagram} \right)$$

Fig. 11-2

$$\theta_B - \overline{\theta}_B = 0 \qquad \text{Fixed End}$$

$$\theta_B = \overline{\theta}_B$$

$$\frac{\overline{M}}{M} = \frac{1}{2} \qquad \text{Carry Over Factor}$$

and

$$\theta = \theta_A - \overline{\theta}_A$$

$$\theta = \frac{ML}{4EI}$$

$$K = \frac{M}{\theta} = \frac{4EI}{L} \qquad \text{Stiffness Factor}$$

Example: 11-1

Determine the reactions for the beam shown in Fig 11-3a.

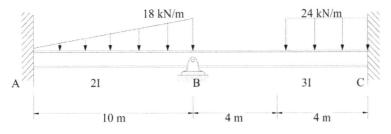

Fig. 11-3a

SOLUTION

The fixed end moments, stiffness factors, and distribution factors for the spans AB
and BC, which are determined using Table 10-1, Equation [11-3] and
Equation [11-5], are listed below.

$$FEM_{AB} = -60 \, kN.m \quad , \quad FEM_{BA} = 90 \, kN.m$$

$$FEM_{BC} = -40 \, kN.m \quad , \quad FEM_{CB} = 88 \, kN.m$$

$$K_{AB} = \frac{2I}{10} \quad , \quad K_{BC} = \frac{3I}{8}$$

$DF_{AB} = 0$ \qquad (The Fixed End is a self balancing joint).

$DF_{BA} = \dfrac{K_{AB}}{K_{AB} + K_{BC}} = 0.348$ ⎫

⎬ Summation of the distribution factors at joint B

$DF_{BC} = \dfrac{K_{BC}}{K_{AB} + K_{BC}} = 0.652$ ⎭

$$\sum = 1.0$$

$DF_{CB} = 0$ \qquad (Fixed End).

Moment Distribution Table

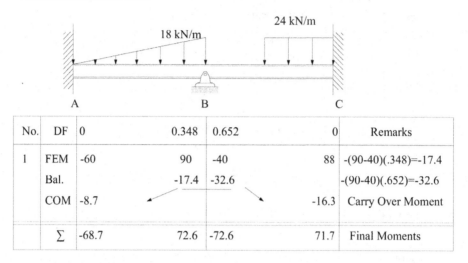

No.	DF	0	0.348	0.652	0	Remarks
1	FEM	-60	90	-40	88	-(90-40)(.348)=-17.4
	Bal.		-17.4	-32.6		-(90-40)(.652)=-32.6
	COM	-8.7			-16.3	Carry Over Moment
	Σ	-68.7	72.6	-72.6	71.7	Final Moments

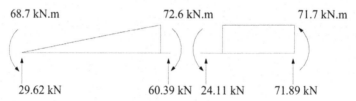

Fig. 11-3c. Member Free-Body diagram

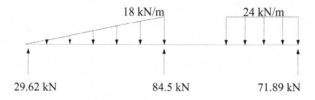

Fig. 11-3d. Applied Loads and Reactions

Note

One cycle is sufficient herein to yield accurate moment values.

Example: 11-2

Draw the shear and moment diagrams for the beam shown in Fig. 11-4a if the support B settles 0.04m.
$[E = 28(10)^6 \text{ N/m}^2, \; I = 200 \times 10^{-6} \text{ m}^4]$

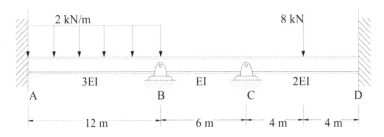

2 kN/m

8 kN

3EI

EI

2EI

A

B

C

D

12 m

6 m

4 m

4 m

Fig. 11-4a

SOLUTION

The fixed end moments due to the external loads and the support settlement as well as the stiffness factors for the spans AB, BC, and CD are listed below.

$$FEM_{AB} = (-24) + \left(\frac{-6E(3I)(0.04)}{12^2} = -0.028 \right) = -24.028 \, kN.m$$

$$FEM_{BA} = (24) + (-0.028) = 23.972 \, kN.m$$

$$FEM_{BC} = 0 + \left(\frac{6EI(0.04)}{6^2} = 0.037 \right) = 0.037 \, kN.m$$

$$FEM_{CB} = 0 + (0.037) = 0.037 \, kN.m$$

$$FEM_{CD} = (-8) + 0 = -8 \, kN.m$$

$$FEM_{DC} = (+8) + 0 = +8 \, kN.m$$

$$K_{AB} = \frac{3I}{12} \quad, \quad K_{BC} = \frac{I}{6} \quad, \quad K_{CD} = \frac{2I}{8}$$

The distribution factors for the spans AB, BC, and CD are listed below.

$DF_{AB} = DF_{DC} = 0$ (FIXED END).

$DF_{BA} = 0.6$, $DF_{BC} = 0.4$

$DF_{CB} = 0.4$, $DF_{CD} = 0.6$

Moment Distribution Table

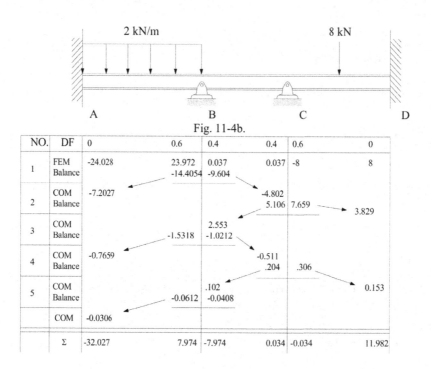

Fig. 11-4b.

NO.	DF	0		0.6	0.4		0.4	0.6		0
1	FEM	-24.028		23.972	0.037		0.037	-8		8
	Balance			-14.4054	-9.604					
2	COM	-7.2027					-4.802			
	Balance						5.106	7.659		3.829
3	COM				2.553					
	Balance			-1.5318	-1.0212					
4	COM	-0.7659					-0.511			
	Balance						.204	.306		0.153
5	COM				.102					
	Balance			-0.0612	-0.0408					
	COM	-0.0306								
	Σ	-32.027		7.974	-7.974		0.034	-0.034		11.982

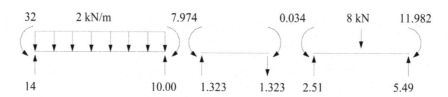

Fig. 11-4c. Member Free Body Diagram

Fig. 11-4d. Applied Loads and Reactions

380

Fig. 11-4e. Shear Diagram (kN)

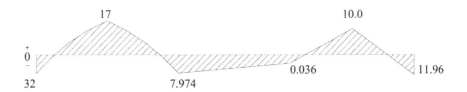

Fig. 11-4f. Moment Diagram (kN.m)

<u>Note</u>
Three cycles are sufficient to give good and accurate moment values.

Example: 11-3

Draw the shear and moment diagrams for the beam shown in Fig 11-5a.

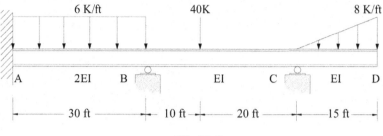

Fig. 11-5a

SOLUTION

      The fixed end moments, the stiffness factors, and the distribution factors for the spans AB, BC, and CD are listed below.

$$FEM_{AB} = -450 \; K\,ft$$

$$FEM_{BA} = \;\;\;450 \; K\,ft$$

$$FEM_{BC} = \;-177.778 \;\; K\,ft$$

$$FEM_{CB} = \;\;88.889 \;\; K\,ft$$

$$M_{CD} = -600 \quad K\,ft, \;\text{Fig.11-5b}$$

$$K_{AB} = \frac{2I}{30} \quad , \quad K_{BC} = \frac{I}{30}\left(\frac{3}{4}\right) \quad [PINNED \; END \; SPAN]$$

$DF_{BA} = 0.728$	$DF_{CB} = 1.0$	[PINNED END]
$DF_{BC} = 0.272$	$DF_{CD} = DF_{AB} = 0$	

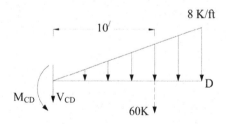

Fig. 11-5b. Over hang moment M_{CD}

Moment Distribution Table

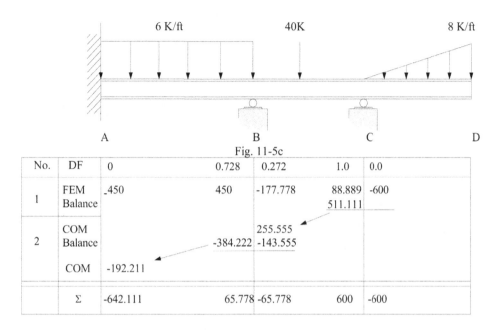

Fig. 11-5c

No.	DF	0		0.728	0.272	1.0	0.0
1	FEM	-450		450	-177.778	88.889	-600
	Balance					511.111	
2	COM				255.555		
	Balance			-384.222	-143.555		
	COM	-192.211					
	Σ	-642.111		65.778	-65.778	600	-600

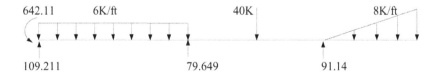

Fig. 11-5d. Member Free Body Diagram

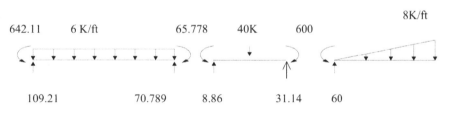

Fig. 11-5e. Loads and Reactions

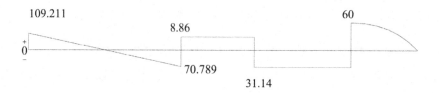

109.211

8.86

60

0 +
  −

70.789

31.14

Fig. 11-5f. Shear Diagram (Kips)

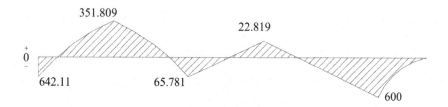

351.809

22.819

0 +
  −

642.11

65.781

600

Fig. 11-5g. Moment Diagram (K.ft)

Example: 11-4

Determine the reactions for the beam shown Fig 11-6a.

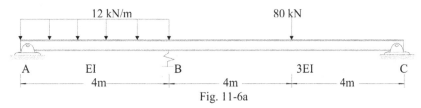

Fig. 11-6a

$[\,E = 200.10^6 \, kN/m^2, \, I = 200.10^6 \, mm^4, \, K \text{ spring} = 400 \, kN/m \,]$

SOLUTION

Because of the spring at B, two beam analyses are needed. The first one considers no settlement, Table 11-1, while the second one considers a settlement effect (Δ) as shown in Table 11-2.

The fixed end moments for the first analysis (without settlement) are listed below.

$$FEM_{AB} = -16 \, kN.m$$
$$FEM_{BA} = \;\;\;16 \, kN.m$$
$$FEM_{BC} = \;-80 kN.m$$
$$FEM_{CB} = \;\;\;80 kN.m$$

The fixed end moments for the second analysis (with settlement) are listed below.

$$FEM_{AB} = -\frac{6\,EI\,\Delta}{L^2} \;\;\; = -15000\,\Delta$$

$$FEM_{BA} = -\frac{6\,EI\,\Delta}{L^2} \;\;\; = -15000\,\Delta$$

$$FEM_{BC} = \frac{6\,E(3I)\,\Delta}{L^2} \;\;\; = 11250\,\Delta$$

$$FEM_{CB} = \frac{6\,E(3I)\,\Delta}{L^2} \;\;\; = 11250\,\Delta$$

The stiffness factors and distribution factors are listed below.

$$K_{AB} = \frac{I}{4}\left(\frac{3}{4}\right) \quad , \quad K_{BC} = \frac{3I}{8}\left(\frac{3}{4}\right) \quad \textbf{Pinned End Spans}$$

$$DF_{BA} = 0.4 \quad , \quad DF_{BC} = 0.6$$

Moment Distribution Tables

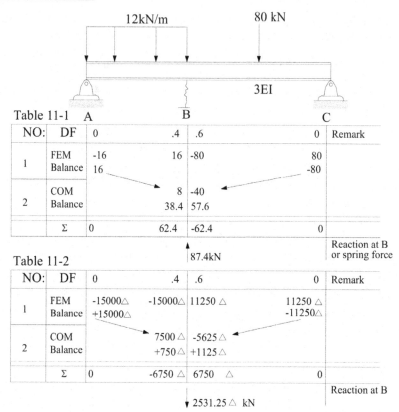

Table 11-1

NO:	DF	0	.4	.6	0	Remark
1	FEM	-16	16	-80	80	
	Balance	16			-80	
2	COM		8	-40		
	Balance		38.4	57.6		
	Σ	0	62.4	-62.4	0	

Reaction at B or spring force

87.4kN

Table 11-2

NO:	DF	0	.4	.6	0	Remark
1	FEM	-15000△	-15000△	11250 △	11250 △	
	Balance	+15000△			-11250△	
2	COM		7500 △	-5625 △		
	Balance		+750△	+1125 △		
	Σ	0	-6750 △	6750 △	0	

Reaction at B

2531.25 △ kN

The summing of the forces at joint B for static equilibrium yields the following equation:

$2531.25\Delta + K_{sp}\,\Delta = 87.4$

$2531.25\Delta + 400\;\Delta = 87.4$

$\therefore \Delta = 0.0298$ m.

Using the value of spring settlement (Δ), the member end moments are determined and listed below.

$M_{AB} = 0 + 0 = 0$

$M_{BA} = 62.4 + (-6750\,\Delta) = -138.75 = 138.75$

$M_{BC} = -62.4 + (6750\,\Delta) = 138.75$

$M_{CB} = 0 + 0 = 0$

The member free body diagram and the beam reactions are shown in Figs. 11-6b and 11-6c, respectively.

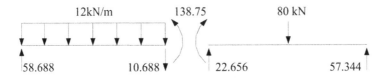

Fig. 11-6b. Member Free Body Diagram

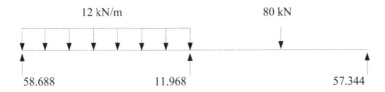

Fig. 11-6c. Loads and Reactions

Example: 11-5

Draw the shear and moment diagrams for the beam shown in Fig 11-7a.

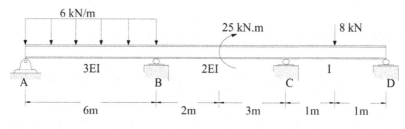

Fig. 11-7a

SOLUTION

The fixed end moments, the stiffness factors, and the distribution factors for spans AB, BC, and CD are listed below.

$FEM_{AB} = -18\,kN.m$

$FEM_{BA} = 18\,kN.m$

$FEM_{BC} = \dfrac{25(3)}{(5)^2}\left[2(2)-3\right] = 3kN.m$ (+25 Clockwise)

$FEM_{CB} = \dfrac{25(2)}{(5)^2}\left[2(3)-2\right] = 8kN.m$ (+25 Clockwise)

$FEM_{CD} = -2\,kN.m$

$FEM_{CD} = 2\,kN.m$

$K_{AB} = \dfrac{3I}{6}\left(\dfrac{3}{4}\right)$, $K_{BC} = \dfrac{2I}{5}$, $K_{CD} = \dfrac{I}{2}\left(\dfrac{3}{4}\right)$

DF$_{BA}$ = 0.484   DF$_{BC}$ = 0.516

DF$_{CB}$ = 0.516,   DF$_{CD}$ = 0.484

Moment Distribution Table

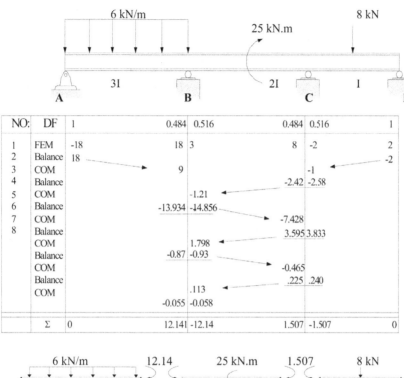

NO:	DF	1	0.484	0.516		0.484	0.516	1
1	FEM	-18	18	3		8	-2	2
2	Balance	18						-2
3	COM			9			-1	
4	Balance					-2.42	-2.58	
5	COM			-1.21				
6	Balance		-13.934	-14.856				
7	COM					-7.428		
8	Balance					3.595	3.833	
	COM			1.798				
	Balance		-0.87	-0.93				
	COM					-0.465		
	Balance					.225	.240	
	COM			.113				
			-0.055	-0.058				
	Σ	0	12.141	-12.14		1.507	-1.507	0

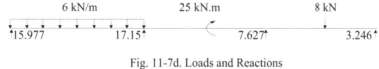

Fig. 11-7c. Member Free Body Diagram

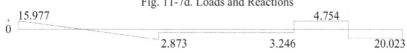

Fig. 11-7d. Loads and Reactions

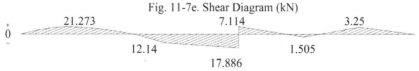

Fig. 11-7e. Shear Diagram (kN)

Fig. 11-7f. Moment Diagram (kN.m))

Example: 11-6

Determine the reactions for the beam shown in Fig. 11-8a if both supports at B and C settle by 0.03 m. [EI = 40000 kN.m^2]

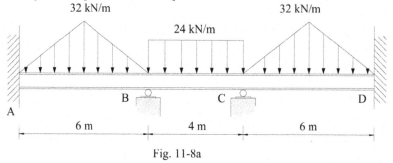

Fig. 11-8a

SOLUTION

The fixed end moments, stiffness factors, and distribution factors for the spans AB, BC, and CD are listed below.

$$FEM_{AB} = -60 + (-200) = -260\,kN.m$$

$$FEM_{BA} = 60 + (-200) = -140\,kN.m$$

$$FEM_{BC} = -32 + 450 - 450 = -32kN.m$$

$$FEM_{CB} = 32 + 450 - 450 = 32kN.m$$

$$FEM_{CD} = -60 + 200 = -140kN.m$$

$$FEM_{CD} = 60 + 200 = 260kN.m$$

$$K_{AB} = K_{CD} = \frac{I}{6} \quad , \quad K_{BC} = \frac{I}{4}\left(\frac{1}{2}\right) \quad [Symmetry]$$

$$DF_{BA} = 0.571 \quad , \quad DF_{BC} = 0.429$$

Chapter 11.   Moment Distribution Method

Moment Distribution Table

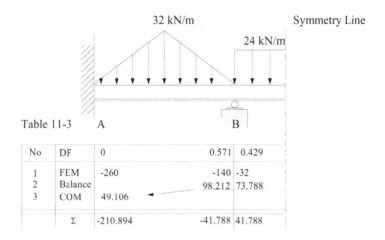

Table 11-3

No	DF	0		0.571	0.429
1	FEM	-260		-140	-32
2	Balance			98.212	73.788
3	COM	49.106			
	Σ	-210.894		-41.788	41.788

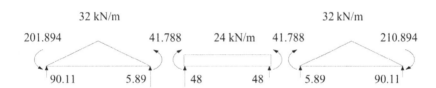

Fig. 11-8c. Member Free Body Diagram

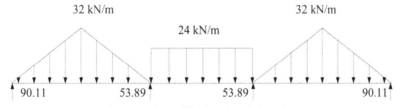

Fig. 11-8d. Applied Loads and Reactions

Note

The solution of this problem is tedious and long without the use of the structure symmetry feature.

391

Example: 11-7

Determine the reaction for the frame shown in Fig 11-9a.

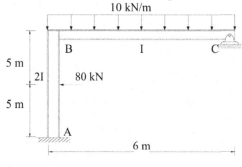

Fig. 11-9a

SOLUTION

The frame is braced against sides-way at its top. Therefore, the moment distribution method procedure is the same as that of beams.

The fixed end moments, stiffness factors, and distribution factors for the spans AB and BC are listed below.

$FEM_{AB} = 100 \, kN.m$

$FEM_{BA} = -100 \, kN.m$

$FEM_{BC} = -30 \, kN.m$

$FEM_{CB} = 30 \, kN.m$

$$K_{AB} = \frac{2I}{10} \quad , \quad K_{BC} = \frac{I}{6}\left(\frac{3}{4}\right) \quad [PINNED \, END]$$

$DF_{BA} = 0.615$, $DF_{BC} = 0.385$

$DF_{AB} = 0$ [FIXED END], $DF_{CB} = 1.0$ [PINNED END]

Moment Distribution Table

	Joint	A	B		C
No.	DF	0	0.615	0.385	1
1	FEM	100	-100	-30	30
2	Bal				-30
	COM				
3				-15	
4	Bal		89.175	55.825	
5	COM	44.588			
	Σ	144.588	-10.825	10.825	0

The member free body diagrams and reactions are shown in Fig. 11-9b.

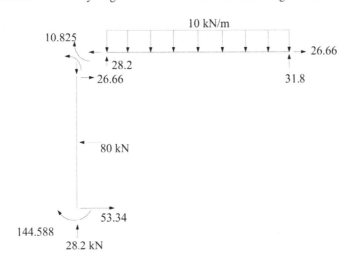

Fig. 11-9b. Member Free Body Diagram

393

Example: 11-8

Determine the reactions for the frame shown in Fig 11-10a.

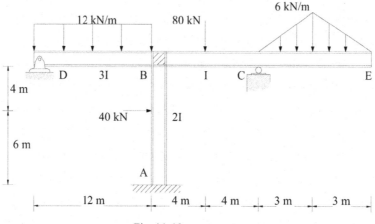

Fig. 11-10a

SOLUTION

The frame has no sides-way. The fixed end moments, stiffness factors, and distribution factors for the spans AB, BC, and CD as well as the over hang moment M_{CE} are listed below.

$FEM_{AB} = -38.4$ kN.m

$FEM_{BA} = 57.6$ kN.m

$FEM_{DB} = -144$ kN.m

$FEM_{BD} = 144$ kN.m

$FEM_{BC} = -80$ kN.m

$FEM_{CB} = 80$ kN.m

$M_{CE} = -54$ kN.m

$$K_{AB} = \frac{2I}{10} \quad , \quad K_{DB} = \frac{3I}{12}\left(\frac{3}{4}\right) \quad , \quad K_{BC} = \frac{I}{8}\left(\frac{3}{4}\right)$$

$DF_{BA} = 0.416$, $DF_{BC} = 0.195$, $DF_{BD} = 0.389$

$DF_{AB} = 0$, $DF_{DB} = DF_{CB} = 1.0$ [PINNED END]

394

Moment Distribution Table

	Joint	D		B		C		A	
No.	Member	DB	BD	BA	BC	CB	CE	BA	
	DF	1	0.389	0.416	0.195	1	NA	0	
1	FEM	-144	144	57.6	-80	80	-54	-38.4	
2	Balance	144				-26			
3	COM		72		-13				
4	Balance		-70.25	-75.13	-35.2				
5	COM							► -37.57	
	Σ	0	145.75	-17.53	128.2	54	-54	-75.97	

The member free body diagrams and reactions are shown in Fig. 11-10c.

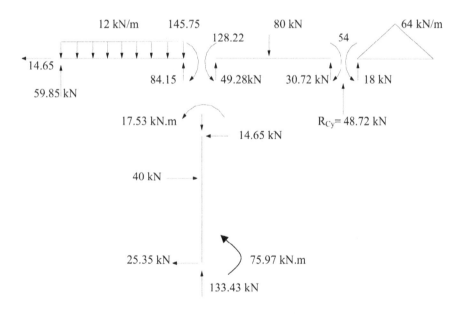

Fig. 11-10c. Member Free Body Diagram

Example: 11-9

Determine the reactions for the frame shown in Fig 11-11a.

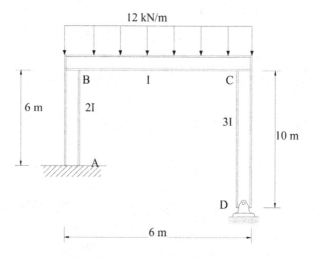

Fig. 11-11a

SOLUTION

Because it is not symmetrical and not prevented from sidesway at its top, the frame is unbraced (i.e., allows for sidesway). The solution of unbraced (sway) frames consists of two analyses: non-sway and sway analyses.

Non-sway Analysis

The fixed end moments, stiffness factors, and distribution factors for the spans AB, BC, and CD are listed below.

$FEM_{AB} = FEM_{BA} = 0$

$FEM_{BC} = -36 \text{ kN.m}$

$FEM_{CB} = 36 \text{ kN.m}$

$FEM_{CD} = FEM_{DC} = 0$

$$K_{AB} = \frac{2I}{6} \quad , \quad K_{BC} = \frac{I}{6} \quad , \quad K_{CD} = \frac{3I}{10}\left(\frac{3}{4}\right)$$

$DF_{BA} = 0.667$, $DF_{BC} = 0.333$, $DF_{CB} = 0.426$

$DF_{AB} = 0$ [FIXED END] , $DF_{CD} = 0.574$, $DF_{DC} = 1$ [PINNED END]

Moment Distribution Table

	Joints	A	B		C		D
No:	Members	AB	BA	BC	CB	CD	DC
	DF	0	0.667	0.333	0.426	0.574	1
1	FEM	0	0	-36	36	0	0
2	Balance		24	12			
3	COM	12			6		
4	Bal				-17.892	-24108	
5	COM			-8.946			
6	Bal		5.967	2.979			
7	COM	2.984			1.49		
8	Bal				-.635	-.855	
9	COM			-0.318			
10	Bal		0.212	0.106			
	Σ	14.984	30.179	-30.179	24.963	-24.963	0

The reactions H_A and H_B (Fig. 11-11b) are computed by considering the free body diagrams of the columns and by taking the moments about the joints B and C.

H_A	7.527 kN →
H_D	2.496 kN ←

To satisfy the horizontal equilibrium of the frame, Fig. 11-11c, F must be equal to 5.031 kN ←

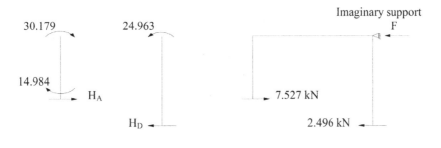

Fig. 11-11b   Fig. 11-11c

Sway Analysis

The frame will sway by an arbitrary Δ that will produce a counter-clockwise moments due to the force F as shown in Fig. 11-11d.

$$\text{FEM}_{AB} \quad : \quad \text{FEM}_{CD}$$

$$\frac{6\,EI_{AB}\,\Delta}{L_{AB}^2} \quad : \quad \frac{6\,EI_{CD}\,\Delta}{L_{CD}^2}$$

$$\frac{\text{FEM}_{AB}}{\text{FEM}_{CD}} = \frac{\left(\dfrac{I_{AB}}{L_{AB}^2}\right)}{\left(\dfrac{I_{CD}}{L_{CD}^2}\right)} = \frac{\left(\dfrac{2I}{36}\right)}{\left(\dfrac{3I}{100}\right)} = \frac{33.333}{18}$$

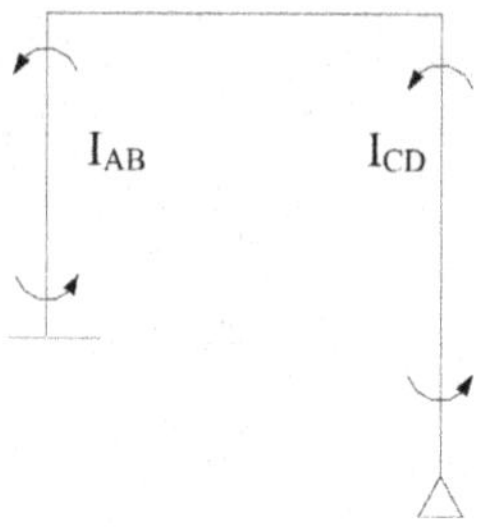

Let us assume the initial end moments.

Fig. 11-11d

$$\text{FEM}_{AB} = \text{FEM}_{BA} = -33.333 \text{ kN.m}$$
$$\text{FEM}_{CD} = \text{FEM}_{DC} = -18 \text{ kN.m}$$

The moment distribution procedure is carried out using these end moments as follows

No:	Joints	A	B		C		D
	Members	AB	BA	BC	CB	CD	DC
	DF	0	0.667	0.333	0.426	0.574	1
1	FEM	-33.333	-33.333	0	0	-18	-18
2	Balance						+18
3	COM					9	
4	Bal				3.834	5.166	
5	COM			1.917			
6	Bal		20.954	10.462			
7	COM	10.477			5.231		
8	Bal				-2.228	-3.003	
9	COM			-1.114			
10	Bal		.743	.371			
	Σ	-22.856	-11.636	11.636	6.837	-6.837	0

The reactions $\bar{H}_A$ and $\bar{H}_D$ are obtained by taking moments about the column top joints B and C as shown in Fig. 11-11e.

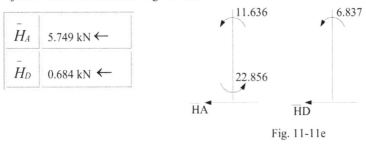

$\bar{H}_A$	5.749 kN ←
$\bar{H}_D$	0.684 kN ←

Fig. 11-11e

To satisfy the horizontal equilibrium of the frame, $F' = 6.433$ kN → as shown in Fig. 11-11f.

The correction factor $CF = \dfrac{F}{F'} = \dfrac{5.031}{6.433} = 0.782$

The final moments are given below.

$$M_{AB} = M_{AB}\,(\text{Load}) + CF\,(M_{AB}^{Sway})$$

$M_{AB} = 14.984 + (0.782)(-22.856) = -2.889$ kN.m
$M_{BA} = 30.179 + (0.782)(-11.636) = 21.08$ kN.m
$M_{BC} = -30.179 + (0.782)(11.636) = -21.08$ kN.m
$M_{CB} = 24.963 + (0.782)(6.837) = 30.31$ kN.m
$M_{CD} = -24.963 + (0.782)(-6.837) = -30.31$ kN.m
$M_{DC} = 0$

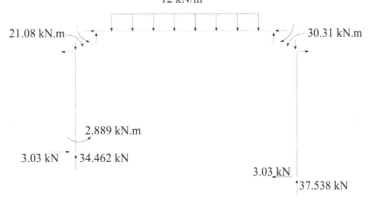

The member free body diagrams and the frame reactions are shown in Fig. 11-11g.

Fig. 11-11g. Member Free Body Diagrams and Reactions

Example: 11-10

Determine the reactions for the frame shown in Fig. 11-12a.

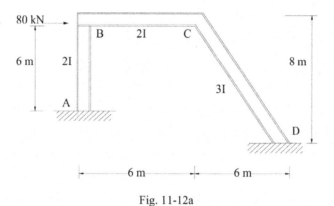

Fig. 11-12a

SOLUTION

The inclined member deflection induces different displacements for the three members as shown in Fig. 11-12 b and 11-12c.

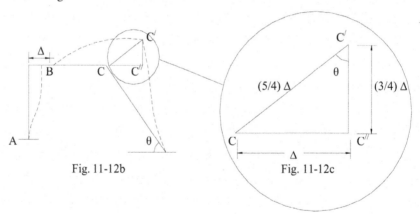

Fig. 11-12b   Fig. 11-12c

Non-Sway Analysis

The fixed end moments, stiffness factors, and distribution factors for the spans AB, BC, and CD are listed below.

$$\text{FEM}_{AB} = \text{FEM}_{BA} = \text{FEM}_{BC} = \text{FEM}_{CB} = \text{FEM}_{CD} = \text{FEM}_{DC} = 0$$

$$K_{AB} = \frac{2I}{6} \quad , \quad K_{BC} = \frac{2I}{6} \quad , \quad K_{CD} = \frac{3I}{10}$$

$DF_{BA} = 0.5$, $DF_{BC} = 0.5$, $DF_{CB} = 0.526$
$DF_{AB} = 0$ [FIXED END] , $DF_{CD} = 0.473$, $DF_{DC} = 0$ [FIXED END]

Since there are no fixed end moments, the final moments are equal to zero, the reactions H_A and H_D are also equal zero, and the sides-way force F is equal to 80 kN ←

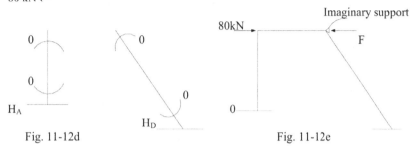

Fig. 11-12d   Fig. 11-12e

Sway Analysis

The arbitrary frame sides-way Δ produces counter-clockwise end moments in the members AB and CD and clockwise end moments in member BC.

FEM_{AB}	:	FEM_{BC}	:	FEM_{CD}
$\dfrac{6\,E(2I)\,\Delta}{(6)^2}$	:	$\dfrac{6\,E(2I)\left(\frac{3}{4}\right)\Delta}{(6)^2}$	:	$\dfrac{6\,E(3I)\left(\frac{5}{4}\right)\Delta}{(10)^2}$
0.333	:	0.25	:	0.225

Let us assume that the initial end moments are as follows.
$FEM_{AB} = FEM_{BA} = -33.3$ kN.m
$FEM_{BC} = FEM_{CB} = 25$ kN.m
$FEM_{CD} = FEM_{DC} = -22.5$ kN.m

Moment Distribution Table

No.	Joints	A	B		C		D
	DF	0	0.5	0.5	0.526	0.473	0
1	FEM	-33.3	-33.3	25	25	-22.5	-22.5
2	Balance		4.15	4.15			
3	COM	2.075			2.075		
4	Bal				-2.406	-2.164	
5	COM			-1.203			-1.1082
6	Bal		0.602	0.602			
7	COM	0.301			0.301		
8	Bal				-0.158	-0.142	
9	COM			-0.079			-0.071
10	Bal		0.0395	0.0395			
	Σ	-30.92	-28.51	28.51	24.81	-24.81	-23.653

The reactions $\bar{H}_A$ and $\bar{H}_D$ are computed by taking moments about the column top joints B and C as shown in Fig. 11-12f.

$\bar{H}_A$	9.905 kN ←
$\bar{H}_D$	12.733 kN ←

In order to satisfy the horizontal equilibrium of the frame, the sway force F′ has to be equal to 22.638 kN → as shown in Fig. 11-12g

The correction factor CF = $\dfrac{F}{F'}$ = 3.534

The final moments are equal to the following:

$M_{AB} = 0$ + 3.534 (-30.92)   = -109.27 kN.m
$M_{BA} = 0$ + 3.534 (-28.51)   = -100.75 kN.m
$M_{BC} = 0$ + 3.534 (28.51)   = 100.75 kN.m
$M_{CB} = 0$ + 3.534 (24.81)   = 87.68 kN.m
$M_{CD} = 0$ + 3.534 (-24.81)   = -87.68 kN.m
$M_{DC} = 0$ + 3.534 (-23.653)   = -83.6 kN.m

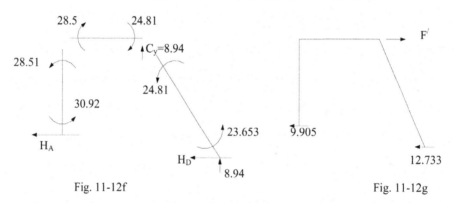

Fig. 11-12f   Fig. 11-12g

The member free body diagrams and reactions are shown in Fig. 11-12h.

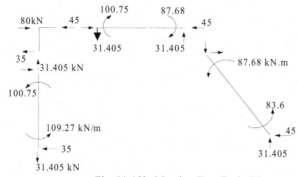

Fig. 11-12h. Member Free Body Diagram

Example: 11-11

Determine the reactions for the gable frame shown in Fig 11-13a. [EI = Constant]

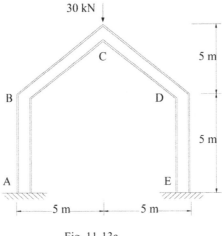

Fig. 11-13a

SOLUTION

The gable frame will have the joint displacements shown in Figs. 11-13b and 11-13c.

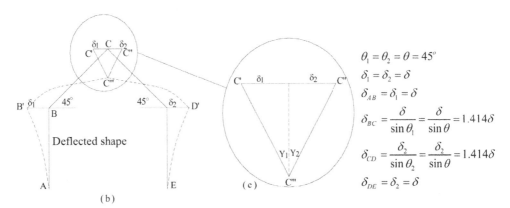

$$\theta_1 = \theta_2 = \theta = 45°$$

$$\delta_1 = \delta_2 = \delta$$

$$\delta_{AB} = \delta_1 = \delta$$

$$\delta_{BC} = \frac{\delta}{\sin\theta_1} = \frac{\delta}{\sin\theta} = 1.414\delta$$

$$\delta_{CD} = \frac{\delta_2}{\sin\theta_2} = \frac{\delta_2}{\sin\theta} = 1.414\delta$$

$$\delta_{DE} = \delta_2 = \delta$$

Non-Sway Analysis

The stiffness factors, and distribution factors for the gable frame spans are listed below.

$$K_{AB} = \frac{I}{5} \quad , \quad K_{BC} = \frac{I}{7.07} \quad , \quad K_{CD} = \frac{I}{7.07} \quad , \quad K_{DE} = \frac{I}{5}$$

$$DF_{BA} = 0.414 \qquad , \qquad DF_{BC} = 0.586 \qquad , \qquad DF_{CB} = 0.5$$

$DF_{DE} = 0.414$, $DF_{CD} = 0.5$, $DF_{DC} = 0.586$

The Fixed end moments are zeros for all spans.

Since the fixed end moments are zeros and there are no horizontal loads, the final

moments are zeros and $V_A = V_D = 15$ kN as shown in Fig. 11-13d.

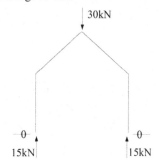

Sway Analysis

The gable frame will sides way by an arbitrary
δ that will produce M^+ in span AB, M^- in
span BC, M^+ in span CD, and M^- in span DE

Fig. 11-13d

FEM_{AB}	:	FEM_{BC}	:	FEM_{CD}
$\dfrac{6\,EI\,\delta}{(5)^2}$	:	$\dfrac{6\,EI\,1.414}{(7.07)^2}$	:	$\dfrac{6\,EI\,\delta}{(5)^2}$
0.24	:	0.17	:	0.24

Let us assume that the initial end moments are equal to the following:

$$FEM_{AB} = FEM_{BA} = 24 \text{ kN.m}$$
$$FEM_{BC} = FEM_{CB} = -17 \text{ kN.m}$$
$$FEM_{CD} = FEM_{DC} = 17 \text{ kN.m}$$
$$FEM_{DE} = FEM_{ED} = -24 \text{ kN.m}$$

Moment Distribution Table

No:	Joints	A	B		C		D		E
	Member	AB	BA	BC	CB	CD	DC	DE	ED
	DF	0	.414	.586	.5	.5	.586	.414	0
1	FEM	24	24	-17	-17	17	17	-24	-24
2	Balance		-2.9	-4.1			4.1		
3	COM	-1.45			-2.05	2.05		2.9	
4	Bal				0	0			1.45
	Σ		22.55	21.1	-21.1	-19.05	19.05	21.1	-21.1	-22.55

Let us consider the free body diagram of the column AB and the section ABC of the gable frame and let us consider take the summation of the moments about B and C as shown Figs. 11-13e and 11-13f .

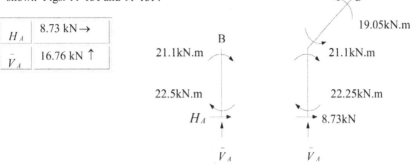

Fig. 11-13e   Fig. 11-13f

The correction Factors are equal to $CF = \dfrac{\bar{V_A}}{\bar{\bar{V_A}}} = \dfrac{15}{16.76} = 0.895$

The final moments, which are computed using Equation [11-6], are equal to the following:

$M_{AB} = 0$ $+ 0.895\ (22.55)$ $= 20.18$ kN.m

$M_{BA} = 0$ $+ 0.895\ (21.1)$ $= 18.88$ kN.m

$M_{BC} = 0$ $+ 0.895\ (-21.1)$ $= -18.88$ kN.m

$M_{CB} = 0$ $+ 0.895\ (-19.05)$ $= -17.05$ kN.m

$M_{CD} = 0$ $+ 0.895\ 19.05)$ $= 17.05$ kN.m

$M_{DC} = 0$ $+ 0.895\ (21.1)$ $= 18.88$ kN.m

$M_{DE} = 0$ $+ 0.895\ (-21.1)$ $= -18.88$ kN.m

$M_{ED} = 0$ $+ 0.895\ (-22.55)$ $= -20.18$ kN.m

The member free body diagrams and the frame reactions are shown in Fig. 11-13g.

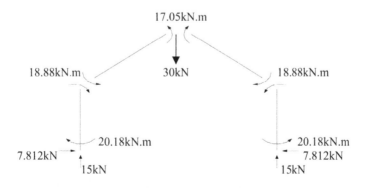

Example: 11-12

Determine the reactions for the two-story frame shown in Fig 11-14a.

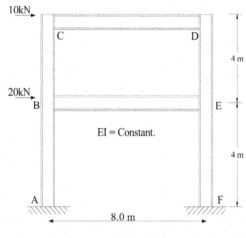

Fig. 11-14a

SOLUTION

The two-story frame has two independent displacements as shown in Fig. 11-14b.

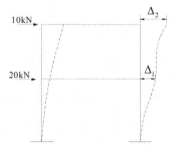

Fig. 11-14b

Non-Sway Analysis

The stiffness factors, and distribution factors for the two story spans are listed below.

$$K_{AB} = K_{BC} = K_{DE} = K_{EF} = \frac{I}{4}$$

$$K_{CD} = K_{BE} = \frac{I}{8}$$

$DF_{BA} = DF_{BC} = DF_{ED} = DF_{EF} = 0.4$
$DF_{BE} = DF_{ED} = 0.2$
$DF_{CB} = DF_{DE} = 0.67$
$DF_{DC} = DF_{CD} = 0.33$
The fixed end moments are zeros for all spans.
Since there is no external load, the final moments are zeros as shown in Figs. 11-14c and 11-14d.

$H_B + H_E = 10$ kN   & $H_A + H_F = 30$ kN

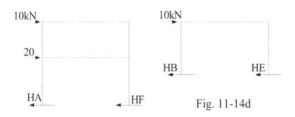

Fig. 11-14d

Fig. 11-14c

Sway Analysis

The sway moments for the lower columns are equal to the following:

$$
\begin{array}{ccc}
FEM_{AB} & : & FEM_{FE} \\
\dfrac{6\,EI\,\Delta_1}{(4)^2} & : & \dfrac{6\,EI\,\Delta_1}{(4)^2} \\
1 & : & 1
\end{array}
$$

Let us assume that the initial end moments are equal to the following:

$$FEM_{AB} = FEM_{BA} = FEM_{FE} = FEM_{EF} = 10 \text{ kN.m}$$

Similarly, the sway moments for the top columns are equal to the following:

$$
\begin{array}{ccc}
FEM_{BC} & : & FEM_{ED} \\
\dfrac{6\,EI\,\Delta_2}{(4)^2} & : & \dfrac{6\,EI\,\Delta_2}{(4)^2} \\
1 & : & 1
\end{array}
$$

Therefore, $FEM_{BC} = FEM_{CB} = FEM_{DE} = FEM_{ED} = 10$ kN.m

Moment Distribution Table (Δ_1)

Frame center and [anti symmetry]

	Joints	A	B			C		D	E	F
No:	Member	AB	BA	BE	BC	CB	CD			
	DF	0	0.4	0.2	0.4	0.67	0.33			
1	FEM	10	10	0	0	0	0			
2	Balance		-4	-2	-4					
3	COM	-2				-2				
4	Bal					1.34	0.66			
	Σ	8	6	-2	-4	-0.66	0.66			

Let us consider the free body diagram of the frame column and Let us consider the summation of the moments about top joints as shown in Figs.11-14e and 11-14f.

$H_B = H_E$	1.165 kN ←
$H_A = H_F$	3.5 kN →

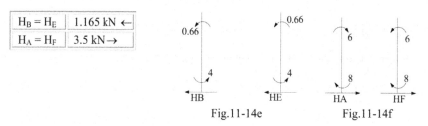

Fig.11-14e   Fig.11-14f

Moment Distribution Table (Δ_2)

Frame center and [anti symmetry]

	Joints	A	B			C		D	E	F
No:	Member	AB	BA	BE	BC	CB	CD			
	DF	0	0.4	0.2	0.4	0.67	0.33			
1	FEM	0	0	0	10	10	0			
2	Balance	0	-4	-2	-4					
3	COM	-2				-2				
4	Bal					-5.36	-2.64			
5	COM			-1	-2.68					
6	Bal		1.472	0.736	1.472					
7	COM					0.736				
8	Bal					-0.493	-.243			
	Σ	-2	-2.528	-2.264	4.792	2.883	-2.883			

408

Chapter 11.   Moment Distribution Method

Let us consider the free body diagram of the frame columns and let us consider the summation of the moments about top joints as shown in Figs. 11-14g and 11-14h.

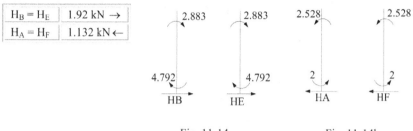

| $H_B = H_E$ | 1.92 kN $\rightarrow$ |
| $H_A = H_F$ | 1.132 kN $\leftarrow$ |

Fig. 11-14g   Fig. 11-14h

The correction factors CF_1 and CF_2 , which are used to satisfy the horizontal equilibrium, are computed as follows.

Δ_1 Δ_2

$$-(H_B + H_E) \, CF_1 \; + \; (H_B + H_E) \, CF_2 \; = \; -10 \qquad [1]$$
$$-2.33 \, CF_1 \quad + \; 3.84 \, CF_2 \quad = -10 \qquad [2]$$

Δ_1 Δ_2

$$(H_A + H_F) \, CF_1 \; + \; (H_A + H_F) \, CF_2 \; = \; -30 \qquad [3]$$
$$7 \, CF_1 \quad - \; 2.264 \, CF_2 \quad = -30 \qquad [4]$$

Solving Equations [2] and [4] yields the following values for CF_1 and CF_2:

$$CF_1 = -6.38$$
$$CF_2 = -6.48$$

The final moments are determined using Equation [11-7] and making use of the anti-symmetry

Final	Load	Δ_1	Δ_2

$$M_{AB} \, (Final) = M_{AB} \, (load) \; + \; CF_1 \, (M_{AB}) \; + \; CF_2 \, (M_{AB})$$

$$M_{AB} = M_{FE} = \; 0 \qquad - \quad 51.04 \quad + \quad 12.96 = 38.08 \text{ kN.m} \quad \curvearrowleft$$

$$M_{BA} = M_{EF} = 21.89 \text{ kN.m} \quad \curvearrowleft$$

$M_{BE} = M_{EB} = 27.43$ kN.m

$M_{BC} = M_{ED} = 5.53$ kN.m

$M_{CB} = M_{DE} = 14.47$ kN.m

$M_{CD} = M_{DC} = 14.47$ kN.m

The member free body diagram and the reactions are shown in Fig. 11-14i.

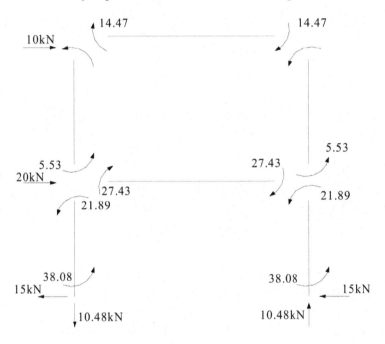

Fig. 11-14i. Member Free Body Diagram

Chapter 11.   Moment Distribution Method

Summary

- The moment distribution method, which is an approximate procedure for analyzing indeterminate beams and frames, eliminates the need to write and solve the simultaneous equations required in the slope-deflection method.
- The analyst begins by assuming that all joints free to rotate are restrained by clamps, producing fixed-end conditions. When loads are applied, fixed-end moments are induced. The solution is completed by unlocking and relocking joints in succession and distributing moments to both ends of all the members framing into the joint until all joints are in equilibrium. The time required to complete the analysis increases significantly if frames are free to side-sway. The method can be extended to non-prismatic members where standard tables of fixed-end moments are available (see Table 13.1).
- Once end moments are established, free bodies of members are analyzed to determine shear forces. After shears are established, axial forces in members are computed using free bodies of joints.
- Although moment distribution provide students with an insight into the behavior of continuous structures, its use is limited in practice because a computer analysis is much faster and more accurate.

Problems

1- Determine, using moment distribution method, the reactions of the continuous beam shown below if the support B settles by 0.05m. [I=200.10^6 mm^4, E = 20.10^6 kN/m^2].

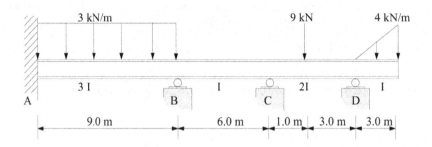

2- Using the moment distribution method, draw the shear and moment diagrams for the continuous beam shown below if the support B settles by 40 mm. [E =200.10^6 kN/m^2, I = 100.10^6 mm^4]

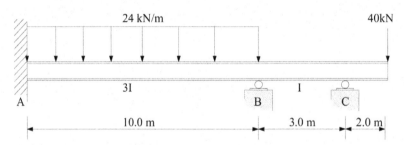

3- Using the moment distribution method, determine the reactions and draw the shear and moment diagrams for the continuous beam shown below.

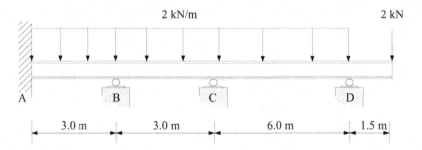

Chapter 11.   Moment Distribution Method

4-   Determine the reactions of the continuous beam shown below using the moment
      distribution method.

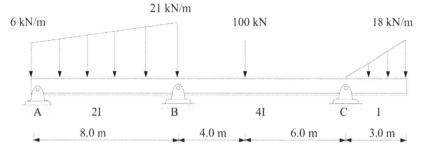

5-   Using the moment distribution method, draw the shear and moment diagrams of the
      continuous beam shown below if the support B settles by 0.03m. [E = 200.10^9 N/m^2,
      I =2 000.10^{-6} m^4]

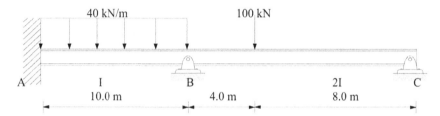

6-   Using the moment distribution method, draw the shear and moment diagrams of the
      continuous beam shown below if the support B settles by 0.05m. [E = 200.10^9 N/m^2,
      I =2000.10^{-6} m^4]

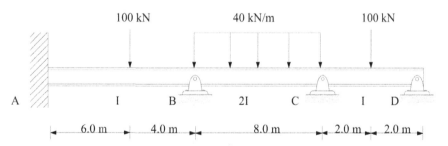

7-   Determine the final moments for the continuous beam shown below using the
      moment distribution method.

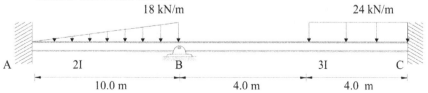

8-   Draw the shear and moment diagrams for the continuous beam shown below using the moment distribution method.

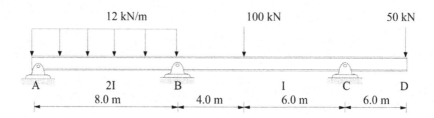

9-   Determine the reactions of the continuous beam shown below using the moment distribution method.

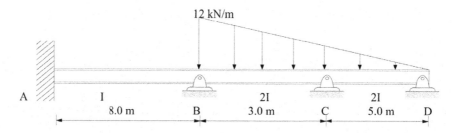

10-   Draw the shear and moment diagrams for the frame shown below using the moment distribution method.

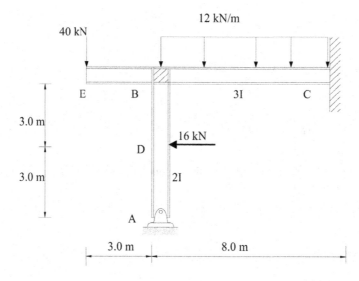

Chapter 11.   Moment Distribution Method

11-   Determine the reactions of the frame shown blow using the moment distribution
       method.

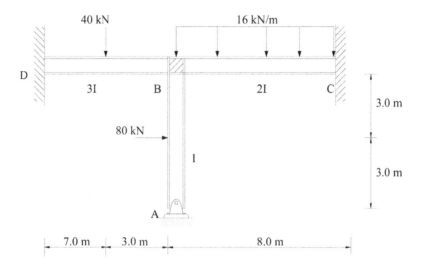

12-   Using the moment distribution method, determine the reactions of the frame shown
       below.

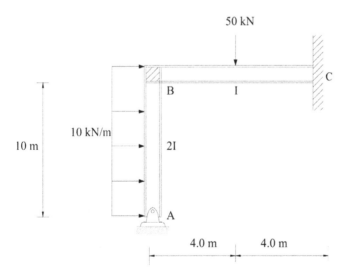

13-   Using the moment distribution method, determine the reactions of the frame shown below.

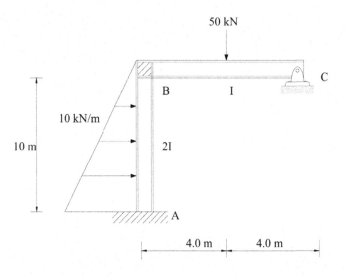

14-   Using the moment distribution method, determine the reactions of the frame shown below.

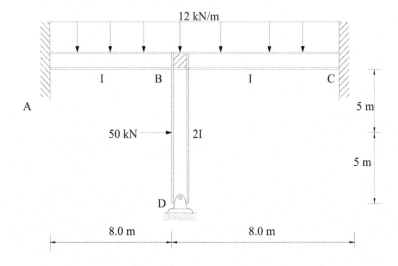

Chapter 11.   Moment Distribution Method

15-   Determine the reactions of the frame shown below using the moment distribution method.

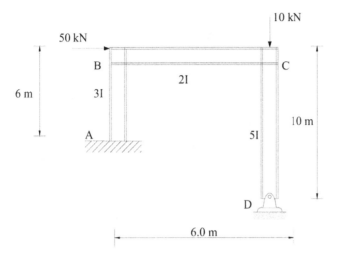

16-   Determine the reactions.

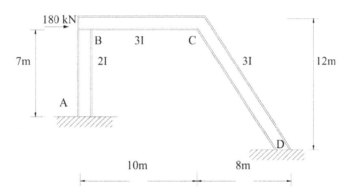

17-   Determine the reactions.

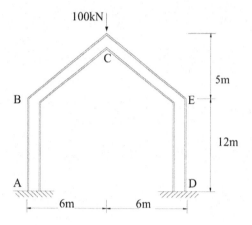

18-   Determine the reactions.

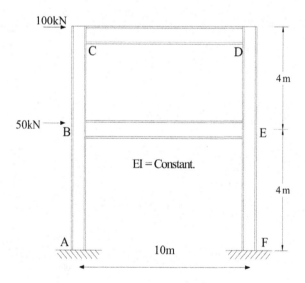

CHAPTER-12
STRUCTURAL MATRIX ANALYSIS

Taj Mahl , India

Chapter 12.   Structural Matrix Analysis

Introduction

The need for the analysis of multi-story buildings with large number of unknowns and the advent of computers has given rise to the matrix method. The flexibility and stiffness methods constitute the two basic methods of structural matrix analysis.

Flexibility Matrix Method

The formulation of a set of equations in matrix form by the consistent deformation method has led to the flexibility method, the compatibility method, or the force method. In this method, the unknowns, which are called redundant forces, are equal in numbers to the degree of static indeterminacy. The joint displacements due to given loads and due to the redundant forces are found through the formulation of the consistent deformation conditions which results in a set of equations in a matrix form. Computer programs, which can nowadays easily solve the set of equations in matrix form, impose no limit on the number of equations (i.e., number of unknowns) to be solved.

Stiffness Matrix Method.

The formulation of a set of equations in the matrix form by the slope deflection method has led to the stiffness method, the equilibrium method, or the displacement method. In this method, the unknowns are joint displacements that are equal in numbers to the degree of kinematic indeterminacy. The formulation of the joint equilibrium equations results in a set of equations in a matrix form that are used to determine joint displacements such as slopes and deflections. The moments and shear forces are then calculated using these joint displacements values.

Degree of Static Indeterminacy.

The number of equations over the static equilibrium equations that are needed to analyze a structure are known as the degree of static indeterminacy as shown Fig. 12-1.

Degree of Kinematic Indeterminacy (Degree of freedom).

A joint displacement due to the structure movement is referred to as kinematic degree of freedom. In the case of kinematic indeterminate structures, the number of unknown joint displacement components is greater than the number of compatibility equations. The number of equilibrium equations needed to find the displacement components of all joints of the structure is known as the degrees of freedom DOF as shown in Fig. 12-1.

Kinematic Indeterminacy (DOF)   = 0
Static Indeterminacy   = 3

Kinematic Indeterminacy (DOF)   = 2
Static Indeterminacy   = 1

Kinematic Indeterminacy (DOF)   = 2
Static Indeterminacy   = 0

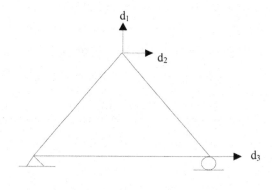

Kinematic Indeterminacy (DOF)   = 3
Static Indeterminacy   = 0

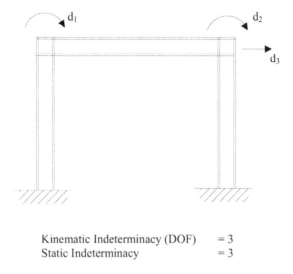

Kinematic Indeterminacy (DOF)   = 3
Static Indeterminacy   = 3

Fig. 12-1. Degrees of Static and Kinematic Indeterminacy

Summary

- The advent of computers and the need for the analysis of multi-story buildings with very large number of unknowns has given rise to the matrix analysis of structures.
- The flexibility and stiffness methods are the two basic methods of structural matrix analysis.
- The formulation in matrix form of a set of equations by the consistent deformation method has led to the flexibility method, compatibility method, or the force method.
- The formulation in matrix form of a set of equations by the slope deflection has led to the stiffness method, equilibrium method, or the displacement method. In this method, the unknowns are the joint displacements, which are equal to the degree of kinematic indeterminacy.
- A set of equations, which can put in a matrix form, are developed by expressing the equilibrium of all of the structure joints. Theses equations can then be easily solved to determine the joint displacements such as slopes and deflections.
- The degree of static indeterminacy is equal to the total number of unknowns minus the number of unknowns that can be solved by statics equations (i.e., three unknown).
- The kinematic indeterminacy (degree of freedom) is the number of unknown joint displacement components.

Chapter 12.   Structural Matrix Analysis

CHAPTER -13
FLEXIBILITY MATRIX METHOD

Kingdom of Saudia Arabia - Jeddah

Introduction

The flexibility of a structure is defined as the displacement caused by a unit force. The forces and the corresponding joint displacements are related to each other by the flexibility matrice $[\delta]$, which relate the displacements $[d]$ to the external loads $[D]$ as shown in Fig. 13 – 1. The displacement is given by

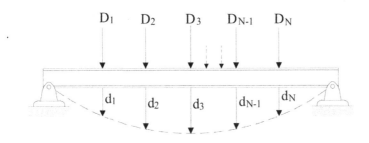

Fig.13-1

$$d_1 = \delta_{11}D_1 + \delta_{11}D_1 + \cdots + \cdots + \delta_{1N}D_N$$
$$d_2 = \delta_{21}D_1 + \delta_{22}D_1 + \cdots + \cdots + \delta_{2N}D_N$$
$$\cdots \quad \cdots \quad \cdots \quad \cdots \quad \cdots \quad \cdots$$
$$\cdots \quad \cdots \quad \cdots \quad \cdots \quad \cdots$$
$$d_N = \delta_{N1}D_1 + \delta_{N2}D_1 + \cdots + \cdots + \delta_{NN}D_N$$

Where d_i = displacement of the ith joint, δ_{ij} =displacement of the ith joint due a unit force at the jth joint j, and D_i = external load at the ith joint.

These expressions can be presented in a matrix form as follows.

$$[d] = [F][D] \qquad\qquad [13\text{-}1]$$

$$
\begin{bmatrix} d_1 \\ d_2 \\ \cdots \\ \cdots \\ d_N \end{bmatrix}
=
\begin{bmatrix}
\delta_{11} & \delta_{12} & \cdots & \cdots & \delta_{1N} \\
\delta_{21} & \delta_{22} & \cdots & \cdots & \delta_{2N} \\
\cdots & \cdots & \cdots & \cdots & \cdots \\
\cdots & \cdots & \cdots & \cdots & \cdots \\
\delta_{N1} & \delta_{N2} & \cdots & \cdots & \delta_{NN}
\end{bmatrix}
\begin{bmatrix} D_1 \\ D_2 \\ \cdots \\ \cdots \\ D_N \end{bmatrix}
$$

Where [d] = displacement matrix, [F] = flexibility matrix, and [D] = external load matrix

The displacement response of a structure is presented as the flexibility matrix $[F]$. As an example, the flexibility coefficients δ_{ij} for the cantilever beam shown in Fig. 13-2 are determined using the moment area method.

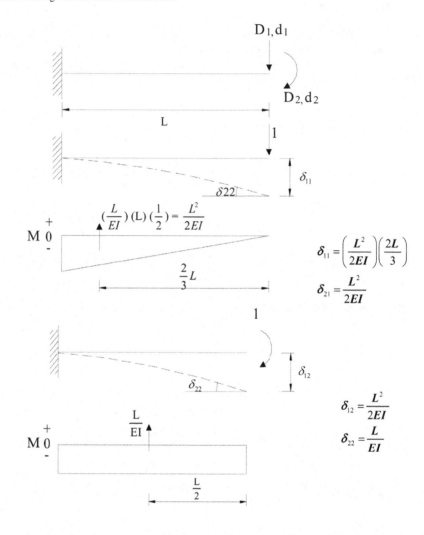

Fig. 13-2

Substituting these flexibility values into Equation $[13-1]$ yields the following equation:

$$\begin{bmatrix} d_1 \\ d_2 \end{bmatrix} = \begin{bmatrix} \dfrac{L^3}{3EI} & \dfrac{L^2}{2EI} \\ \dfrac{L^2}{2EI} & \dfrac{L}{EI} \end{bmatrix} \begin{bmatrix} D_1 \\ D_2 \end{bmatrix}$$

This method of constructing the flexibility matrix is extremely impractical for large and complex structures because of the difficulty of obtaining deflections. Another method is used in theses cases for the determination of the flexibility matrix, where the structure is subdivided into several members. The flexibility matrix of the structure is obtained by combining the flexibility matrices of the individual members. As an example, the construct of the flexibility matrix for the cantilever beam shown in Fig. 13-3 consists of the following steps:

1. The member flexibility coefficients are determined using the moment area method.
2. The flexibility matrices for each structure member are formed as shown below.

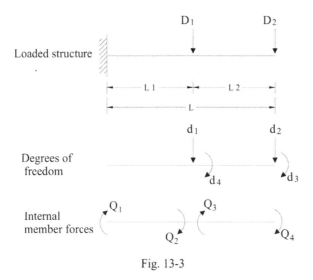

Fig. 13-3

$$\begin{bmatrix} \overset{member1}{f} \end{bmatrix} = \frac{1}{6EI}\begin{bmatrix} 2L_1 & -L_1 \\ -L_1 & 2L_1 \end{bmatrix}, \quad \begin{bmatrix} \overset{member2}{f} \end{bmatrix} = \frac{1}{6EI}\begin{bmatrix} 2L_2 & -L_2 \\ -L_2 & 2L_2 \end{bmatrix}$$

The combination of the individual member matrices yields the following structure flexibility matrix:

$$[f] = \frac{1}{6EI}\begin{bmatrix} 2L_1 & -L_1 & 0 & 0 \\ -L_1 & 2L_1 & 0 & 0 \\ 0 & 0 & 2L_2 & -L_2 \\ 0 & 0 & -L_1 & 2L_2 \end{bmatrix} \qquad [13\text{-}2]$$

If the axial effects need to be included, the structure flexibility matrix becomes as follows.

$$[f] = \begin{bmatrix} \dfrac{2L_1}{6EI} & \dfrac{-L_1}{6EI} & 0 & 0 & 0 & 0 \\[2mm] \dfrac{-L_1}{6EI} & \dfrac{2L_1}{6EI} & 0 & 0 & 0 & 0 \\[2mm] 0 & 0 & \dfrac{L_1}{EA_1} & 0 & 0 & 0 \\[2mm] 0 & 0 & 0 & \dfrac{2L_2}{6EI} & \dfrac{-L_2}{6EI} & 0 \\[2mm] 0 & 0 & 0 & \dfrac{-L_2}{6EI} & \dfrac{2L_2}{6EI} & 0 \\[2mm] 0 & 0 & 0 & 0 & 0 & \dfrac{L_2}{EA_2} \end{bmatrix} \qquad [13\text{-}3]$$

The force transformation matrix $[b]$, which relates the member internal force matrix $[Q]$ to the structure external force matrix $[D]$, represents the internal forces caused by applied unit load forces as shown in Fig. 13-2. The relationship between the member internal force matrix $[Q]$ and the structure external force matrix $[D]$ is expressed by the following equation:

$$[Q] = [b][D]$$

$$\begin{bmatrix} Q^1 \\ Q^2 \\ Q^3 \\ Q^4 \end{bmatrix} = \begin{bmatrix} \overset{D_1=1}{-L_1} & \overset{D_2=1}{-L} & \overset{D_3=1}{-1} & \overset{D_4=1}{-1} \\ 0 & L_2 & 1 & 1 \\ 0 & -L_2 & -1 & 0 \\ 0 & 0 & 1 & 0 \end{bmatrix} \begin{bmatrix} D_1 \\ D_2 \\ D_3 \\ D_4 \end{bmatrix} \qquad [13\text{-}4]$$

The structure flexibility matrix $[F]$ is obtained using the following equation:

$$[F] = [b^T][f][b] \qquad [13\text{-}5]$$

$$[F] = \frac{1}{EI}
\begin{bmatrix}
\dfrac{L^3_1}{3} & \dfrac{L^2_1(2L+L_2)}{6} & \dfrac{L^2_1}{2} & \dfrac{L^2_1}{2} \\[3mm]
\dfrac{L^2_1(2L+L_2)}{6} & \dfrac{L^2L_1 + LL_1L_2 + L^2_2L_1 + L^3_2}{3} & \dfrac{L_2L_1 + LL_1 + L^2_2}{2} & \dfrac{L_1L + L_1L_2}{2} \\[3mm]
\dfrac{L^2_1}{2} & \dfrac{L_2L_1 + LL_1 + L^2_2}{2} & L_1 + L_2 & L_1 \\[3mm]
\dfrac{L^2_1}{2} & \dfrac{L_1L + L_1L_2}{2} & L_1 & L_1
\end{bmatrix}$$

The structure displacements are calculated using Equation [13-1] when the flexibility matrix is determined.

Flexibility matrix analysis for indeterminate structures

The flexibility method for the analysis of indeterminate structures, which is essentially a matrix formulation of the force method of Chapter 9, consists of the following steps:

1. Define the external load matrix $[D]$ and displacement matrix $[d]$.

2. Define the internal member force matrix $[Q]$ and the redundant matrix $[R]$.

3. Construct the force transformation matrix $[b]$ and partition it to the following sub-matrices:

$$[b_D \vdots b_R].\tag{13-6}$$

4. Construct the element flexibility matrix $[f]$ using Equation [13-3]

5. construct the matrices [FRD] and [FRR] using the following equations:

$$[F_{RD}] = [b_R^{\ T}][f][b_D]\tag{13-7}$$

$$[F_{RR}] = [b_R^{\ T}][f][b_R]\tag{13-8}$$

6. Determine the redundant matrix $[R]$ using the following equations:

$$[R] = -[F_{RR}^{\ -1}][F_{RD}][D] \qquad \text{for rigid supports}\tag{13-9}$$

$$[R] = [\Delta - [F_{RD}][D]][F_{RR}^{\ -1}] \qquad \text{for supports with settlement } \Delta\tag{13-10}$$

7. Obtain the member force matrix $[Q]$ from equilibrium using the following equation

$$[Q] = [b_D \vdots b_R]\begin{bmatrix} D \\ R \end{bmatrix}\tag{13-11}$$

8. Obtain the matrix [B] using the following equation:

$$[B] = [b_D] - [b_R][F_{RR}^{\ -1}][F_{RD}]\tag{13-12}$$

431

9. Construct the structure flexibility matrix $[F]$ using the following equation:

$$[F] = [b_D{}^T][f][B] \qquad\qquad [13\text{-}13]$$

10. Obtain the displacement matrix $[d_D]$ using the following equation:

$$[d_D] = [F][D] \qquad\qquad [13\text{-}14]$$

11. Alternatively the member forces matrix $[Q]$ can be obtained using the following equation:

$$[Q] = [B][D] \qquad\qquad [13\text{-}15]$$

Example: 13-1

Determine the reactions, the member forces, and the vertical displacement at B, for the beam shown in Fig.13-4a. [$E = 200.10^6 \, kN/m^2$, $I = 300.10^6 \, mm^4$]

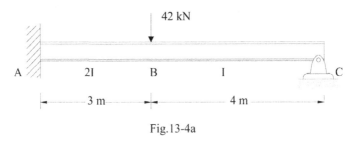

Fig.13-4a

SOLUTION

The cantilever beam is indeterminate to the 1st degree statically. The redundant can be either the moment at the left end or the reaction at the right end. The load matrix [D], the displacement matrix [d], the internal member force matrix [Q], and redundant matrix [R] are defined in Figs.13-4b, 13-4c, 13-4d, and 13-4e, respectively.

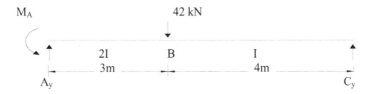

Fig. 13-4b. Structural Model

Fig. 13-4c. DOF and Redundant

Fig. 13-4d. Applied Loads

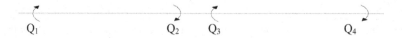

Fig. 13-4e. Internal Member Forces

The force transformation matrix [b] is constructed using Equations [13-16] and [13-11], respectively.

$$
\begin{bmatrix} Q_1 \\ Q_2 \\ Q_3 \\ Q_4 \end{bmatrix} =
\left[
\begin{array}{c|c}
\overset{D=1}{\begin{array}{c} -L_1 = -3 \\ 0 \\ 0 \\ 0 \end{array}} &
\overset{R=1}{\begin{array}{c} L_1 + L_2 = 7 \\ -L_2 = -4 \\ +L_2 = +4 \\ 0 \end{array}}
\end{array}
\right]
\begin{bmatrix} D \\ \cdots \\ R \end{bmatrix}
$$
$$\;\;\;\;\;\;\;\;\;\;\;\;\;\;\;\;\;\; \mathbf{b_D} \;\;\;\;\;\;\;\;\; \mathbf{b_R}$$

The element flexibility matrix [f], which is constructed using Equation [13-3], is equal to the following:

$$
[f] = \frac{1}{6E}
\begin{bmatrix}
\dfrac{2(L_1 = 3)}{2I} & -\dfrac{(L_1 = 3)}{2I} & 0 & 0 \\[2mm]
-\dfrac{3}{2I} & -\dfrac{2(3)}{2I} & 0 & 0 \\[2mm]
0 & 0 & -\dfrac{2(L_2 = 4)}{I} & \dfrac{-(L_2 = 4)}{I} \\[2mm]
0 & 0 & \dfrac{-4}{I} & \dfrac{2(4)}{I}
\end{bmatrix}
$$

The matrices [F$_{RD}$] and [F$_{RR}$], which are computed using Equations [13-7] and [13-8], respectively, are equal to the following:

$$[F_{RD}] = [-2.25 \times 10^{-4}]$$

$$[F_{RR}] = [1.131 \times 10^{-3}]$$

The redundant matrix [R], which is determined using Equation [13-9], is equal to the following:

$$[R] = -[F_{RR}]^{-1}[F_{RD}][D] = +8.359 = 8.359 \text{ kN} \uparrow$$

Where $[D] = [42 \text{ kN @ d}]$

434

Note: The load is positive because it is in the assumed direction of d.

Once the redundant is determined, the other reactions are calculated using the principles of statics as shown in Fig.13-4f.

67.487 kN.m   42 kN

33.641 kN   8.359 kN

Fig. 13-4f. Load and Reactions

The member force matrix [Q], which is determined using Equation [13-11], is shown in Fig.13-4g.

$$Q = \begin{bmatrix} -67.487 \\ -33.436 \\ 33.436 \\ 0 \end{bmatrix}$$

33.436

67.487

Fig. 13-4g. Member Forces

The force transformation matrix [B], which is determined using Equation [13-12], is as follows.

$$B = \begin{bmatrix} -1.607 \\ -0.796 \\ 0.796 \\ 0 \end{bmatrix}$$

The flexibility matrix [F], which is determined using Equation [13-12] and [13-13], is as follows.

$$[F] = [3.022 \times 10^{-5}]$$

The displacement matrix [d_B] at B, which is obtained using Equation [13-14], is as follows.

$$[d] = [1.269 \times 10^{-3}] \text{ m} = 1.269 \text{ mm} \downarrow$$

The positive sign of the displacement means that the assumed displacement direction is correct.

Example: 13-2

Determine the reactions, member forces, and the rotational displacement at B for the beam shown in Fig.13-5a. [$E = 200.10^6$ kN/m^2, $I = 200.10^6$ mm^4]

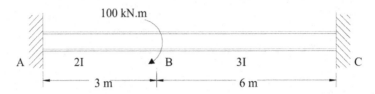

100 kN.m

A   2I   B   3I   C

3 m   6 m

Fig.13-5a

SOLUTION

The beam is statically indeterminate to the 2nd degree because axial deformations are neglected. The reaction and moment on the right support are the selected redundants. The external load matrix [D], the displacement matrix [d], the internal member force matrix [Q], and the redundant matrix [R] are defined in Figs.13-5 b, 13-5c, 13-5d, and 13-5e, respectively.

M_A 100 kN.m M_C

A_y 3 m   B   4 m C_y

Fig. 13-5b. Structural model

d_1 R_2

R_1

Fig. 13-5c. Displacements and Redundants

100 kN.m

Fig. 13-5d. Applied Loads

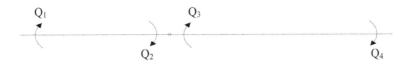

Fig. 13-5e. Internal Member Forces

The force transformation matrix [B], which is constructed using Equations [13-6] and [13-11], is as follows.

$$
\begin{bmatrix} Q_1 \\ Q_2 \\ Q_3 \\ Q_4 \end{bmatrix} =
\begin{array}{ccc}
{\scriptstyle D=1} & {\scriptstyle R_1=1} & {\scriptstyle R_2=1}
\end{array}
\left[\begin{array}{c:cc}
-1 & 9 & -1 \\
1 & -6 & 1 \\
0 & 6 & -1 \\
0 & 0 & 1 \\
\mathbf{b_D} & \mathbf{b_R} &
\end{array} \right]
\left[\begin{array}{c}
D \\ \hline R_1 \\ R_2
\end{array} \right]
$$

The element flexibility matrix [f], which is constructed using Equation [13-3], is as follows.

$$
[f] = \frac{1}{6E}
\begin{bmatrix}
\dfrac{6}{2I} & \dfrac{-3}{2I} & 0 & 0 \\[2mm]
\dfrac{-3}{2I} & \dfrac{6}{2I} & 0 & 0 \\[2mm]
0 & 0 & \dfrac{12}{3I} & \dfrac{-6}{3I} \\[2mm]
0 & 0 & \dfrac{-6}{3I} & \dfrac{12}{3I}
\end{bmatrix}
$$

The matrices [F_RD] and [F_RR], which are constructed using Equations [13-7] and [13-8], respectively, are as follows.

$$
[F_{RD}] = \begin{bmatrix} -2.813 \times 10^{-4} \\ 3.75 \times 10^{-5} \end{bmatrix}
$$

$$
[F_{RR}] = \begin{bmatrix} -2.738 \times 10^{-3} & -4.313 \times 10^{-4} \\ -4.313 \times 10^{-4} & 8.75 \times 10^{-5} \end{bmatrix}
$$

The redundant matrix [R], which is computed using Equation [13-9], is as shown below.

$$[R] = -[F_{RR}^{-1}] [F_{RD}][D] = \begin{Bmatrix} 15.755 \ kN \\ 34.792 \ kN \ m \end{Bmatrix}$$

Where [D] = [100 kN.m @ d]

Note : The moment is positive because it is in the assumed direction of d.

Once the redundant is determined, the other reactions are calculated using the principles of statics as shown in Fig.13-5f.

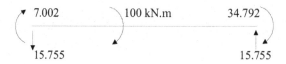

7.002   100 kN.m   34.792

15.755   15.755

Fig. 13-5f. Applied Loads and Reactions

The member force matrix [Q], which is determined using Equation [13-11], is shown in Fig.13-5g.

$$[Q] = \begin{bmatrix} 7.002 \\ 40.263 \\ 59.737 \\ 34.792 \end{bmatrix}$$

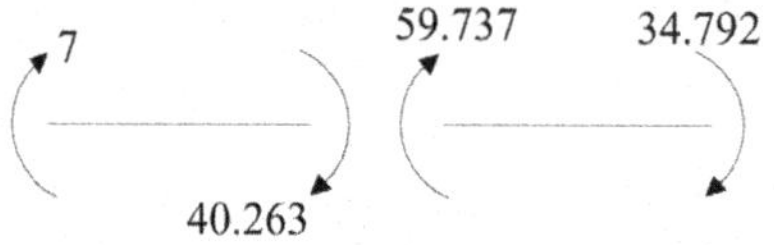

7   59.737   34.792

40.263

Fig. 13-5g. Member Forces

To determine the rotational displacement at B, the matrices [B] and [F] have to be determined first.

The force transformation matrix [B], which is determined using Equation [13-12], is listed below.

$$B = \begin{pmatrix} 0.07 \\ 0.403 \\ 0.597 \\ 0.348 \end{pmatrix}$$

The flexibility matrix [F], which is determined using Equation [13-13], is listed below.

$$[F] = [\ 6.236 \times 10^{-6}\ ]$$

The rotational displacement at B [d_B] is obtained using equation [13-14] as follows.

$$[d] = [6.236 \times 10^{-4} \ rad \ \curvearrowleft \ ]$$

Example: 13-3

Determine the reactions for the two-span beam shown in Fig. 13-6a.
[E = 200.10^6 kN/m², I = 200.10^6 mm⁴]

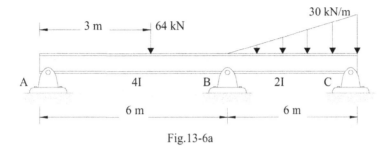

Fig.13-6a

SOLUTION

The reaction at B is chosen as a redundant. The redundant matrix [R], the load matrix [D], the displacement matrix [d], and the internal member force matrix [Q] are defined in Figs.13-6b,13-6c,13-6d, and 13-6e, respectively.

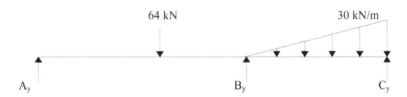

Fig. 13-6b. Structural model

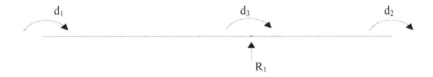

Fig. 13-6c. Joint displacements (DOF)

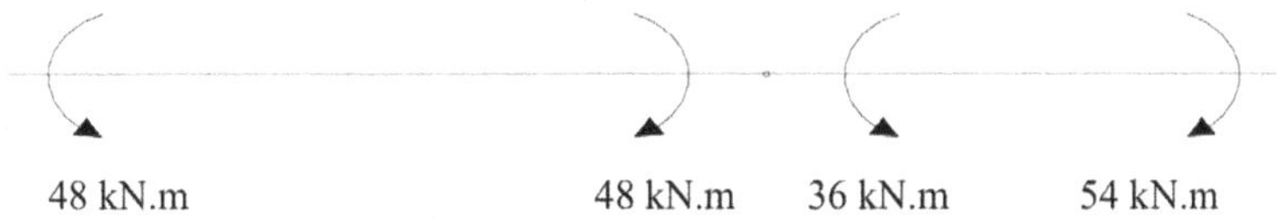

48 kN.m   48 kN.m   36 kN.m   54 kN.m

Fig. 13-6d. Fixed End Moments

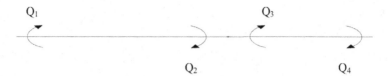

Fig. 13-6e. Internal Member Forces

The force transformation matrix [B] is constructed using Equations [13-6] and [13-11] as follows.

$$
\begin{bmatrix} Q^1 \\ Q^2 \\ Q^3 \\ Q^4 \end{bmatrix}
\begin{array}{cccc} D_1{=}1 & D_2{=}1 & D_3{=}1 & R_1{=}1 \\ \end{array}
\left[\begin{array}{ccc|c} 1 & 0 & 0 & 0 \\ 0 & 0 & 1 & 3 \\ 0 & 0 & 0 & -3 \\ 0 & 1 & 0 & 0 \end{array}\right]
= \begin{bmatrix} D_1 \\ D_2 \\ D_3 \\ \hline R_1 \end{bmatrix}
$$

$$\mathbf{b_D} \qquad\qquad \mathbf{b_R}$$

The element flexibility matrix [f], which is constructed using Equation[13-3], is as follows.

$$
f = \begin{bmatrix}
1.25 \times 10^{-5} & -6.25 \times 10^{-6} & 0 & 0 \\
-6.25 \times 10^{-6} & 1.25 \times 10^{-5} & 0 & 0 \\
0 & 0 & 2.5 \times 10^{-5} & -1.25 \times 10^{-5} \\
0 & 0 & -1.25 \times 10^{-5} & 2.5 \times 10^{-5}
\end{bmatrix}
$$

The matrices $[F_{RD}]$ and $[F_{RR}]$ are constructed using Equations [13-7] and [13-8], respectively, as follows.

$$[F_{RD}] = [-1.875 \times 10^{-5} \quad 3.75 \times 10^{-5} \quad 3.75 \times 10^{-5}]$$
$$[F_{RR}] = [3.375 \times 10^{-4}]$$

The redundant matrix [R], which is determined using Equation [13-9], is as follows.

$$[R] = -[F_{RR}^{-1}] \, [F_{RD}][D] = [10] = 10 \text{ kN} \uparrow$$

where the displacement matrix [D] is equal to the following:

$$[D] = \begin{bmatrix} -(48) = 48^{kN \cdot m} \ @ \ d_1 \\ -(54) = -54^{kN \cdot m} \ @ \ d_2 \\ -(48-36) = -12^{kN \cdot m} \ @ \ d_3 \end{bmatrix}$$

The minus sign is for joint equilibrium (equal and opposite.) .

The magnitude of the redundant R does not equal to the reaction B_y due to span loading. In order to get the final reaction, the internal member force matrix [Q] is determined using Equation [13-11].

$$\begin{bmatrix} Q_1 \\ Q_2 \\ Q_3 \end{bmatrix} = \begin{bmatrix} 48 \\ 18 \\ -30 \\ -54 \end{bmatrix}$$

The final moment matrix [M] is determined using the following equation:
$$[M] = [Q_0] + [Q]$$
where [M] = final moment matrix and $[Q_0]$ = fixed end moment matrix ,Fig.13-6d.

The final moment matrix [M] is therefore equal to the following:

$$\begin{bmatrix} M_1 \\ M_2 \\ M_3 \\ M_4 \end{bmatrix} = \begin{bmatrix} -48 \\ 48 \\ -36 \\ 54 \end{bmatrix} + \begin{bmatrix} 48 \\ 18 \\ -30 \\ -54 \end{bmatrix} = \begin{bmatrix} 0 \\ 66 \\ -66 \\ 0 \end{bmatrix}$$

Once the final moments are determined the reactions are easily computed using the principles of statics as shown in Figs. 13-6f, 13-6g, and 13-6h.

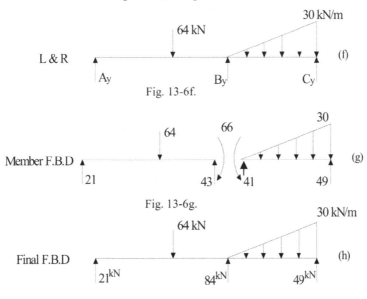

Fig. 13-6f.

Fig. 13-6g.

Fig. 13-6h.

441

Example: 13-4

Determine for the beam shown in Fig. 13-7a the reaction, slope, and deflection at B.
$[E = 200.10^6 \text{ kN/m}^2, I = 200.10^6 \text{ mm}^4, K_{Spring} = 200 \text{ kN/m}]$

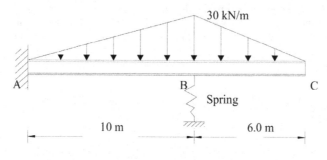

Fig. 13-7a

SOLUTION

The beam is statically indeterminate to the first degree. The spring force B_y is the chosen redundant as shown in Figs.13-7b, 13-7c, 13-7d, and 13-7e.

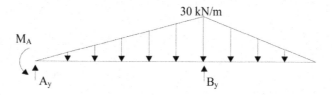

Fig. 13-7b. Structural Model

Fig. 13-7c. Displacements (DOF) and Redundant

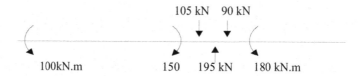

Fig. 13-7d. Fixed End Moments

Fig. 13-7e. Internal Member Forces

The force transformation matrix [B], which is constructed using Equations [13-6] and [13-11], is as follows.

$$\begin{bmatrix} Q^1 \\ Q^2 \\ Q^3 \end{bmatrix} = \begin{array}{c} D_1=1 \quad D_2=1 \quad R_1=1 \\ \begin{bmatrix} -10 & -1 & | & 10 \\ 0 & 1 & | & 0 \\ 0 & 0 & | & 1 \end{bmatrix} \end{array} \begin{bmatrix} D_1 \\ D_2 \\ \hline R_1 \end{bmatrix}$$

$$\qquad\qquad \mathbf{b_D} \qquad\qquad \mathbf{b_R}$$

The element flexibility matrix [f], which is constructed using Equation [13-3], is as follows.

$$f = \begin{bmatrix} \dfrac{2L}{6EI} & -\dfrac{L}{6EI} & 0 \\[2ex] -\dfrac{L}{6EI} & \dfrac{2L}{6EI} & 0 \\[2ex] 0 & 0 & \dfrac{1}{K_s} \end{bmatrix} = \begin{bmatrix} 5.556\times10^{-5} & -2.778\times10^{-5} & 0 \\ -2.778\times10^{-5} & 5.556\times10^{-5} & 0 \\ 0 & 0 & 5\times10^{-3} \end{bmatrix}$$

where the spring flexibility is given by the following equation:

$$f_{spring} = \frac{1}{K_{Spring}}$$

The matrices [F_{RD}] and [F_{RR}], which are determined using Equations [13-7] and [13-8], respectively, are as follows.

$$[F_{RD}] = [-5.556 \times 10^{-5} \quad -8.333 \times 10^{-4}]$$
$$[F_{RR}] = [0.011]$$

The displacement matrix [D] is as follows.

$$[D] = \begin{bmatrix} -195\ \text{kN} & @d_1 \\ -(150-180) = +30\ \text{kN.m} & @d_2 \end{bmatrix}$$

443

The redundant matrix [R], which is computed using Equation [13-9], is as follows.

The redundant matrix [R], which is computed using Equation [13-9], is as follows.

$$[R] = -[F_{RR}^{-1}] [F_{RD}][D] = [105] = 105 \text{ kN} \uparrow$$

The internal member force matrix [Q], which is determined using Equation [13-11], is as follows.

$$\begin{bmatrix} Q_1 \\ Q_2 \\ Q_3 \end{bmatrix} = \begin{bmatrix} -930 \\ 30 \\ 105 \end{bmatrix}$$

where Q_3 is the redundant spring force.

The final moment matrix [M] is determined using Equation [13-16].

$$[M] = \begin{bmatrix} -100 \\ 150 \end{bmatrix} + \begin{bmatrix} -930 \\ 30 \end{bmatrix} = \begin{bmatrix} -1030 \\ 180 \end{bmatrix}$$

The beam reactions are shown in Fig.13-7f.

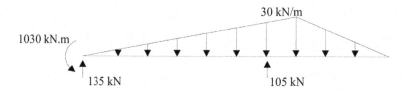

Fig. 13-7f. Applied Loads and Reactions

The force transformation matrix [B], which is determined using Equation [13-12], is equal to the following:

$$B = \begin{bmatrix} -4.737 & -0.211 \\ 0 & 1 \\ 0.526 & 0.079 \end{bmatrix}$$

The flexibility matrix [F], which is determined using Equation [13-13], is equal to the following:

$$[F] = \begin{bmatrix} 2.632 \times 10^{-3} & 3.947 \times 10^{-4} \\ 3.947 \times 10^{-4} & 1.009 \times 10^{-4} \end{bmatrix}$$

The displacement matrix [d] (at the spring), which is obtained using equation [13-14], is as follows.

$$[d] = \begin{bmatrix} 0.525 \\ 0.08 \end{bmatrix} = \begin{bmatrix} 0.525 \text{ m} \downarrow \\ 0.08 \text{ rad} \curvearrowleft \end{bmatrix}$$

Example: 13-5

Determine for the frame shown in Fig. 13-8 the displacement at B.

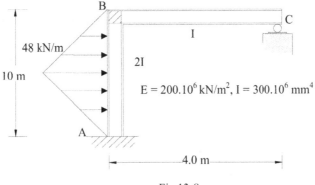

Fig.13-8a

SOLUTION

The frame is statically indeterminate to the first degree. The reaction C_y is the chosen redundant. The total number of unknown displacements (DOF) for the frame is equal to three. However, only two DOF are selected for the solution as shown in Figs. 13-8b, 13-8c, 13-8d, and 13-8e.

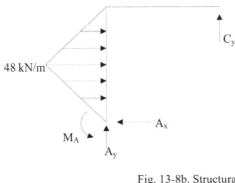

Fig. 13-8b. Structural model

Fig. 13-8c. DOF and Redundant

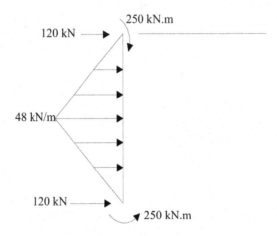

Fig. 13-8d. Fixed End Moment

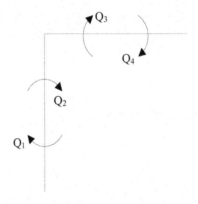

Fig. 13-8e. Internal Member Forces

The force transformation matrix [B], which is constructed using Equation [13-11], is as follows.

$$
\begin{bmatrix} Q_1 \\ Q_2 \\ Q_3 \\ Q_4 \end{bmatrix} = \begin{matrix} D_1=1 \quad D_2=1 \quad R_1=1 \\ \begin{bmatrix} -1 & -10 & \vdots & 4 \\ 1 & 0 & \vdots & -4 \\ 0 & 0 & \vdots & 4 \\ 0 & 0 & \vdots & 0 \end{bmatrix} \end{matrix} \begin{bmatrix} D_1 \\ D_2 \\ \hdashline R_1 \end{bmatrix}
$$

$$\mathbf{b_D} \qquad\qquad \mathbf{b_R}$$

446

The element flexibility matrix [f], which is built using Equation [13-3], is as follows.

$$
f = \begin{bmatrix}
2.778 \times 10^{-5} & -1.389 \times 10^{-5} & 0 & 0 \\
-1.389 \times 10^{-5} & 2.778 \times 10^{-5} & 0 & 0 \\
0 & 0 & 2.222 \times 10^{-5} & -1.111 \times 10^{-5} \\
0 & 0 & -1.111 \times 10^{-5} & 2.222 \times 10^{-5}
\end{bmatrix}
$$

The matrices [F_{RD}] and [F_{RR}], which are computed using Equations [13-7] and [13-8], are as follows.

$$[F_{RD}] = [-3.333 \times 10^{-4} \quad -1.667 \times 10^{-3}]$$
$$[F_{RR}] = [1.689 \times 10^{-3}]$$

The redundant matrix [R], which is computed using Equation [13-9], is as follows.

$$[R] = -[F_{RR}^{-1}] \, [F_{RD}][D] = [69] = 69 \text{ kN} \uparrow$$

Where the matrix displacement D is given by

$$[D] = \begin{bmatrix} -(250+0) = -250kN \, .m & @d_1 \\ 120kN & @d_2 \end{bmatrix}$$

The internal member forces matrix[Q], which is computed using Equation [13-11], is equal to

$$[Q] = \begin{bmatrix} Q_1 \\ Q_2 \\ Q_3 \\ Q_4 \end{bmatrix} = \begin{bmatrix} -673.684 \\ -526.316 \\ 276.316 \\ 0 \end{bmatrix}$$

The final moments matrix [M], which is computed using Equation [13-16], is equal to

$$[M] = \begin{bmatrix} -250 \\ 250 \\ 0 \\ 0 \end{bmatrix} + \begin{bmatrix} -673.684 \\ -526.316 \\ 276.316 \\ 0 \end{bmatrix} = \begin{bmatrix} -923.684 \\ -276.316 \\ 276.316 \\ 0 \end{bmatrix}$$

The frame reactions are shown in Fig.13-8f.

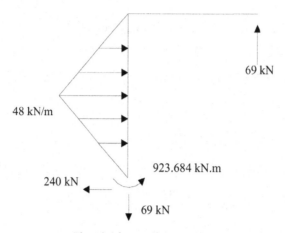

Fig. 13-8f. Applied Loads and Reactions

To determine the displacement at B the flexibility matrix [F] is constructed using Equations [13-12] and [13- 13] as follows.

$$B = \begin{bmatrix} -0.211 & -6.053 \\ 0.211 & -3.947 \\ 0.789 & 3.947 \\ 0 & 0 \end{bmatrix}$$ [13-12]

$$[F] = \begin{bmatrix} 1.754 \times 10^{-5} & 8.772 \times 10^{-5} \\ 8.772 \times 10^{-5} & 1.133 \times 10^{-3} \end{bmatrix}$$ [13-13]

The displacement matrix [d] at B is obtained using Equation [13-14] as follows

$$[d] = \begin{bmatrix} 0.00614 \\ 0.114 \end{bmatrix} = \begin{bmatrix} 0.00614 \ rad \ \curvearrowright \\ 0.114 \ m \ \rightarrow \end{bmatrix}$$

Example: 13-6

Determine the displacement under the load for the frame shown in Fig.13-9a.
[$E = 200.10^6$ kN/m^2, $I = 400.10^6$ mm^4]

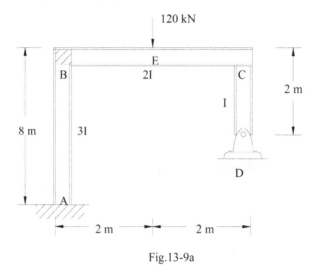

120 kN

E

B   2I   C

2 m

I

8 m   3I

D

A

2 m   2 m

Fig.13-9a

SOLUTION

The frame is statically indeterminate to the second degree. The reactions D_X and Dy are the chosen redundants as shown in Figs. 13-9b, 13-9c, and 13-9d.

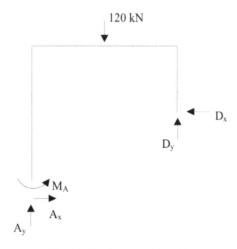

120 kN

D_x

D_y

M_A

A_x

A_y

Fig. 13-9b. Structural Model

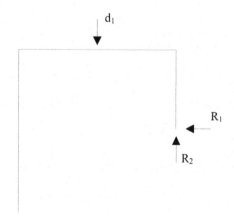

Fig. 13-9c. DOF and Redundants

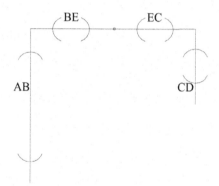

(No Loads in the specified members => Zero Moments)

Fig. 13-9d. Fixed End Moments

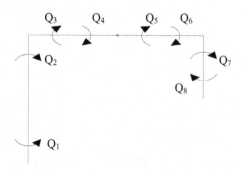

Fig. 13-9e. Internal Member Forces

The force transformation matrix [b], which is built using Equation[13-11], is equal to the following:

$$[b] = \begin{bmatrix} Q_1 \\ Q_2 \\ Q_3 \\ Q_4 \\ Q_5 \\ Q_6 \\ Q_7 \\ Q_8 \end{bmatrix} = \begin{bmatrix} -3 & 6 & 4 \\ 3 & 2 & -4 \\ -3 & -2 & 4 \\ 0 & 2 & -2 \\ 0 & -2 & 2 \\ 0 & 2 & 0 \\ 0 & -2 & 0 \\ 0 & 0 & 0 \end{bmatrix} \begin{bmatrix} D_1 \\ R_1 \\ R_2 \end{bmatrix}$$

with column headings $D_1=1 \quad R_1=1 \quad R_2=1$

The element flexibility matrix [f], which is constructed using Equation [13-13], is given by

$$f = \begin{bmatrix} \dfrac{2.L_{AB}}{6E.3I} & \dfrac{-L_{AB}}{6E.3I} & 0 & 0 & 0 & 0 & 0 & 0 \\[2ex] \dfrac{-L_{AB}}{6E.3I} & \dfrac{2.L_{AB}}{6E.3I} & 0 & 0 & 0 & 0 & 0 & 0 \\[2ex] 0 & 0 & \dfrac{2.L_{BE}}{6E.2I} & \dfrac{-L_{BE}}{6E.2I} & 0 & 0 & 0 & 0 \\[2ex] 0 & 0 & \dfrac{-L_{BE}}{6E.2I} & \dfrac{2.L_{BE}}{6E.2I} & 0 & 0 & 0 & 0 \\[2ex] 0 & 0 & 0 & 0 & \dfrac{2.L_{EC}}{6E.2I} & \dfrac{-L_{EC}}{6E.2I} & 0 & 0 \\[2ex] 0 & 0 & 0 & 0 & \dfrac{-L_{EC}}{6E.2I} & \dfrac{2.L_{EC}}{6E.2I} & 0 & 0 \\[2ex] 0 & 0 & 0 & 0 & 0 & 0 & \dfrac{2.L_{CD}}{6EI} & \dfrac{-L_{CD}}{6EI} \\[2ex] 0 & 0 & 0 & 0 & 0 & 0 & \dfrac{-L_{CD}}{6EI} & \dfrac{2.L_{CD}}{6EI} \end{bmatrix}$$

Where L_{AB} = 8.0 m, L_{BE} = 2.0 m, L_{EC} = 2.0 m, and L_{CD} = 2.0 m.

The matrices [F_{RD}] and [F_{RR}], which are computed using Equations [13-7] and [13-8], are equal to

$$[F_{RD}] = \begin{bmatrix} -1.083 \times 10^{-4} \\ -3.083 \times 10^{-4} \end{bmatrix}$$

$$[F_{RR}] = \begin{bmatrix} 4.444 \times 10^{-4} & 1.667 \times 10^{-4} \\ 1.667 \times 10^{-4} & 6.667 \times 10^{-4} \end{bmatrix}$$

The redundant matrix [R], which is determined using Equation [13-9], is as follows.

$$R = \begin{bmatrix} 9.31 \\ 53.172 \end{bmatrix} = \begin{matrix} 9.31 \text{ kN} \leftarrow \\ 53.172 \text{ kN} \uparrow \end{matrix}$$

The matrix [D] is equal to [D] = [120 kN @ d]

The Internal member force matrix [Q], which is built using Equation [13-11], is as follows.

$$Q = \begin{bmatrix} 28.552 \\ 45.931 \\ -45.931 \\ -87.724 \\ 87.724 \\ 18.621 \\ -18.621 \\ 0 \end{bmatrix}$$

The final moment matrix [M] is computed using Equation [13-16] as follows.

$$M = \begin{bmatrix} 28.552 \\ 45.931 \\ -45.931 \\ -87.724 \\ 87.724 \\ 18.621 \\ -18.621 \\ 0 \end{bmatrix} + \begin{bmatrix} 0 \\ 0 \\ 0 \\ 0 \\ 0 \\ 0 \\ 0 \\ 0 \end{bmatrix} = \begin{bmatrix} 28.552 \\ 45.931 \\ -45.931 \\ -87.724 \\ 87.724 \\ 18.621 \\ -18.621 \\ 0 \end{bmatrix}$$

The frame reactions are shown in Fig.13-9f.

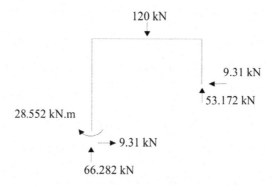

Fig. 13-9f. Applied Loads and Reactions

To determine the vertical displacement under the concentrated load, the flexibility matrix [F] is constructed using Equations [13-12] and [13-13].

$$[B] = \begin{bmatrix} 0.238 \\ 0.383 \\ -0.383 \\ -0.731 \\ 0.731 \\ 0.155 \\ -0.155 \\ 0 \end{bmatrix} \qquad\qquad [13\text{-}12]$$

$$[F] = [4.971 \times 10^{-6}] \qquad\qquad [13\text{-}13]$$

The displacement under the concentrated load, which is computed using Equation [13-14], is equal to

$$[d] = [0.001] = [0.001 \text{ m}\downarrow]$$

Example: 13-7

Analyze the frame shown in Fig.13-10a. $[E = 200.10^6 \, kN/m^2, I = 400.10^6 \, mm^4]$

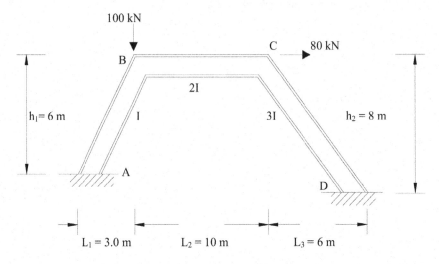

100 kN

C   80 kN

B

2I

$h_1 = 6$ m   I   3I $h_2 = 8$ m

A

D

$L_1 = 3.0$ m $L_2 = 10$ m $L_3 = 6$ m

Fig.13-10a

SOLUTION

The inclined frame is statically indeterminate to the third degree. The reactions at D are the chosen redundants, as shown in Figs.13-10b, 13-10c, 13-10d, and 13-10e.

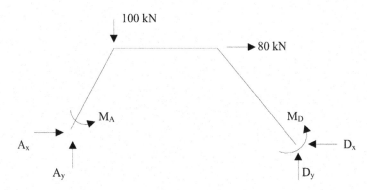

100 kN

80 kN

M_A

A_x M_D D_x

A_y D_y

Fig. 13-10b. Structure Free Body Diagram

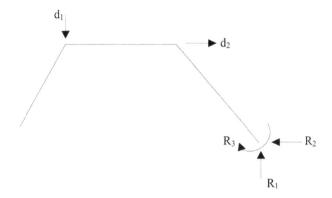

Fig. 13-10c, DOF and Redundants

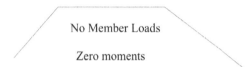

No Member Loads

Zero moments

Fig. 13-10d. Fixed End Moments

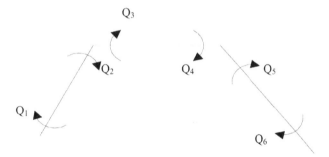

Fig. 13-10e. Internal Member Forces

The force transformation matrix [b], which is constructed using Equations [13-6] and [13-11], is as follows.

$$[Q] = \begin{bmatrix} Q^1 \\ Q^2 \\ Q^3 \\ Q^4 \\ Q^5 \\ Q^6 \end{bmatrix} = \begin{matrix} D_1=1 & D_2=1 & R_1=1 & & R_2=1 & R_3=1 \\ \begin{bmatrix} -L_1 & -h_1 & (L_1+L_2+L_3) & -(h_2-h_1) & -1 \\ 0 & 0 & -(L_2+L_3) & h_2 & 1 \\ 0 & 0 & (L_2+L_3) & -h_2 & -1 \\ 0 & 0 & -L_3 & h_2 & 1 \\ 0 & 0 & L_3 & -h_2 & -1 \\ 0 & 0 & 0 & 0 & 1 \end{bmatrix} \end{matrix} \begin{bmatrix} D_1 \\ D_2 \\ R_1 \\ R_2 \\ R_3 \end{bmatrix}$$

$\mathbf{b_D}$ $\qquad\qquad$ $\mathbf{b_R}$

Substituting the values of L_1, L_2, L_3, H_1 and H_2, the matrix [b] becomes

$$[b] = \begin{bmatrix} -3 & -6 & 19 & -2 & -1 \\ 0 & 0 & -16 & 8 & 1 \\ 0 & 0 & 16 & -8 & -1 \\ 0 & 0 & -6 & 8 & 1 \\ 0 & 0 & +6 & -8 & -1 \\ 0 & 0 & 0 & 0 & 1 \end{bmatrix}$$

$\mathbf{b_D}$ $\qquad\quad$ $\mathbf{b_R}$

The flexibility matrix [f], which is built using Equation [13-3], is as follows.

$$[f] = \begin{bmatrix} \dfrac{2.L_{AB}}{6EI} & \dfrac{-L_{AB}}{6EI} & 0 & 0 & & 0 & 0 \\[2ex] \dfrac{-L_{AB}}{6EI} & \dfrac{2.L_{AB}}{6EI} & 0 & 0 & & 0 & 0 \\[2ex] 0 & 0 & \dfrac{2.L_{BC}}{6E.2I} & \dfrac{-L_{BC}}{6E.2I} & & 0 & 0 \\[2ex] 0 & 0 & \dfrac{-L_{BC}}{6E.2I} & \dfrac{2.L_{BC}}{6E.2I} & & 0 & 0 \\[2ex] 0 & 0 & 0 & 0 & & \dfrac{2.L_{CD}}{6E.3I} & \dfrac{-L_{CD}}{6E.3I} \\[2ex] 0 & 0 & 0 & 0 & & \dfrac{-L_{CD}}{6E.3I} & \dfrac{2L_{CD}}{6E.3I} \end{bmatrix}$$

Where L_{AB} = 6.708 m, L_{BC} = 10.0 m, L_{CD} = 10.0m

The matrices [F_{RD}] and [F_{RR}], which are computed using Equations [13-7] and [13-8], respectively, are equal to the following:

456

$$[F_{RD}] = \begin{bmatrix} -2.264 \times 10^{-3} & -4.528 \times 10^{-3} \\ 5.031 \times 10^{-4} & 1.006 \times 10^{-3} \\ 1.258 \times 10^{-4} & 2.516 \times 10^{-4} \end{bmatrix}$$

$$[F_{RR}] = \begin{bmatrix} 0.034 & -0.013 & -2.28 \times 10^{-3} \\ -0.013 & 7.237 \times 10^{-3} & 1.086 \times 10^{-3} \\ -2.28 \times 10^{-3} & 1.086 \times 10^{-3} & 1.88 \times 10^{-4} \end{bmatrix}$$

The redundant matrix [R], which is determined using Equation [13-9], is equal to:

$$R = \begin{bmatrix} 30.861 \\ 66.901 \\ -186.107 \end{bmatrix} \Rightarrow \begin{bmatrix} 30.861 \; kN \; \uparrow \\ 66.901 \; kN \; \leftarrow \\ 186.107 \; kN.m \; \curvearrowleft \end{bmatrix}$$

The matrix [D] is equal to:

$$[D] = \begin{bmatrix} 100 \; kN & @d_1 \\ 80 \; kN & @d_2 \end{bmatrix}$$

The internal member force matrix [Q], which is computed using Equation [13-11], is equal to

$$[Q] = \begin{bmatrix} -141.336 \\ -144.675 \\ 144.675 \\ 163.935 \\ -163.935 \\ -186.107 \end{bmatrix}$$

The final moment matrix [M] is built using Equation [13-16] as follows.

$$M = \begin{bmatrix} -141.336 \\ -144.675 \\ 144.675 \\ 163.935 \\ -163.935 \\ -186.107 \end{bmatrix} + \begin{bmatrix} 0 \\ 0 \\ 0 \\ 0 \\ 0 \\ 0 \end{bmatrix} = \begin{bmatrix} -141.336 \\ -144.675 \\ 144.675 \\ 163.935 \\ -163.935 \\ -186.107 \end{bmatrix}$$

The frame reactions are shown in Fig. 13-10f.

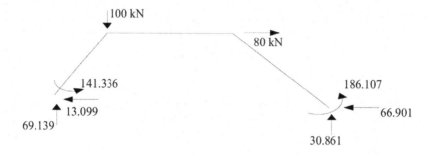

Fig. 13-10f. Applied Loads and Reactions

To determine the frame displacements, the flexibility matrix [F] is constructed using Equations [13-12] and [13-13], as shown below.

$$[B] = \begin{bmatrix} -0.544 & -1.087 \\ -0.556 & -1.113 \\ 0.556 & 1.113 \\ 0.631 & 1.261 \\ -0.631 & -1.261 \\ -0.716 & -1.432 \end{bmatrix} \quad\quad [13\text{-}12]$$

$$[F] = \begin{bmatrix} 2.225 \times 10^{-5} & 4.45 \times 10^{-5} \\ 4.45 \times 10^{-5} & 8.901 \times 10^{-5} \end{bmatrix} \quad\quad [13\text{-}13]$$

The displacement matrix [d], which is computed using Equation [13-14], is equal to

$$[d] = \begin{bmatrix} 5.786 \times 10^{-3} \\ 0.012 \end{bmatrix} \Rightarrow \begin{bmatrix} 5.786 \times 10^{-3} & m \downarrow \\ 0.012 & m \rightarrow \end{bmatrix}$$

Example: 13-8

Analyze the truss shown in Fig.13-11a.   [$E = 200.10^6$ kN/m², $A_{AB} = 200$ mm², $A_{BD} = 300$ mm² , $A_{CD} = 150$ mm²]

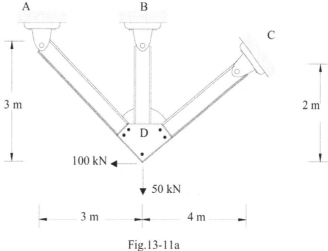

Fig.13-11a

SOLUTION

The truss is indeterminate to the first degree. The internal force F_{BD} in the member BD is chosen as redundant (Figs.13-11b, 13-11c, and 13-11d).

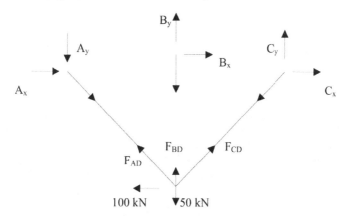

Fig. 13-11b. Structure Free Body Diagram

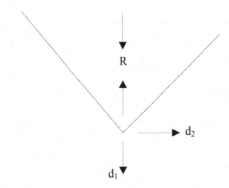

Fig. 13-11c. DOF and Redundants

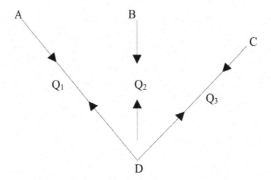

Fig. 13-11d. Internal Member Forces

The force transformation matrix [b] is constructed as follows (see Figs. 13-11e,13-11f, and 13-11g).

1. A unit load is applied in the direction of each degree of freedom.
2. The member forces, which represent the elements of the matrix [b], are determined.

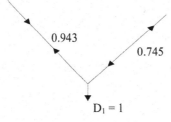

Fig. 13-11e

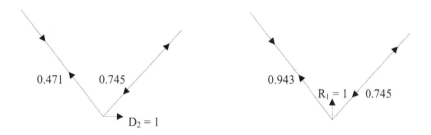

Fig. 13-11f   Fig. 13-11g

The force transformation matrix [b], which is computed using Equations [13-6] and [13-11], is equal to

$$[b] = \begin{bmatrix} Q_1 \\ Q_2 \\ Q_3 \end{bmatrix} = \begin{array}{c} \begin{array}{cc} D_1=1 & D_2=1 \end{array} \quad R_1=1 \\ \begin{bmatrix} 0.943 & 0.745 & -0.943 \\ 0 & 0 & 1 \\ 0.745 & -0.745 & -0.745 \end{bmatrix} \end{array} \begin{bmatrix} D_1 \\ D_2 \\ R \end{bmatrix}$$

$$\qquad\qquad\qquad\qquad b_D \qquad\qquad b_R$$

The flexibility matrix [f], which is built using Equation [13-3], is equal to

$$[f] = \begin{bmatrix} \dfrac{L_{AD}}{A_1 E} & 0 & 0 \\ 0 & \dfrac{L_{BD}}{A_2 E} & 0 \\ 0 & 0 & \dfrac{L_{CD}}{A_3 E} \end{bmatrix} = \begin{bmatrix} 1.061 \times 10^{-4} & 0 & 0 \\ 0 & 5 \times 10^{-5} & 0 \\ 0 & 0 & 1.491 \times 10^{-4} \end{bmatrix}$$

The matrices [F_{RD}] and [F_{RR}], which are computed using [13-7] and [13-8], respectively, are equal to

$$[F_{RD}] = \begin{bmatrix} -1.771 \times 10^{-4} & 3.562 \times 10^{-5} \end{bmatrix}$$

$$[F_{RR}] = \begin{bmatrix} 2.271 \times 10^{-4} \end{bmatrix}$$

The redundant matrix [R] is equal to the following

$$R = [54.679] \Rightarrow [F_{BD} = 54.679 \text{ kN} \quad T]$$

The matrix [D] is equal to

$$[D] = \begin{bmatrix} 50 \\ -100 \end{bmatrix}$$

The member force matrix [Q], which is computed using Equation [13-11], is as follows.

$$[Q] = \begin{bmatrix} Q_1 \\ Q_2 \\ Q_3 \end{bmatrix} = \begin{bmatrix} -51.512 \\ 54.679 \\ 71.014 \end{bmatrix} \Rightarrow \begin{bmatrix} F_{AD} = 51.512 \text{ kN} & Compression \\ F_{BD} = 54.679 \text{ kN} & Tension \\ F_{CD} = 71.014 \text{ kN} & Tension \end{bmatrix}$$

The reactions shown in Fig. 13-11h are easily computed once the element forces are determined

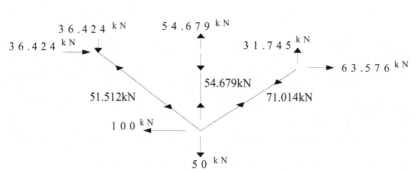

Fig. 13-11h. Applied Loads ,Reactions and Bar forces

To determine the truss displacement matrix [d], the flexibility matrix [F] needs first to be determined using Equations [13-12] and [13-13] as shown below.

$$[B] = \begin{bmatrix} 0.208 & 0.619 \\ 0.78 & -0.157 \\ 0.164 & -0.628 \end{bmatrix} \tag{13-12}$$

$$[F] = \begin{bmatrix} 3.899 \times 10^{-5} & -7.844 \times 10^{-6} \\ -7.844 \times 10^{-6} & 1.007 \times 10^{-4} \end{bmatrix} \tag{13-13}$$

The truss displacement matrix [d], which is determined using Equation [13-14], is equal to:

$$d = \begin{bmatrix} 2.734 \times 10^{-3} \\ -0.01 \end{bmatrix} = \begin{bmatrix} 2.734 \text{ mm} & \downarrow \\ 0.01 \text{ mm} & \leftarrow \end{bmatrix}$$

Summary

- The analysis procedure of indeterminate structures using the flexibility method is a matrix formulation of the force method presented in Chapter 9.
- In order to meet the requirements of statics and geometry (deformation), the given structure is resolved into two components:
    1. **Basic system**, obtained from the initial system by removing the redundants (release of redundants), but retaining other causes, such as the applied loads and the change in volume.
    2. **Complementary system**, obtained from the initial system by removing the applied loads and the change in volume, but retaining the redundants.
- The basic and the complementary systems must satisfy (each one independently) the conditions of static equilibrium and must be geometrical stable.
- The superposition of both systems must also satisfy the conditions of compatibility; namely, the algebraic sum of deformations at the sections of release must be either zero or equal to the prescribed displacement. In matrix form,

$$\Delta_x + \Delta_o = \left\{ \begin{array}{c} 0 \\ \text{or} \\ \Delta \end{array} \right\}$$

Where Δ_x = deformation vector, complementary system, $\{n \times 1\}$,
Δ_o = deformation vector, basic system, $\{n \times 1\}$
Δ = prescribed displacement vector, initial system, $\{n \times 1\}$,
n = number of redundants

- The selection of the possible basic and complementary systems for a proposed structure is not exclusive. Usually there is available a large number of choices some of which offer a simpler solution, others a more numerically accurate solution.
- The vector matrices Δ_x and Δ_o can always be expressed as follows

$\Delta_x = f X$ and $\Delta_o = l W$

where f = redundant effect flexibility matrix, $\{n \times n\}$,
l = load effect flexibility matrix, $\{n \times m\}$,
X = redundant vector matrix, $\{n \times 1\}$,
m = number of loads.

- The matrices f and l are formed respectively by flexibility coefficients.
- The flexibility is defined as the deformation due to a unit cause. Since the unit cause can be a force or a moment and the deformation can be measured at the point and in the direction of the cause or at another point or in another direction, the flexibilities are further designated as direct (near) or indirect (far), respectively.
- The direct flexibility δ_{jj} is the deformation at the point of application of $X_j = +1$ due to X_j and along the line of action of X_j.
- The indirect flexibility δ_{ji} is the deformation at the point of application of $X_j = +$ along the line of action of X_j, due to $X_i = +1$ acting at another point i.
- The load flexibility $l_{j_o,k}$ is the deformation at the point of application of $X_j = +1$ along the line of action of X_j, due to $P_k = +1$ or $Q_k = +1$, acting at another point k. Obviously, the load flexibility and the direct or indirect flexibility (under certain conditions) may be the same value.

Problems

1-   Determine the reactions for the beam shown below using the flexibility method.

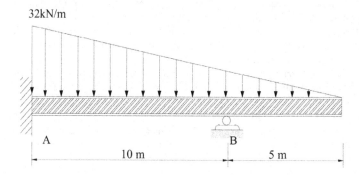

2-   Using the flexibility method draw the shear and moment diagrams for the beam shown below when its support B settles by 0.03-meter. [E = 200.10^9 N/m^2, I =2 000.10^{-6} m^4]

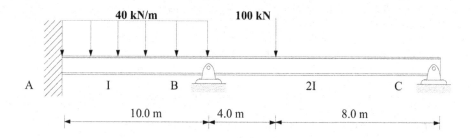

3-   Determine the reactions of the beam shown below using the flexibility method.

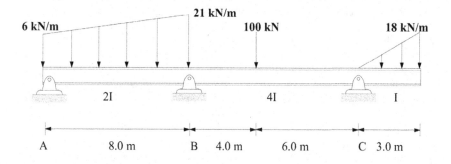

4-Using the flexibility method draw the shear and moment diagrams for the beam shown below.

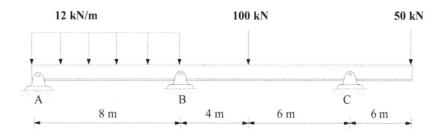

5-   Using the flexibility method determine the reactions of the frame shown below.

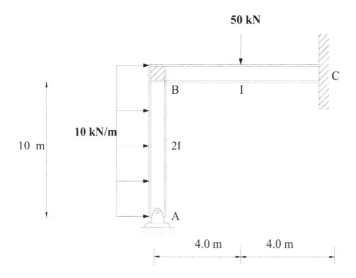

6-   Using the flexibility method draw the shear and moment diagrams for the the frame.

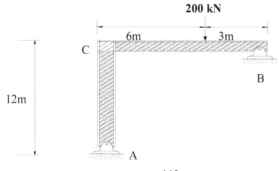

7-   Using the flexibility method analyze the frame shown below.

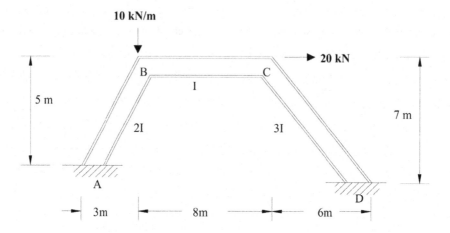

8-   Using the flexibility method analyze the truss shown below. [EA = Constant].

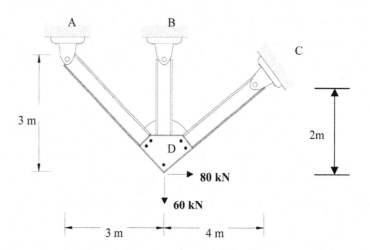

9-   Analyze problem No. 7 by having support D as a pinned end.

10-   Analyze problem No. 7 by having both supports D & A as a pinned end.

CHAPTER- 14
STIFFNESS MATRIX METHOD

Cairo - Egypt

Introduction

The stiffness of a structure is defined as the force developed due to a unit displacement. The stiffness matrix $[k]$ relates the displacement matrix $[d]$ to the external load matrix $[D]$ as shown below (see Fig. 13-1).

$$D_1 = k_{11}d_1 + k_{12}d_1 + \cdots + \cdots + k_{1N}d_N$$

$$D_2 = k_{21}d_1 + k_{22}d_1 + \cdots + \cdots + k_{2N}d_N$$

$$\cdots \quad \cdots \quad \cdots \quad \cdots \quad \cdots \quad \cdots$$

$$\cdots \quad \cdots \quad \cdots \quad \cdots \quad \cdots$$

$$D_N = k_{N1}d_1 + k_{N2}d_1 + \cdots + \cdots + k_{NN}d_N$$

Where d_i = displacement of the ith point , k_{ij} = force at point i due to a displacement at a point j, and D_i = external load at the ith point

These expressions can be written in matrix form as follows.

$$[D] = [K][d] \tag{14-1}$$

$$
\begin{bmatrix} D_1 \\ D_2 \\ \cdots \\ \cdots \\ D_N \end{bmatrix}
=
\begin{bmatrix}
k_{11} & k_{12} & \cdots & \cdots & k_{1N} \\
k_{21} & k_{22} & \cdots & \cdots & k_{2N} \\
\cdots & \cdots & \cdots & \cdots & \cdots \\
\cdots & \cdots & \cdots & \cdots & \cdots \\
k_{N1} & k_{N2} & \cdots & \cdots & k_{NN}
\end{bmatrix}
\begin{bmatrix} d_1 \\ d_2 \\ \cdots \\ \cdots \\ d_N \end{bmatrix}
\tag{14-2}
$$

Where $[d]$ = displacement matrix, $[K]$ = stiffness matrix, and $[D]$ = external load matrix

The force response of a structure is represented by the stiffness matrix $[K]$. For a beam element subjected to clockwise moments at both ends as shown in Fig. 14-1, the stiffness coefficients are determined using slope deflection method and are equal to:

$$k_{11} = \frac{4EI}{L} \quad , \quad k_{12} = \frac{2EI}{L} \quad , \quad k_{21} = \frac{2EI}{L} \quad , \quad k_{22} = \frac{4EI}{L} \tag{14-3}$$

Equation [14-1] becomes

$$
\begin{bmatrix} D_1 \\ D_2 \end{bmatrix}
=
\begin{bmatrix}
\dfrac{4EI}{L} & \dfrac{2EI}{L} \\
\dfrac{2EI}{L} & \dfrac{4EI}{L}
\end{bmatrix}
\begin{bmatrix} d_1 \\ d_2 \end{bmatrix}
\tag{14-4}
$$

The matrix displacement matrix $[d]$ can be computed using the following equation:

$$[d] = [K]^{-1}[D] \tag{14-5}$$

For a structure of two members or more as shown in Fig. 14-2, the stiffness matrix of the structure is obtained by combining the stiffness matrices of the individual members. The stiffness matrices for each member of the structure are as follows.

$$
\begin{bmatrix} {}^{member\,1}k \end{bmatrix} = \begin{bmatrix} \dfrac{4EI}{L_1} & \dfrac{2EI}{L_1} \\[2ex] \dfrac{2EI}{L_1} & \dfrac{4EI}{L_1} \end{bmatrix} \;,\quad \begin{bmatrix} {}^{member\,2}k \end{bmatrix} = \begin{bmatrix} \dfrac{4EI}{L_2} & \dfrac{2EI}{L_2} \\[2ex] \dfrac{2EI}{L_2} & \dfrac{4EI}{L_2} \end{bmatrix}
\qquad [14\text{-}6]
$$

Combining the individual member matrices yields the following structure stiffness matrix :

$$
[k] = \begin{bmatrix}
\dfrac{4EI}{L_1} & \dfrac{2EI}{L_1} & 0 & 0 \\[2ex]
\dfrac{2EI}{L_1} & \dfrac{4EI}{L_1} & 0 & 0 \\[2ex]
0 & 0 & \dfrac{4EI}{L_2} & \dfrac{2EI}{L_2} \\[2ex]
0 & 0 & \dfrac{2EI}{L_2} & \dfrac{4EI}{L_2}
\end{bmatrix}
\qquad [14\text{-}7]
$$

If the axial effects are to be included, the structure stiffness matrix $[k]$ becomes

$$
[k] = \begin{bmatrix}
\dfrac{4EI}{L_1} & \dfrac{2EI}{L_1} & 0 & 0 & 0 & 0 \\[2ex]
\dfrac{2EI}{L_1} & \dfrac{4EI}{L_1} & 0 & 0 & 0 & 0 \\[2ex]
0 & 0 & \dfrac{EA_1}{L_1} & 0 & 0 & 0 \\[2ex]
0 & 0 & 0 & \dfrac{4EI}{L_2} & \dfrac{2EI}{L_2} & 0 \\[2ex]
0 & 0 & 0 & \dfrac{2EI}{L_2} & \dfrac{4EI}{L_2} & 0 \\[2ex]
0 & 0 & 0 & 0 & 0 & \dfrac{EA_2}{L_2}
\end{bmatrix}
\qquad [14\text{-}8]
$$

The member internal force matrix $[Q]$ is related to the structure external displacement matrix $[d]$ by the displacement transformation matrix $[a]$. The elements of the matrix $[a]$ are generated by: 1) setting each nodal displacement equals to 1 unit and 2) determining the related member deformation for the applied force of unit load as shown in Fig. 14-2. Thus, the member internal force matrix $[Q]$ is obtained using the following equation:

$$[Q] = [k][a][d]$$

$$
\begin{bmatrix} Q^1 \\ Q^2 \\ Q^3 \\ Q^4 \end{bmatrix} =
\begin{bmatrix}
\dfrac{4EI}{L_1} & \dfrac{2EI}{L_1} & 0 & 0 \\
\dfrac{2EI}{L_1} & \dfrac{4EI}{L_1} & 0 & 0 \\
0 & 0 & \dfrac{4EI}{L_2} & \dfrac{2EI}{L_2} \\
0 & 0 & \dfrac{2EI}{L_2} & \dfrac{4EI}{L_2}
\end{bmatrix}
\begin{bmatrix}
0 & 0 & \dfrac{-1}{L1} \\
1 & 0 & \dfrac{-1}{L1} \\
1 & 0 & \dfrac{1}{L2} \\
0 & 1 & \dfrac{1}{L2}
\end{bmatrix}
\begin{bmatrix} d_1 \\ d_2 \\ d_3 \end{bmatrix}
\qquad [14\text{-}10]
$$

The structure stiffness matrix $[K]$ that relates nodal forces to the nodal displacements can be constructed directly using the following equation:

$$[K] = [a^T][k][a] \qquad [14\text{-}11]$$

The structure stiffness matrix $[K]$ for the structure shown in Fig. 14-3 is as follows.

$$
[K] = EI
\begin{bmatrix}
4\dfrac{(L_2 + L_1)}{(L_1 L_2)} & \dfrac{2}{L_2} & -6\dfrac{(L_2^2 - L_1^2)}{(L_1^2 L_2^2)} \\
\dfrac{2}{L_2} & \dfrac{4}{L_2} & \dfrac{6}{L_2^2} \\
-6\dfrac{(L_2^2 - L_1^2)}{(L_1^2 L_2^2)} & \dfrac{6}{L_2^2} & 12\dfrac{(L_2^3 + L_1^3)}{(L_1^3 L_2^3)}
\end{bmatrix}
\qquad [14\text{-}12]
$$

Once the structure stiffness matrix is determined, the displacements are computed using Equation [14-5].

Stiffness method analysis of indeterminate structures

The analysis procedure of indeterminate structures using the stiffness method is essentially a matrix formulation of the slope deflection, which is presented in Chapter 10. The steps of the solution are as follows.

1. Define the external load matrix $[D]$ and the displacement matrix $[d]$.

2. Define the internal member force matrix $[Q]$.

3. Build the displacement transformation matrix $[a]$.

4. Build the element stiffness matrix $[k]$.

5. Build the structure stiffness matrix $[K]$ using Equation [14-11]
6. Obtain the displacement matrix $[d]$ using Equation [14-5]
7. Obtain the member force matrix $[Q]$ using Equation [14-10]

The following illustrative examples will employ this procedure of analysis.

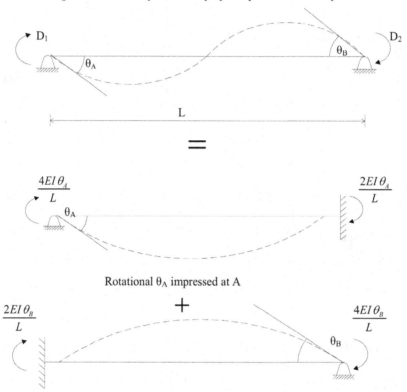

Rotational θ_A impressed at A

Rotational θ_B impressed at B

Fig. 14-1

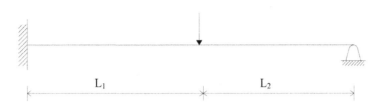

Single-Span Beam Structure

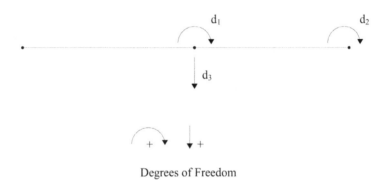

Degrees of Freedom

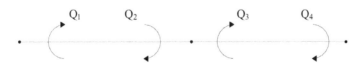

Member Internal Forces

Fig. 14-2

Example: 14-1

Determine the bar forces for the truss shown in Fig 14-3a. [E = 200.10^6 kN/m^2]

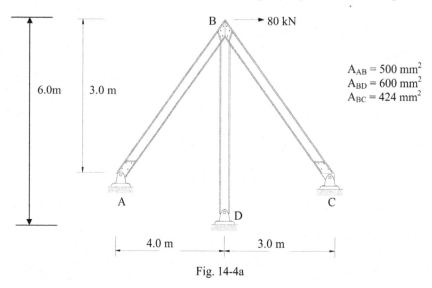

$A_{AB} = 500$ mm^2
$A_{BD} = 600$ mm^2
$A_{BC} = 424$ mm^2

Fig. 14-4a

SOLUTION

The truss reactions, the external load matrix [**D**], the displacement matrix [**d**] and the internal force matrix [**Q**] are shown in Figs. 14-3b,14-3c, and 14-3d, respectively.

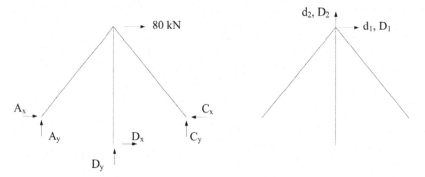

Fig. 14-3b. Structure Free Body Diagram   Fig. 14-3c. Degrees of Freedom

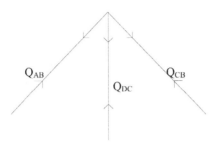

Fig. 14-3d. Internal Member Forces

The displacement transformation matrix [**a**] is built as shown in Figs. 14-3e, 14-3f, and 14-3g.

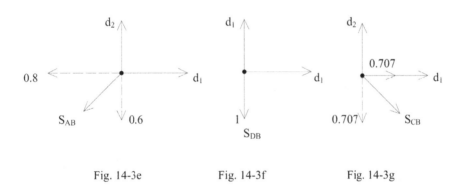

Fig. 14-3e   Fig. 14-3f   Fig. 14-3g

The relationship between the member total deformation matrix [**S**] (one unit deformation) and the structural displacement matrix [**d**] is given by the following equation:

$$[S] = [a][d]$$

[14-13]

$$[S] = \begin{bmatrix} S_{AB} \\ S_{DB} \\ S_{CB} \end{bmatrix} = \begin{bmatrix} 0.8 & 0.6 \\ 0 & 1 \\ -0.707 & 0.707 \end{bmatrix} \begin{bmatrix} d_1 \\ d_2 \end{bmatrix}$$

The element stiffness matrix $[k]$, which is built using Equation [14-8], is equal to:

$$[k] = \begin{bmatrix} \dfrac{EA_{AB}}{L_{AB}} & 0 & 0 \\ 0 & \dfrac{EA_{DB}}{L_{DB}} & 0 \\ 0 & 0 & \dfrac{EA_{CB}}{L_{CB}} \end{bmatrix}$$

where $A_{AB} = 500.10^{-6}$ m^2, $A_{DB} = 600.10^{-6}$ m^2, $A_{CB} = 424.10^{-6}$ m^2, $L_{AB} = 5.0$ m, $L_{DB} = 6.0$ m, $L_{CB} = 4.24$ m

$$[k] = \begin{bmatrix} 2 \times 10^4 & 0 & 0 \\ 0 & 2 \times 10^4 & 0 \\ 0 & 0 & 2 \times 10^4 \end{bmatrix}$$

The structure stiffness matrix $[K]$, which is built using Equation [14-11], is equal to:

$$[K] = [a^T]\,[k]\,[a] = \begin{bmatrix} 2.28 \times 10^4 & -396.98 \\ -396.98 & 3.72 \times 10^4 \end{bmatrix}$$

The displacement matrix $[d]$, which is determined using Equation [14-5], is equal to:

$$[d] = [K]^{-1}\,[D] = \begin{bmatrix} 3.5 \times 10^{-3} \\ 3.746 \times 10^{-5} \end{bmatrix} = \begin{bmatrix} 3.5 \times 10^{-3}\ m\ \rightarrow \\ 3.746 \times 10^{-5}\ m\ \uparrow \end{bmatrix}$$

where $[D] = \begin{bmatrix} 80 \\ 0 \end{bmatrix}$

Notes:
- The 80-kN load has a positive sign because it has the same direction as the displacement d_1.
- There is no load in the direction of the displacement d_2.

The element (bar) force matrix $[Q]$, which is determined using Equation [14-10], is equal to:

$$[Q] = \begin{bmatrix} 56.608 \\ 0.749 \\ -49.1 \end{bmatrix} = \begin{bmatrix} 56.608\ kN & Tension \\ 0.749\ kN & Tension \\ 49.1\ kN & Compression \end{bmatrix}$$

The reactions are easily be computed using principles of equilibrium once the bar forces are determined as shown in Fig. 14-4h.

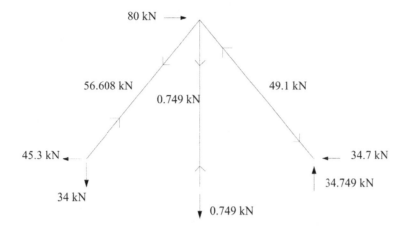

Fig. 14-3h. Final Free Body Diagram

Example: 14-2

Draw the shear and moment diagrams for the continuous beam shown in Fig 14-4a.
$[E= 200.10^6 \text{ kN/m}^2 , I = 400.10^6 \text{ mm}^4]$

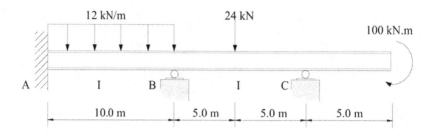

12 kN/m   24 kN

100 kN.m

A   I   B   I   C

10.0 m   5.0 m   5.0 m   5.0 m

Fig. 14-4a

SOLUTION

The beam reactions, the external load matrix [D], the displacement matrix [d], and the internal force matrix [Q] are shown in Figs. 14-4b, 14-4c, 14-4d, and 14-4e, respectively. The fixed end moments are summarized in Table 10-1.

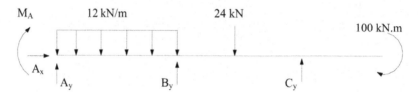

M_A 12 kN/m   24 kN   100 kN.m

A_x

A_y B_y C_y

Fig. 14-4b. Structural Model

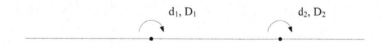

d_1, D_1 d_2, D_2

Fig. 14-4c. Degrees of Freedom

100 kN.m   100 kN.m   30 kN.m   30 kN.m   100 kN.m

Fig. 14-4d. Fixed End Moments

Fig. 14-4e. Internal Member Forces

The displacement transformation matrix [a] is equal to

$$[S] = \begin{array}{c} d_1 \ d_2 \\ \begin{bmatrix} 0 & 0 \\ 1 & 0 \\ 1 & 0 \\ 0 & 1 \end{bmatrix} \begin{bmatrix} d_1 \\ d_2 \end{bmatrix} \end{array}$$

The element stiffness matrix [k], which is built using Equation [14-3], is equal to

$$[k] = \begin{bmatrix} 3.2 \times 10^4 & 1.6 \times 10^4 & 0 & 0 \\ 1.6 \times 10^4 & 3.2 \times 10^4 & 0 & 0 \\ 0 & 0 & 3.2 \times 10^4 & 1.6 \times 10^4 \\ 0 & 0 & 1.6 \times 10^4 & 3.2 \times 10^4 \end{bmatrix}$$

The structure stiffness matrix [K], which is built using Equation [14-11], is equal to

$$[K] = [a^T] [k] [a] = \begin{bmatrix} 6.4 \times 10^4 & 1.6 \times 10^4 \\ 1.6 \times 10^4 & 3.2 \times 10^4 \end{bmatrix}$$

The displacement matrix [d], which is determined using Equation [14-5], is equal to

$$[d] = [K]^{-1} [D] = \begin{bmatrix} -1.875 \times 10^{-3} & rad \\ 3.125 \times 10^{-3} & rad \end{bmatrix}$$

where $[D] = \begin{bmatrix} -(100 - 30) = -70 \\ -(30 - 100) = 70 \end{bmatrix}$

Note
    Clock wise moment is positive but the negative sign in front is for equal and
    opposite moment, to balance the moments at the joint.

The structure force matrix [Q], which is obtained using Equation [14-10], is equal to

$$[Q] = \begin{bmatrix} -30 \\ -60 \\ -10 \\ 70 \end{bmatrix}$$

The final member force matrix $\left[\overline{Q}\right]$ due to span loading is computed using the following equation:

$$\left[\overline{Q}\right] = [Q] + [Q_0] \qquad\qquad [14\text{-}13]$$

where $[Q_0]$ = fixed end moment matrix (Fig. 14-4d).

The final member force matrix $\left[\overline{Q}\right]$ is equal to

$$\left[\overline{Q}\right] = \begin{bmatrix} -30 \\ -60 \\ -10 \\ 70 \end{bmatrix} + \begin{bmatrix} -100 \\ 100 \\ -30 \\ 30 \end{bmatrix} = \begin{bmatrix} -130 \\ 40 \\ -40 \\ 100 \end{bmatrix}$$

The beam free body diagram, shear and moment diagrams are shown in Fig. 14-4f, 14-4g, and 14-4h, respectively.

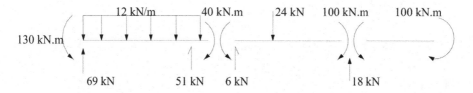

Fig. 14-4f. Beam Free Body Diagram

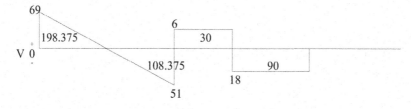

Fig. 14-4g. Shear Diagram (kN)

480

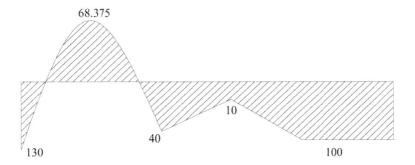

Fig. 14-4h. Moment Diagram (kN.m)

Example: 14-3

Determine the reactions for the structure shown in Fig 14-5a.
[E= 200.10^6 kN/m^2 , I = 100.10^6 mm^4, K_{Sp} = 200 kN/m]

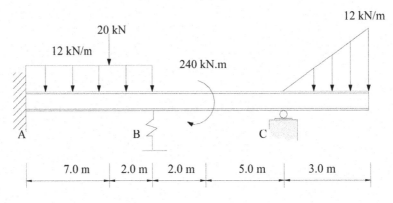

Fig. 14-5a

SOLUTION

The structure reactions, the external load matrix [D], the displacement matrix [d], the fixed-end forces, and the internal force matrix [Q] are shown in Figs. 14-5b, 14-5c, 14-5d, and 14-5e, respectively.

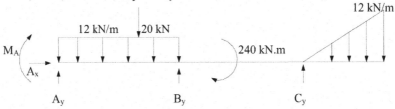

Fig. 14-5b. Structural Model

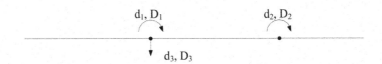

Fig. 14-5c. Degrees of Freedom

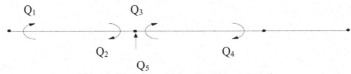

Fig. 14-5d. Internal Member Force Matrix

482

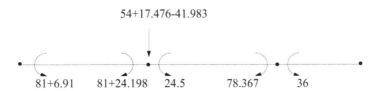

Fig. 14-5e. Fixed End Forces

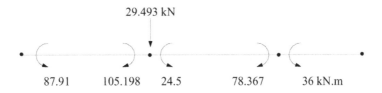

Fig. 14-5f. Total Fixed End Forces

The displacement transformation matrix [*a*] is equal to

$$[S] = \begin{bmatrix} 0 & 0 & -\dfrac{1}{9} \\ 1 & 0 & -\dfrac{1}{9} \\ 1 & 0 & \dfrac{1}{7} \\ 0 & 1 & \dfrac{1}{7} \\ 0 & 0 & 1 \end{bmatrix} \begin{bmatrix} d_1 \\ d_2 \\ d_3 \end{bmatrix}$$

The element stiffness matrix [*k*], which is obtained using Equation [14-8], is as shown below.

$$[k] = \begin{bmatrix} 8.889\times10^3 & 4.444\times10^3 & 0 & 0 & 0 \\ 4.444\times10^3 & 8.889\times10^3 & 0 & 0 & 0 \\ 0 & 0 & 1.143\times10^4 & 5.714\times10^3 & 0 \\ 0 & 0 & 5.714\times10^3 & 1.143\times10^4 & 0 \\ 0 & 0 & 0 & 0 & 200 \end{bmatrix}$$

The structure stiffness matrix $[K]$, which is built using Equation [14-11], is equal to

$$[K] = [a^T][k][a] = \begin{bmatrix} 2.032 \times 10^4 & 5.714 \times 10^3 & 967.498 \\ 5.714 \times 10^3 & 1.143 \times 10^4 & 2.449 \times 10^3 \\ 967.498 & 2.449 \times 10^3 & 1.229 \times 10^3 \end{bmatrix}$$

The displacement matrix $[d]$, which is determined using Equation [14-5], is equal to

$$[d] = [K]^{-1}[D] = \begin{bmatrix} -2.616 \times 10^{-3} & rad \\ -0.014 & rad \\ 0.054 & m \end{bmatrix}$$

$$\text{where} \quad [D] = \begin{bmatrix} -(105.198 - 24.5) = -80.698 \\ -(78.367 - 36) \quad = -42.367 \\ 29.493 \end{bmatrix}$$

The member force matrix $[Q]$, which is determined using Equation [14-10]

$$[Q] = \begin{bmatrix} -91.366 \\ -102.994 \\ 22.296 \\ -42.367 \\ 10.765 \end{bmatrix}$$

⌐—— Spring Force

The final member force matrix $\left[\overline{Q}\right]$ due to span loading is computed using Equation [14-13] as follows.

$$\left[\overline{Q}\right] = [Q] + [Q_0] \qquad\qquad [14\text{-}13]$$

Thus $\left[\overline{Q}\right]$ is equal to

$$\left[\overline{Q}\right] = \begin{bmatrix} -91.366 \\ -102.994 \\ 22.296 \\ -42.367 \end{bmatrix} + \begin{bmatrix} -87.91 \\ 105.198 \\ -24.5 \\ 78.367 \end{bmatrix} = \begin{bmatrix} -179.276 \\ 2.204 \\ -2.204 \\ 36 \end{bmatrix}$$

The beam reactions, which are easily determined after the computation of the final member forces, are shown in Fig. 14-5g.

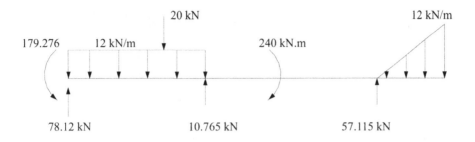

Fig. 14-5g. Applied Loads and Reactions

Example: 14-4

Determine the reactions of the structure shown in Fig. 14-6a if the supports at A and B settle by 40 mm.

$I_{Beam} = 800 . 10^6 \text{ mm}^4$ $A_{cable} = 800 \text{ mm}^2$ $E_{Beam} = E_{cable} = 200 . 10^6 \text{ kN/m}^2$

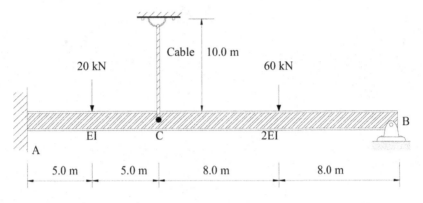

Fig. 14-6a

SOLUTION

The structure reactions, external loads, displacements, internal forces, fixed end forces due to loadings, fixed end forces due to support settlements at A, fixed end forces due to support settlement at B, and total fixed end forces are summarized in Figs. 14-6b, 14-6c, 14-6d, 14-6e, 14-6f, 14-6g, and 14-6h, respectively.

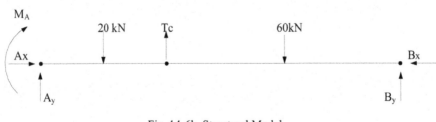

Fig. 14-6b. Structural Model

Fig. 14-6c. Degrees of Freedom

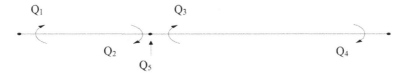

Fig. 14-6d. Member Internal Forces

10+30 = 40 kN

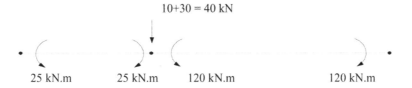

25 kN.m   25 kN.m   120 kN.m   120 kN.m

Fig. 14-6e. Fixed End Forces due to Applied Loads

76.8 kN

384 kN.m

384 kN.m = (6)(200.10^6)(800.10^{-6})(0.04)/(10^2)

Fig. 14-6f. Fixed End Forces due to Support Settlement at A

37.5 kN

300 kN.m

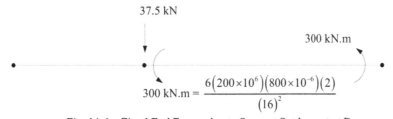

$$300 \text{ kN.m} = \frac{6\left(200\times10^6\right)\left(800\times10^{-6}\right)(2)}{(16)^2}$$

Fig. 14-6g. Fixed End Forces due to Support Settlement at B

154 kN

359 kN.m   180 kN.m

409 kN.m   420 kN.m

Fig. 14-6h. Total Fixed End Forces

The displacement transformation matrix [a] is equal to

$$
[S] = \begin{bmatrix} S_1 \\ S_2 \\ S_3 \\ S_4 \\ S_5 \end{bmatrix} = \begin{array}{ccc} d_1 & d_2 & d_3 \\ \begin{bmatrix} 0 & 0 & -\dfrac{1}{10} \\ 1 & 0 & -\dfrac{1}{10} \\ 1 & 0 & \dfrac{1}{16} \\ 0 & 1 & \dfrac{1}{16} \\ 0 & 0 & 1 \end{bmatrix} \end{array} \begin{bmatrix} d_1 \\ d_2 \\ d_3 \end{bmatrix}
$$

The element stiffness matrix [k], which is computed using Equation [14-8], is equal to

$$
[k] = \begin{bmatrix} \dfrac{4EI}{L_{AB}} & \dfrac{2EI}{L_{AB}} & 0 & 0 & 0 \\ \dfrac{2EI}{L_{AB}} & \dfrac{4EI}{L_{AB}} & 0 & 0 & 0 \\ 0 & 0 & \dfrac{8EI}{L_{BC}} & \dfrac{4EI}{L_{BC}} & 0 \\ 0 & 0 & \dfrac{4EI}{L_{BC}} & \dfrac{8EI}{L_{BC}} & 0 \\ 0 & 0 & 0 & 0 & \dfrac{AE}{L_{Cable}} \end{bmatrix} = \begin{bmatrix} 6.4 \times 10^4 & 3.2 \times 10^4 & 0 & 0 & 0 \\ 3.2 \times 10^4 & 6.4 \times 10^4 & 0 & 0 & 0 \\ 0 & 0 & 8 \times 10^4 & 4 \times 10^4 & 0 \\ 0 & 0 & 4 \times 10^4 & 8 \times 10^4 & 0 \\ 0 & 0 & 0 & 0 & 1.6 \times 10^4 \end{bmatrix}
$$

The structure stiffness matrix [**K**], which is built using Equation [14-11], is equal to

$$
[K] = \begin{bmatrix} 1.44 \times 10^5 & 4 \times 10^4 & -2.1 \times 10^3 \\ 4 \times 10^4 & 8 \times 10^4 & 7.5 \times 10^3 \\ -2.1 \times 10^3 & 7.5 \times 10^3 & 1.886 \times 10^4 \end{bmatrix}
$$

The structure displacement matrix [**d**] is determined using Equation [14-5]

$$
[d] = [K^{-1}][D] = \begin{bmatrix} -2.843 \times 10^{-4} \ \text{rad} \\ 1.691 \times 10^{-3} \ \text{rad} \\ 7.478 \times 10^{-3} \ \text{m} \end{bmatrix}
$$

Where $[D] = \begin{bmatrix} D_1 \\ D_2 \\ D_3 \end{bmatrix} = \begin{bmatrix} -(40 - 420) = 11 \ \text{kN.m} \quad @d_1 \\ -(-180) = 180 \ \text{kN.m} \quad @d_2 \\ 154 \ \text{kN} \qquad\qquad\quad @d_3 \end{bmatrix}$

The structure internal force matrix [Q], which is built using Equation [14-10], is equal to

$$[Q] = \begin{bmatrix} -80.888 \text{ kN.m} \\ -89.986 \text{ kN.m} \\ 100.986 \text{ kN.m} \\ 180 \quad\; \text{kN.m} \\ 119.651 \;\text{ kN} \end{bmatrix}$$
_____ Cable Tension T_c

The final internal force matrix $\left[\overline{Q}\right]$ due to external loadings and settlements is equal

$$\left[\overline{Q}\right] = [Q] + [Q_0] = \begin{bmatrix} 278.112 \\ 319.014 \\ -319.014 \\ 0 \end{bmatrix}$$

where $\qquad [Q_0] = \begin{bmatrix} 359 \\ 409 \\ -420 \\ -180 \end{bmatrix}$ (see Fig. 14-6h)

The reactions are easily computed once the final structure forces are determined as shown in Fig. 14-6k.

278.112 kN.m

20 kN   60 kN

49.713 kN   119.561 kN   10.062 kN

Fig. 14-6k. Applied Loads and Reactions

Example: 14-5

Determine the reactions for the structure shown in Fig 14-7a. [EI = Constant]

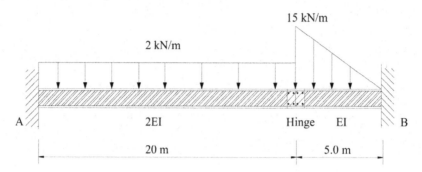

Fig. 14-7a

SOLUTION

The structure reactions, the external loads, the displacements, the fixed end moments, and the internal forces are shown in, Fig. 14-7b, 14-7c, 14-7d, 14-7e, respectively.

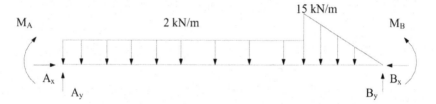

Fig. 14-7b. Structural Model

Fig. 14-7c. Degrees of Freedom

Fig. 14-7d. Internal Member Forces

Fig. 14-7e. Fixed End Forces

The displacement transformation matrix [a] is equal to

$$[S] = \begin{array}{ccc} d_1 & d_2 & d_3 \\ \begin{bmatrix} 0 & 0 & -\dfrac{1}{20} \\ 1 & 0 & -\dfrac{1}{20} \\ 0 & 1 & \dfrac{1}{5} \\ 0 & 0 & \dfrac{1}{2} \end{bmatrix} \end{array} \begin{bmatrix} d_1 \\ d_2 \\ d_3 \end{bmatrix}$$

The element stiffness matrix [k], which is computed using Equation [14-7], is equal to

$$[k] = EI \begin{bmatrix} 0.4 & 0.2 & 0 & 0 \\ 0.2 & 0.4 & 0 & 0 \\ 0 & 0 & 0.8 & 0.4 \\ 0 & 0 & 0.4 & 0.8 \end{bmatrix}$$

The structure stiffness matrix [K], which is computed using Equation [14-11] is equal to

$$[K] = [a^T] [k] [a] = EI \begin{bmatrix} 0.4 & 0 & -0.03 \\ 0 & 0.8 & 0.24 \\ -0.03 & 0.24 & 0.099 \end{bmatrix}$$

The displacement matrix [d], which is built using Equation [14-5], is equal to

$$[d] = [K]^{-1}[D] = \dfrac{1}{EI} \begin{bmatrix} -58.713 & rad \\ -408.38 & rad \\ 1.439 \times 10^3 & m \end{bmatrix}$$

491

where $[D] = \begin{bmatrix} -66.67 \text{ kN.m} & @d_1 \\ 18.75 \text{ kN.m} & @d_2 \\ 46.25 \text{ kN} & @d_3 \end{bmatrix}$

The internal member force matrix [Q], which is built using Equation [14-10], is equal to

$$[Q] = \begin{bmatrix} -54.924 \\ -66.667 \\ 18.75 \\ 182.102 \end{bmatrix}$$

The final member force matrix $\left[\overline{Q}\right]$ due to external loading, which is computed using Equation [14-13], is equal to

$$\left[\overline{Q}\right] = \begin{bmatrix} -54.924 \\ -66.667 \\ 18.75 \\ 182.102 \end{bmatrix} + \begin{bmatrix} -66.667 \\ 66.667 \\ -18.75 \\ 12.5 \end{bmatrix} = \begin{bmatrix} -121.591 \\ 0 \\ 0 \\ 194.602 \end{bmatrix}$$

It is worth noting that the moment at the hinge is equal to zero.

The reactions, which are computed using the final member forces, are shown in Fig. 14-7f.

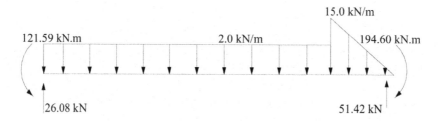

Fig. 14-7f. Applied Loads and Reactions

Example: 14-6

Determine the reactions for the structure shown in Fig 14-8a.
[$E=200.10^6$ kN/m^2, $I=100.10^6$ mm^4, $\alpha = 0.000012 / ^0$C]

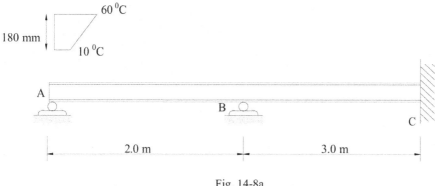

Fig. 14-8a

SOLUTION

The structure reactions, the external loads, the displacements, the internal forces, the
fixed end forces due to external loadings, and the fixed end forces due to temperature
change are shown in Fig. 14-8b,14-8c, and 14-8d, respectively.

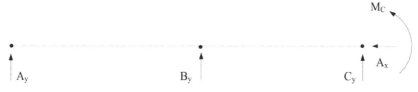

Fig. 14-8b. Structural Model

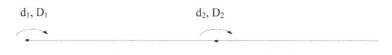

Fig. 14-8c. Degrees of Freedom

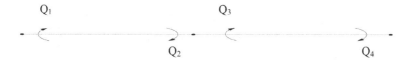

Fig. 14-8d. Member Internal Forces

$\Delta T = T_{bottom} - T_{Top} = 10 - 60 = -50^0 C$

$$FEM_{AB} = -\frac{EI\ \alpha\ \Delta T}{h} = -\frac{200(100)(0.000012)(-50)}{(0.18)} = 66.7 \text{ kN.m}$$

$$FEM_{BA} = \frac{EI\ \alpha\ \Delta T}{h} = \frac{200(100)(0.000012)(-50)}{(0.18)} = -66.7 \text{ kN.m}$$

$FEM_{BC} = FEM_{AB} = -66.7 \text{ kN.m}$

$FEM_{CB} = FEM_{BA} = 66.7 \text{ kN.m}$

Fig. 14-8e. Fixed End Forces due to Temperature Change

The displacement transformation matrix [a] is equal to

$$[S] = \begin{bmatrix} S_1 \\ S_2 \\ S_3 \\ S_4 \end{bmatrix} = \begin{matrix} d_1 & d_2 \\ \begin{bmatrix} 1 & 0 \\ 0 & 1 \\ 0 & 1 \\ 0 & 0 \end{bmatrix} \end{matrix} \begin{bmatrix} d_1 \\ d_2 \end{bmatrix}$$

The element stiffness matrix [k], which is computed using Equation [14-7], is equal to

$$[k] = EI \begin{bmatrix} 2 & 1 & 0 & 0 \\ 1 & 2 & 0 & 0 \\ 0 & 0 & 1.333 & 0.667 \\ 0 & 0 & 0.667 & 1.333 \end{bmatrix}$$

The structure stiffness matrix [K], which is determined using Equation [14-11], is equal to

$$[K] = EI \begin{bmatrix} 2 & 1 \\ 1 & 3.333 \end{bmatrix}$$

The structure displacement matrix [d], which is determined using Equation [14-5], is equal to

$$[d] = [K^{-1}] [D] = \begin{bmatrix} -1.96 \times 10^{-3} \text{ rad} \\ 5.885 \times 10^{-4} \text{ rad} \end{bmatrix}$$

where $[D] = \begin{bmatrix} D_1 \\ D_2 \end{bmatrix} = \begin{bmatrix} -66.7 \text{ kN.m} & @d_1 \\ 0 \text{ kN.m} & @ d_2 \end{bmatrix}$

Compute the structure internal forces [Q], Equation [14-5]

$$[Q] = \begin{bmatrix} -66.7 \\ -15.694 \\ 15.694 \\ 7.847 \end{bmatrix}$$

The final internal force matrix $\left[\overline{Q} \right]$ due to span loading and temperature change, which is determined using Equation [14-13], is equal to

$$\left[\overline{Q} \right] = [Q] + [Q_0] = \begin{bmatrix} 0 \\ -82.4 \\ 82.4 \\ -58.9 \end{bmatrix}$$

Where $[Q_0] = \begin{bmatrix} 66.7 \\ -66.7 \\ 66.7 \\ -66.7 \end{bmatrix}$ (see Fig. 14-8e)

The reactions, which are easily computed once the final structure forces are determined,
Are shown in Fig. 14-8f.

58.9 kN.m

41.2   49.033   7.833

Fig. 14-8f. Applied Loads and Reactions

Example: 14-7

Determine the reactions for the frame shown in Fig 14-9a. [$E = 200.10^6$ kN/m^2 and $I = 400.10^{-6}$ m^4]

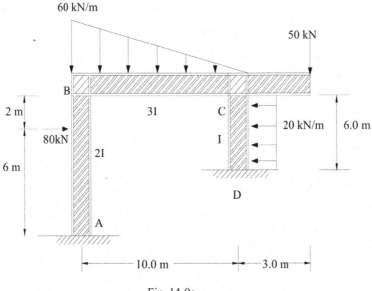

Fig. 14-9a

SOLUTION

The frame reactions, external loads, displacements, internal member forces, and fixed end forces are shown in Fig. 14-9b,14-9c, 14-9d, and 14-9e, respectively.

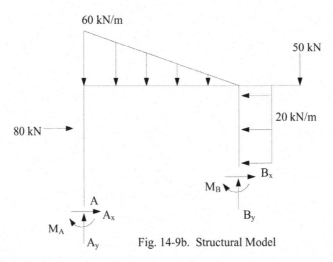

Fig. 14-9b. Structural Model

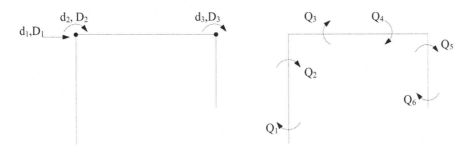

Fig. 14-9c. Degree of Freedom   Fig. 14-9d. Internal Member Forces

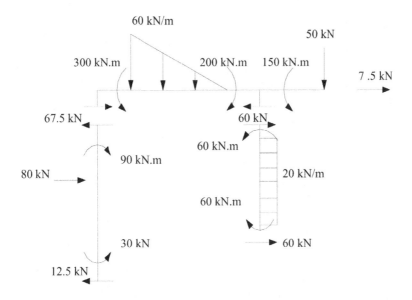

Fig. 14-9e. Fixed End Forces

The displacement transformation matrix [a] is equal to

$$[S] = \begin{bmatrix} S_1 \\ S_2 \\ S_3 \\ S_4 \\ S_5 \\ S_6 \end{bmatrix} = \begin{bmatrix} -\dfrac{1}{8} & 0 & 0 \\ -\dfrac{1}{8} & 1 & 0 \\ 0 & 1 & 0 \\ 0 & 0 & 1 \\ -\dfrac{1}{6} & 0 & 1 \\ -\dfrac{1}{6} & 0 & 0 \end{bmatrix} \begin{bmatrix} d_1 \\ d_2 \\ d_3 \end{bmatrix}$$

The element stiffness matrix [k], which is computed using Equation [14-7], is equal to

$$[k] = EI \begin{bmatrix} 1 & 0.5 & 0 & 0 & 0 & 0 \\ 0.5 & 1 & 0 & 0 & 0 & 0 \\ 0 & 0 & 1.2 & 0.6 & 0 & 0 \\ 0 & 0 & 0.6 & 1.2 & 0 & 0 \\ 0 & 0 & 0 & 0 & 0.667 & 0.333 \\ 0 & 0 & 0 & 0 & 0.333 & 0.667 \end{bmatrix}$$

The structure stiffness matrix [K], which is computed using Equation [14-11], is equal to

$$[K] = EI \begin{bmatrix} 0.102 & -0.188 & -0.167 \\ -0.188 & 2.2 & 0.6 \\ -0.167 & 0.6 & 1.867 \end{bmatrix}$$

The structure displacement matrix [d], which is computed using are determined using Equation [14-5], is equal to

$$[d] = [K]^{-1}[D] = \frac{1}{EI} \begin{Bmatrix} 282.279 \\ 121.86 \\ -8.609 \end{Bmatrix}$$

$$\text{where } [D] = \begin{Bmatrix} D_1 \\ D_2 \\ D_3 \end{Bmatrix} = \begin{bmatrix} 67.5\text{-}60 = 7.5 \text{ kN} & @d_1 \\ -(90-300) = 210 \text{ kN.m} & @d_2 \\ -(200\text{-}60\text{-}150) = 10 \text{ kN.m} & @d_3 \end{bmatrix}$$

498

The structure internal force matrix [Q], which is determined using Equation [14-10], is equal to

$$[Q] = \begin{Bmatrix} 8.003 \\ 68.933 \\ 141.067 \\ 62.786 \\ -52.786 \\ -49.916 \end{Bmatrix}$$

The final member force matrix $\left[\overline{Q} \right]$, which is determined using Equation [14-13], is equal to

$$\left[\overline{Q} \right] = [Q] + [Q_0] = \begin{Bmatrix} -21.997 \\ 158.933 \\ -158.933 \\ 262.786 \\ -112.786 \\ 10.084 \end{Bmatrix}$$

where $[Q_0] = \begin{Bmatrix} -30 \\ 90 \\ -300 \\ 200 \\ -60 \\ 60 \end{Bmatrix}$ (see Fig. 14-9e)

The reactions are shown in Fig. 14-9f.

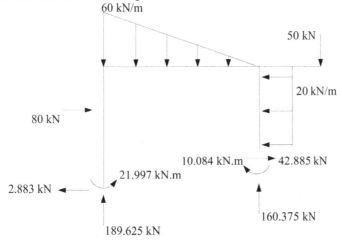

Fig. 14-9f. Applied Loads and Reactions

Example: 14-8

Determine the reactions of the frame shown in Fig. 14-10a if the supports A and D are displaced. $[E = 200.10^6 \text{ kN/m}^2 \quad I = 400.10^{-6} \text{ m}^4]$

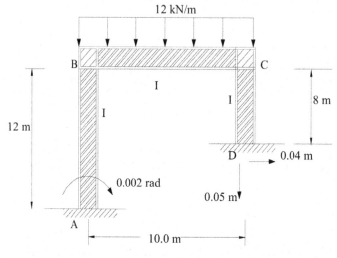

Fig. 14-10a

SOLUTION

The frame reactions, the external loads, the displacements, the internal member forces, the fixed end forces due to a rotational displacement $(\theta = 0.002 \text{ rad})$ at support A, the fixed end forces due to a horizontal displacement $(\Delta_H = 0.04 \text{ m})$ and a vertical displacement $(\Delta_V = 0.05 \text{ m})$ at D, and the total fixed end forces are shown in Fig. 14-10b, 14-10c, 14-10d, 14-10e, 14-10f, 14-10g, 14-10h, and 14-10i, respectively.

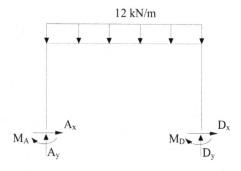

Fig. 14-10b. Structural Model

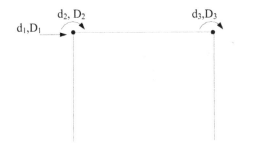

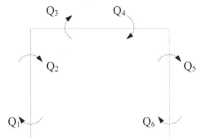

Fig. 14-10c. Degrees of Freedom

Fig. 14-10d. Internal Member Forces

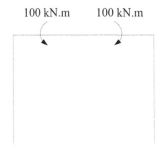

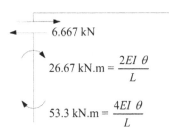

Fig. 14-10e. Fixed End Forces
Due to External Loadings

Fig. 14-10f. Fixed End Forces due to
Rotational displacement at A

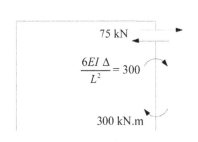

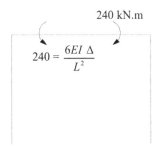

Fig. 14-10g. Fixed End Forces due to
Horizontal displacement at D)

Fig. 14-10h. Fixed End Forces due to
vertical displacement at D

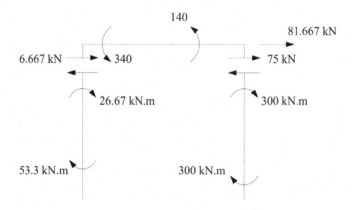

Fig. 14-10i. Total Fixed End Forces

The displacement transformation matrix [a] is equal to

$$[S] = \begin{bmatrix} S_1 \\ S_2 \\ S_3 \\ S_4 \\ S_5 \\ S_6 \end{bmatrix} = \begin{matrix} d_1 & d_2 & d_3 \\ \begin{bmatrix} -\dfrac{1}{12} & 0 & 0 \\ -\dfrac{1}{12} & 1 & 0 \\ 0 & 1 & 0 \\ 0 & 0 & 1 \\ -\dfrac{1}{8} & 0 & 1 \\ -\dfrac{1}{8} & 0 & 0 \end{bmatrix} \end{matrix} \begin{bmatrix} d_1 \\ d_2 \\ d_3 \end{bmatrix}$$

The element stiffness matrix [k], which is computed using Equation [14-7], is equal to

$$[k] = EI \begin{bmatrix} 0.333 & 0.167 & 0 & 0 & 0 & 0 \\ 0.167 & 0.333 & 0 & 0 & 0 & 0 \\ 0 & 0 & 0.4 & 0.2 & 0 & 0 \\ 0 & 0 & 0.2 & 0.4 & 0 & 0 \\ 0 & 0 & 0 & 0 & 0.5 & 0.25 \\ 0 & 0 & 0 & 0 & 0.25 & 0.5 \end{bmatrix}$$

The structure stiffness matrix [K], which is computed using Equation [14-11], is equal to

$$[K] = EI \begin{bmatrix} 0.03 & -0.042 & -0.094 \\ -0.042 & 0.733 & 0.2 \\ -0.094 & 0.2 & 0.9 \end{bmatrix}$$

The frame displacement matrix [d], which is determined using Equation [14-5], is equal to

$$[d] = [K]^{-1}[D] = \frac{1}{EI} \begin{Bmatrix} 3.78 \times 10^3 \\ 620.755 \\ 78.007 \end{Bmatrix}$$

where $[D] = \begin{Bmatrix} D_1 \\ D_2 \\ D_3 \end{Bmatrix} = \begin{bmatrix} 6.667+75 = 81.667 \text{ kN} & @d_1 \\ -(26.67 - 340) = 313.33 \text{ kN.m} & @d_2 \\ -(-140+300) = -160 & @d_3 \end{bmatrix}$

The member internal force matrix [Q], which is computed using Equation [14-10], is equal to

$$[Q] = \begin{Bmatrix} -54.033 \\ 49.426 \\ 263.904 \\ 155.354 \\ -315.354 \\ -334.855 \end{Bmatrix}$$

The final frame internal force matrix $\left[\overline{Q}\right]$, which is computed using Equation [14-13], is equal to

$$\left[\overline{Q}\right] = [Q] + [Q_0] = \begin{Bmatrix} -0.733 \\ 76.096 \\ -76.096 \\ 15.354 \\ -15.354 \\ -34.855 \end{Bmatrix}$$

where $[Q_0] = \begin{Bmatrix} 53.3 \\ 26.67 \\ -340 \\ -140 \\ 300 \\ 300 \end{Bmatrix}$ (see Fig. 14-10i)

503

The reactions are shown in Fig. 14-10k.

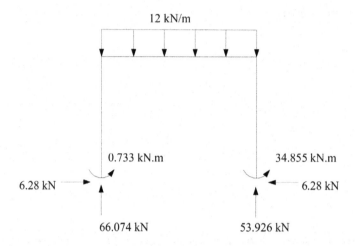

Fig. 14-10k. Applied Loads and Reactions

Example: 14-9

Determine the reactions of the inclined frame shown in Fig 14-11a [EI = Constant].

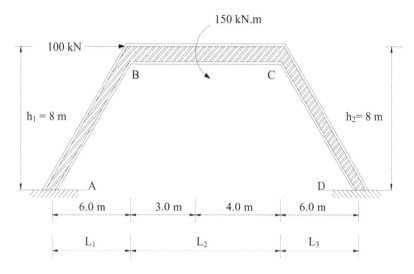

Fig. 14-11a

SOLUTION

The structure reactions, external loads, displacements, internal member forces, and fixed end forces are shown in Fig. 14-11b, 14-11c, 14-11d, and 14-11e, respectively.

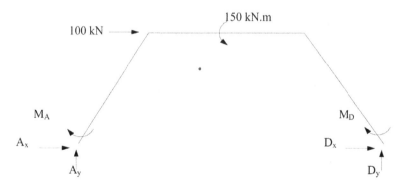

Fig. 14-11b. Structural Model

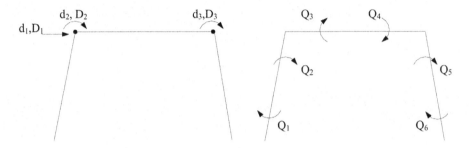

Fig. 14-11c. Degrees of Freedom   Fig. 14-11d. Internal Member Forces

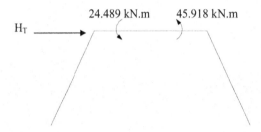

Fig. 14-11e. Fixed End Forces

Notes

$$FEM_{BC} = \frac{(-150)(4)}{7^2}\left[2(3)-4\right] = -24.489 \qquad \text{(-) counter clock wise moment}$$

$$FEM_{CB} = \frac{(-150)(3)}{7^2}\left[2(4)-3\right] = -45.918$$

H_T = Total Horizontal Forces

There is a beam effect from the horizontal member BC to the two inclined members AB and BC, as shown in Figs. 14-11f and 14-11g. The horizontal displacement d_1, which is derived as shown in Fig. 14-11f, yields Equation [14-14]. The horizontal force that will have the same effect as the total vertical load coming from the beam is derived as shown in Fig. 14-11g yielding Equation [14-15].

506

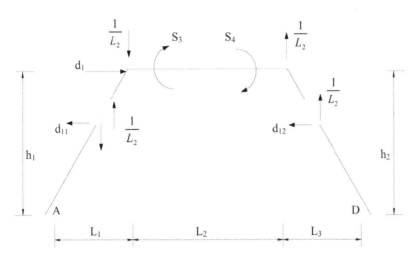

Fig. 14-11f

The displacement transformation matrix [a] is equal to

$$\sum M_A = 0 + \circlearrowright \qquad d_{11} = \frac{1}{L_2}\left[\frac{L_1}{h_1}\right]$$

$$\sum M_D = 0 + \circlearrowright \qquad d_{12} = \frac{1}{L_2}\left[\frac{L_3}{h_2}\right]$$

Therefore, $d_1 = d_{11} + d_{12} = \dfrac{1}{L_2}\left[\dfrac{L_1}{h_1} + \dfrac{L_3}{h_2}\right]$ \hfill [14-14]

$$[S] = \begin{bmatrix} S_1 \\ S_2 \\ S_3 \\ S_4 \\ S_5 \\ S_6 \end{bmatrix} = \begin{array}{ccc} d_1 & d_2 & d_3 \end{array} \begin{bmatrix} -\dfrac{1}{h_1} & 0 & 0 \\[2ex] -\dfrac{1}{h_1} & 1 & 0 \\[2ex] \dfrac{1}{L_2}\left(\dfrac{L_1}{h_1}+\dfrac{L_3}{h_2}\right) & 1 & 0 \\[2ex] \dfrac{1}{L_2}\left(\dfrac{L_1}{h_1}+\dfrac{L_3}{h_2}\right) & 0 & 1 \\[2ex] -\dfrac{1}{h_2} & 0 & 1 \\[2ex] -\dfrac{1}{h_2} & 0 & 0 \end{bmatrix} \begin{bmatrix} d_1 \\ d_2 \\ d_3 \end{bmatrix}$$

507

$$[S] = \begin{bmatrix} S_1 \\ S_2 \\ S_3 \\ S_4 \\ S_5 \\ S_6 \end{bmatrix} = \begin{matrix} d_1 & d_2 & d_3 \\ \begin{bmatrix} -0.125 & 0 & 0 \\ -0.125 & 1 & 0 \\ 0.214 & 1 & 0 \\ 0.214 & 0 & 1 \\ -0.125 & 0 & 1 \\ -0.125 & 0 & 0 \end{bmatrix} \end{matrix} \begin{bmatrix} d_1 \\ d_2 \\ d_3 \end{bmatrix}$$

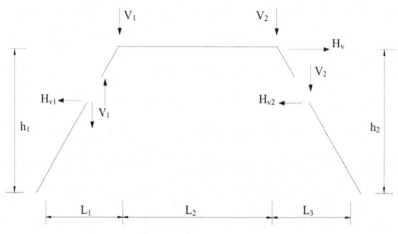

Fig. 14-11g

$$\sum M_A = 0 + \circlearrowleft \quad H_{v1} = \frac{V_1 \, L_1}{h_1}$$

$$\sum M_D = 0 + \circlearrowleft \quad H_{v2} = - \frac{V_2 \, L_3}{h_2}$$

Therefore, $\quad H_v = H_{v1} + H_{v2} = \left[\dfrac{V_1 \, L_1}{h_1} \right] + \left[\dfrac{-V_2 \, L_3}{h_2} \right]$ [14-11]

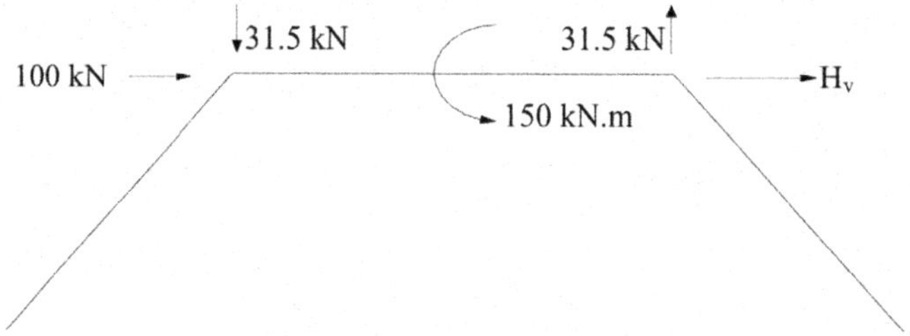

Fig. 14-12h. Structural Loads

The total horizontal force is equal to

$$H_T = 100 + \left[\frac{31.5(6)}{8}\right] + \left[\frac{-(-31.5)(6)}{8}\right] = 147.25 \text{ kN}$$

The element stiffness matrix [k], which is built using Equation [14-4], is equal to

$$[k] = EI \begin{bmatrix} 0.4 & 0.2 & 0 & 0 & 0 & 0 \\ 0.2 & 0.4 & 0 & 0 & 0 & 0 \\ 0 & 0 & 0.571 & 0.286 & 0 & 0 \\ 0 & 0 & 0.286 & 0.571 & 0 & 0 \\ 0 & 0 & 0 & 0 & 0.4 & 0.2 \\ 0 & 0 & 0 & 0 & 0.2 & 0.4 \end{bmatrix}$$

The structure stiffness matrix [K], which is determined using Equation [14-11], is equal to

$$[K] = EI \begin{bmatrix} 0.116 & 0.109 & 0.109 \\ 0.109 & 0.971 & 0.286 \\ 0.109 & 0.286 & 0.971 \end{bmatrix}$$

The frame displacement matrix [d], which is determined using Equation [14-5], is equal to

$$[d] = [K]^{-1}[D] = \frac{1}{EI} \left\{ \begin{array}{c} 1.449 \times 10^3 \\ -112.859 \\ -81.624 \end{array} \right\}$$

$$\text{where } [D] = \left\{ \begin{array}{c} D_1 \\ D_2 \\ D_3 \end{array} \right\} = \left[\begin{array}{cc} 147.25 \text{ kN} & @d_1 \\ -(0 - 24.489) = 24.489 \text{ kN.m} & @d_2 \\ -(-45.918+0) = 45.918 \text{ kN.m} & @d_3 \end{array} \right]$$

The member internal force matrix [Q], which is computed using Equation [14-10], is equal to

$$[Q] = \left\{ \begin{array}{c} -131.238 \\ -153.81 \\ 178.31 \\ 187.234 \\ -141.316 \\ -124.991 \end{array} \right\}$$

The final member internal force matrix $\left[\overline{Q}\right]$, which is computed using Equation [14-12], is equal to

$$\left[\overline{Q}\right] = [Q] + [Q_0] = \begin{Bmatrix} -131.238 \\ -153.81 \\ 153.81 \\ 141.316 \\ -141.316 \\ -124.991 \end{Bmatrix}$$

where $\qquad [Q_0] = \begin{Bmatrix} 0 \\ 0 \\ -24.489 \\ -45.918 \\ 0 \\ 0 \end{Bmatrix}$ (Fig. 14-11e)

The reactions are shown in Fig. 14-11i.

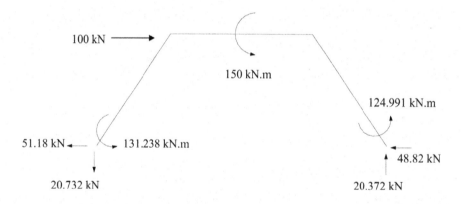

Fig. 14-11i. Applied Loads and Reactions

Summary

- The general stiffness method is the basis of the majority of computer programs used to analyze all types of determinate and indeterminate structures including planar structure and three-dimensional trusses, frames, and shells. The stiffness method eliminates the need to select redundants and a released structure, as required by the flexibility method.
- In the general stiffness method, joint displacements are the unknown. With all joints initially artificially restrained, unit displacements are introduced at each joint and the forces associated with the unit displacements (known as stiffness coefficients) computed.
- The number of unknown displacements will be equal to the degree of kinematic indeterminacy.
- For two dimensional structures with rigid joints, joints, three unknown displacements (two linear and one rotational) are possible at each unrestrained joint.
- For three-dimensional structures with rigid joints, six unknown displacements (three linear and three rotational) are possible at each unrestrained joint. For these situations the torsional stiffness as well as the axial and bending stiffness of members must be considered when evaluating stiffness coefficients.
- In a typical computer program, a coordinate system must be selected to establish the location of joints, specify member properties (such as area, moment of inertia, and modulus of elasticity), and specify the type of loading.

Problems

1-   Using the stiffness method determine the reactions. [EI= const.]

(a)

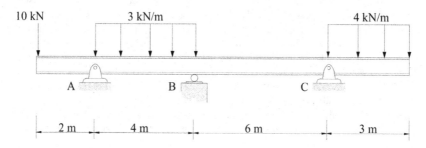

(b)

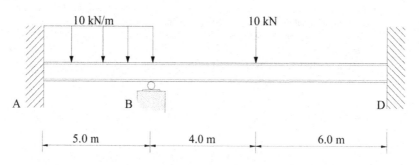

(c)

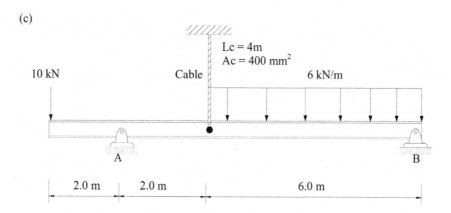

(d)

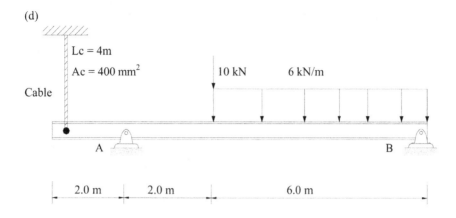

Cable

Lc = 4m

Ac = 400 mm^2

10 kN   6 kN/m

A   B

2.0 m | 2.0 m | 6.0 m

(e)

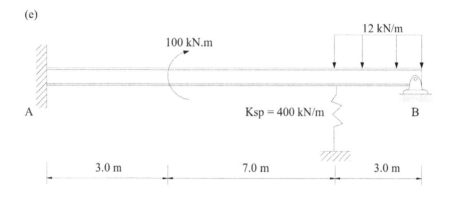

100 kN.m   12 kN/m

A   Ksp = 400 kN/m   B

3.0 m | 7.0 m | 3.0 m

(f)

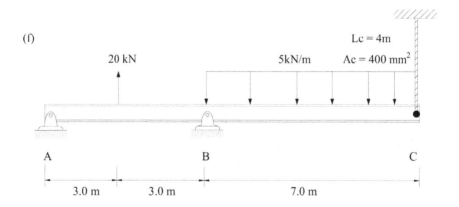

20 kN   5kN/m   Lc = 4m

Ac = 400 mm^2

A   B   C

3.0 m | 3.0 m | 7.0 m

(g)

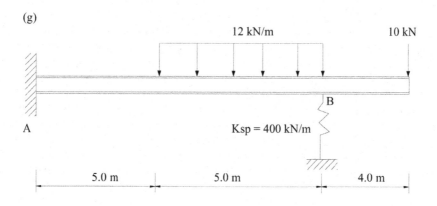

(h)

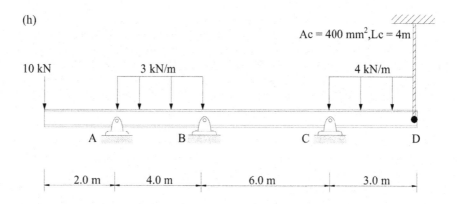

(I)

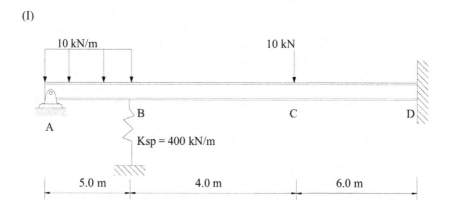

(J)

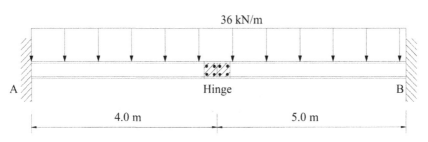

36 kN/m

A

Hinge

B

4.0 m   5.0 m

(k)

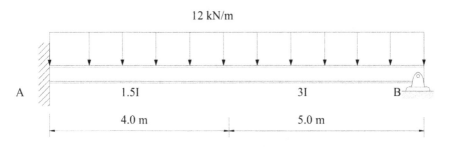

12 kN/m

A   1.5I   3I   B

4.0 m   5.0 m

(L)

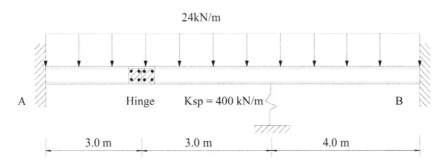

24kN/m

A   Hinge   Ksp = 400 kN/m   B

3.0 m   3.0 m   4.0 m

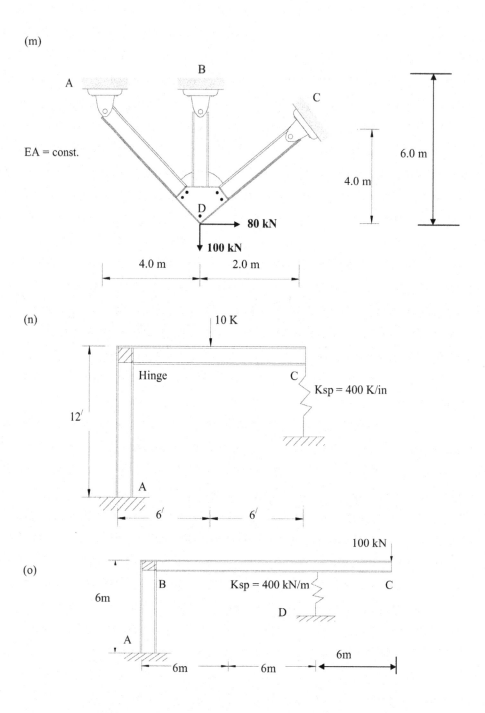

(m)

A

B

C

EA = const.

6.0 m

4.0 m

D

80 kN

100 kN

4.0 m

2.0 m

(n)

10 K

Hinge

C

Ksp = 400 K/in

12'

A

6'

6'

100 kN

(o)

B

Ksp = 400 kN/m

C

6m

D

A

6m

6m

6m

(p)

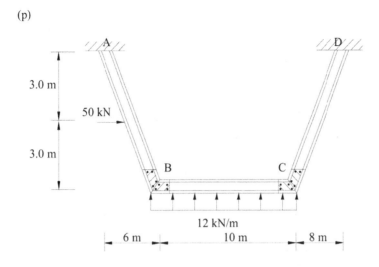

(q)

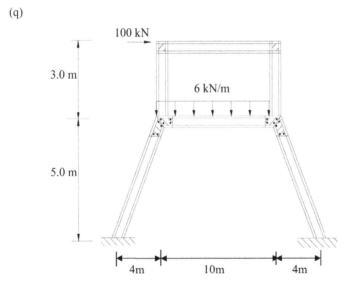

(r)

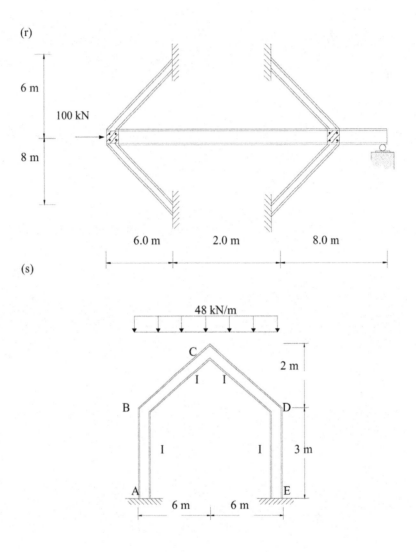

6 m

100 kN

8 m

6.0 m   2.0 m   8.0 m

(s)

48 kN/m

C

2 m

I   I

B   D

I   I

3 m

A   E

6 m   6 m

hint: make use of symmetry.

2-   Determine the reactions of the following frames.
      [E = 200.10^6 kN/m^2, I = 400.10^{-6} m^4, h = 0.263m, α = 0.000012]

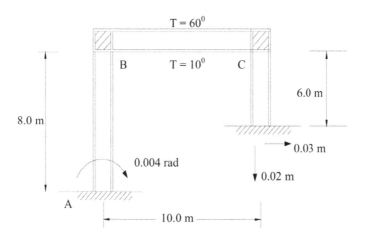

3-   Draw the shear and moment diagrams of the following frame

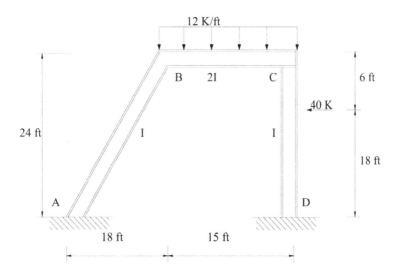

4-   Draw the shear and moment diagrams of the following frame

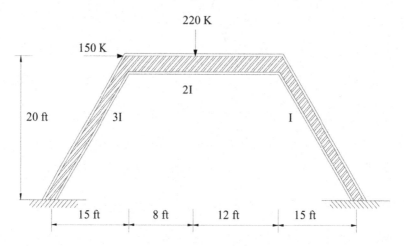

5-   Determine the cable tension. [$E = 200.10^6$ kN/m^2, $I = 150.10^{-6}$ m^4, $A_c = 50.10^{-6}$ m^2]

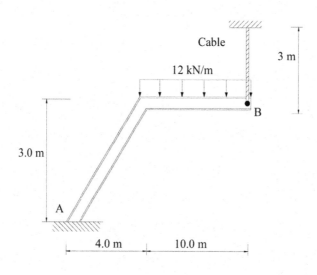

CHAPTER-15

DIRECT STIFFNESS MATRIX METHOD

RIYADH – KINGDOM OF SAUDIA ARABIA

Introduction

Large and complex structures may require huge stiffness matrices. To simply the generation of the structure stiffness, the element stiffness matrices $[k]$ of all members are computed first. Then, the entire structure stiffness matrix $[K]$ is constructed by superimposing the stiffness matrices of the members. This method, which is called the direct stiffness method, provides the basis for most structural computer programs and can be applied to all types of structures including trusses, continuous beams, frames, plates, and shells. It is sometimes called the finite element method when applied to plates, shells and other two or three-dimensional members. The member forces and the corresponding end displacements are related to each other by member stiffness matrices $[k]$. The joint displacements and the external loads are related to each other by the following system of equations (Fig. 15–1):

$$
\begin{aligned}
D_1 &= k_{11}d_1 + k_{12}d_1 + \cdots + \cdots + k_{1N}d_N \\
D_2 &= k_{21}d_1 + k_{22}d_1 + \cdots + \cdots + k_{2N}d_N \\
\cdots \quad & \cdots \quad \cdots \quad \cdots \quad \cdots \quad \cdots \\
\cdots \quad & \cdots \quad \cdots \quad \cdots \quad \cdots \\
D_N &= k_{N1}d_1 + k_{N2}d_1 + \cdots + \cdots + k_{NN}d_N
\end{aligned}
\qquad [15\text{-}1]
$$

Where d_i = displacement of the ith joint, k_{ij} = force at joint i due to a unit displacement at joint j, D_i = external load at the ith joint.

These expressions can be expressed in a matrix form as follows.

$$
[D] = [K][d]
\qquad [15\text{-}2]
$$

$$
[D] = \begin{bmatrix} D \\ \cdots \\ R \end{bmatrix} = \begin{bmatrix} K_{DD} & \vdots & K_{DR} \\ \cdots & \cdots & \cdots \\ K_{RD} & \vdots & K_{RR} \end{bmatrix} \begin{bmatrix} d_D \\ \cdots \\ d_R \end{bmatrix}
\qquad [15\text{-}3]
$$

where $[d_D]$ = unknown joint displacement matrix, $[d_R]$ = known displacement matrix, $[K]$ = partitioned stiffness matrix, $[D]$ = known external load matrix, and $[R]$ = unknown external load matrix.

To construct the stiffness matrix for the truss member consider the truss element in Fig. 15-1 where

$$
c = \cos\theta = \frac{x_j - x_i}{L}
\qquad [15\text{-}4]
$$

$$
s = \sin\theta = \frac{y_j - y_i}{L}
\qquad [15\text{-}5]
$$

The transformation matrix **[a]** that transforms the structure displacements in the global coordinates into member displacements in the local coordinates.

$$[a] = \begin{bmatrix} c & s & 0 & 0 \\ 0 & 0 & c & s \end{bmatrix}$$

[15-6]

The local member stiffness matrix $\left[\overline{k}\right]$ is equal to

$$[\overline{k}] = \frac{AE}{L} \begin{bmatrix} 1 & -1 \\ -1 & 1 \end{bmatrix}$$

[15-7]

The truss member stiffness matrix $[k]$ in the global coordinates is equal to

$$[k] = a^T \overline{k} a$$

[15-8]

$$[k] = \frac{AE}{L} \begin{bmatrix} c^2 & cs & -c^2 & -cs \\ cs & s^2 & -cs & -s^2 \\ -c^2 & -cs & c^2 & cs \\ -cs & -s^2 & cs & s^2 \end{bmatrix}$$

[15-9]

To construct the stiffness matrix for a beam member, let us consider the beam element shown in Fig. 15-1. Neglecting the axial effects $(d_1 \text{ and } d_4)$ and imposing a unit displacement at each degree of freedom $(d_2, d_3, d_5, \text{and } d_6)$, as shown in Fig. 15-2, the beam member global stiffness matrix $[k]$ is found to be equal to

$$[k] = \begin{bmatrix} \dfrac{12EI}{L^3} & \dfrac{6EI}{L^2} & -\dfrac{12EI}{L^3} & \dfrac{6EI}{L^2} \\ \dfrac{6EI}{L^2} & \dfrac{4EI}{L} & -\dfrac{6EI}{L^2} & \dfrac{2EI}{L} \\ -\dfrac{12EI}{L^3} & -\dfrac{6EI}{L^2} & \dfrac{12EI}{L^3} & -\dfrac{6EI}{L^2} \\ \dfrac{6EI}{L^2} & \dfrac{2EI}{L} & -\dfrac{6EI}{L^2} & \dfrac{4EI}{L} \end{bmatrix}$$

[15-10]

If the axial effects are to be considered, then the beam member global stiffness matrix $\left[\overline{k}\right]$ is equal to

$$[\bar{k}] = \begin{bmatrix} \dfrac{AE}{L} & 0 & 0 & -\dfrac{AE}{L} & 0 & 0 \\[2mm] 0 & \dfrac{12EI}{L^3} & \dfrac{6EI}{L^2} & 0 & -\dfrac{12EI}{L^3} & \dfrac{6EI}{L^2} \\[2mm] 0 & \dfrac{6EI}{L^2} & \dfrac{4EI}{L} & 0 & -\dfrac{6EI}{L^2} & \dfrac{2EI}{L} \\[2mm] -\dfrac{AE}{L} & 0 & 0 & \dfrac{AE}{L} & 0 & 0 \\[2mm] 0 & -\dfrac{12EI}{L^3} & -\dfrac{6EI}{L^2} & 0 & \dfrac{12EI}{L^3} & -\dfrac{6EI}{L^2} \\[2mm] 0 & \dfrac{6EI}{L^2} & \dfrac{2EI}{L} & 0 & -\dfrac{6EI}{L^2} & \dfrac{4EI}{L} \end{bmatrix}$$ 　　　[15-11]

To construct the stiffness matrix for a frame member, let us consider the frame element shown in Fig. 15-1 and make use of the beam stiffness matrix $[\bar{k}]$ and the displacement transformation matrix $[t]$, which transforms the six global displacements into local displacements. The frame member global stiffness matrix $[k]$ is obtained using the following equation:

$$[k] = t^T \bar{k}\, t$$ 　　　[15-12]

$$k = \begin{bmatrix} \left(\dfrac{AE}{L}c^2 + \dfrac{12EI}{L^3}s^2\right) & \left(\dfrac{AE}{L} - \dfrac{12EI}{L^3}\right)cs & -\dfrac{6EI}{L^2}s & -\left(\dfrac{AE}{L}c^2 + \dfrac{12EI}{L^3}s^2\right) & \left(\dfrac{AE}{L} - \dfrac{12EI}{L^3}\right)cs & -\dfrac{6EI}{L^2}s \\[3mm] \left(\dfrac{AE}{L} - \dfrac{12EI}{L^3}\right)cs & \left(\dfrac{AE}{L}s^2 + \dfrac{12EI}{L^3}c^2\right) & \dfrac{6EI}{L^2}c & -\left(\dfrac{AE}{L} - \dfrac{12EI}{L^3}\right)cs & -\left(\dfrac{AE}{L}s^2 + \dfrac{12EI}{L^3}c^2\right) & \dfrac{6EI}{L^2}c \\[3mm] -\dfrac{6EI}{L^2}s & \dfrac{6EI}{L^2}c & \dfrac{4EI}{L} & \dfrac{6EI}{L^2}s & -\dfrac{6EI}{L^2}c & \dfrac{2EI}{L} \\[3mm] -\left(\dfrac{AE}{L}c^2 + \dfrac{12EI}{L^3}s^2\right) & -\left(\dfrac{AE}{L} - \dfrac{12EI}{L^3}\right)cs & \dfrac{6EI}{L^2}s & \left(\dfrac{AE}{L}c^2 + \dfrac{12EI}{L^3}s^2\right) & \left(\dfrac{AE}{L} - \dfrac{12EI}{L^3}\right)cs & \dfrac{6EI}{L^2}s \\[3mm] -\left(\dfrac{AE}{L} - \dfrac{12EI}{L^3}\right)cs & -\left(\dfrac{AE}{L}s^2 + \dfrac{12EI}{L^3}c^2\right) & -\dfrac{6EI}{L^2}c & \left(\dfrac{AE}{L} - \dfrac{12EI}{L^3}\right)cs & \left(\dfrac{AE}{L}s^2 + \dfrac{12EI}{L^3}c^2\right) & -\dfrac{6EI}{L^2}c \\[3mm] -\dfrac{6EI}{L^2}s & \dfrac{6EI}{L^2}c & \dfrac{2EI}{L} & \dfrac{6EI}{L^2}s & -\dfrac{6EI}{L^2}c & \dfrac{4EI}{L} \end{bmatrix}$$

where

$$[t] = \begin{bmatrix} c & s & 0 & 0 & 0 & 0 \\ -s & c & 0 & 0 & 0 & 0 \\ 0 & 0 & 1 & 0 & 0 & 0 \\ 0 & 0 & 0 & c & s & 0 \\ 0 & 0 & 0 & -s & c & 0 \\ 0 & 0 & 0 & 0 & 0 & 1 \end{bmatrix}$$

[15-13]

Direct stiffness method for the analysis of indeterminate structures

The analysis of indeterminate structures using the direct stiffness method is outlined in the following steps:

1- Identify the structure members and its nodes (or joint). Nodes are interconnection points, supporting points, points of concentrated loads, points of change in moments of inertia, corners, joints, and points of interest where displacements need to be determined.

2- Label the nodes with unknown displacements (matrix $[d_D]$).

3- Identify the external loads (matrix $[D]$).

4- Construct the stiffness matrix of each member $[k]$.

5- Construct the structure stiffness matrix $[K]$ by superimposing the member stiffness matrices.

6- Obtain the displacement matrix $[d_D]$ using the following equation:

$$[d_D] = [K_{DD}]^{-1}[D]$$

[15-14]

7- Obtain the unknown nodal force matrix $[R]$ using the following equation:

$$[R] = [k_{RD}][d_D]$$

[15-15]

8- Finally, obtain the member internal force matrix $[\bar{D}]$ using the following equation (Fig.15-3):

$$[\bar{D}] = \bar{k} t d$$

[15-16]

The following illustrative examples will employ this procedure of analysis.

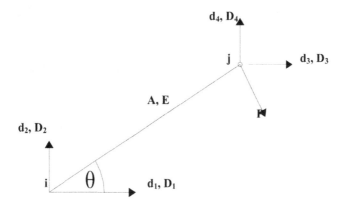

Truss Element

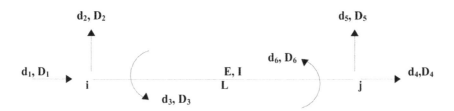

Beam Element

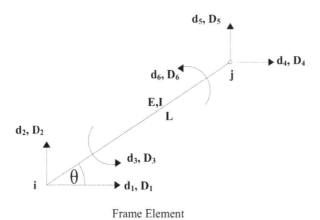

Frame Element

Fig. 15-1. Structural Element Free Body Diagram

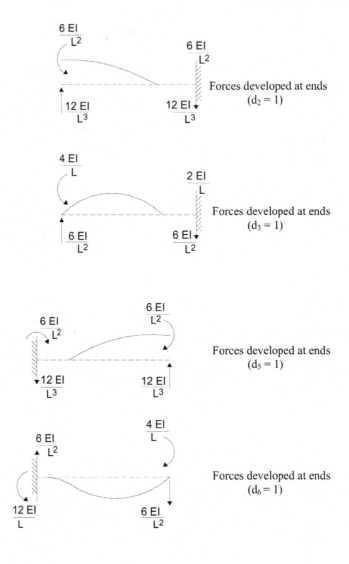

$$\frac{6\,EI}{L^2}$$

$$\frac{6\,EI}{L^2}$$

Forces developed at ends
$(d_2 = 1)$

$$\frac{12\,EI}{L^3}$$

$$\frac{12\,EI}{L^3}$$

$$\frac{4\,EI}{L}$$

$$\frac{2\,EI}{L}$$

Forces developed at ends
$(d_3 = 1)$

$$\frac{6\,EI}{L^2}$$

$$\frac{6\,EI}{L^2}$$

$$\frac{6\,EI}{L^2}$$

$$\frac{6\,EI}{L^2}$$

Forces developed at ends
$(d_5 = 1)$

$$\frac{12\,EI}{L^3}$$

$$\frac{12\,EI}{L^3}$$

$$\frac{6\,EI}{L^2}$$

$$\frac{4\,EI}{L}$$

Forces developed at ends
$(d_6 = 1)$

$$\frac{12\,EI}{L}$$

$$\frac{6\,EI}{L^2}$$

Fig. 15-2. Joint Displacements and Corresponding Internal Forces

528

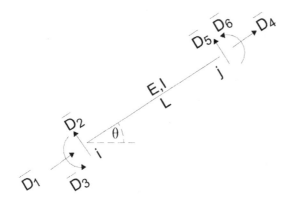

Fig. 15-3. Member Internal Forces

Example: 15-1

Determine the support reactions and the bar forces for the truss shown in Fig 15-4a.
$E = 200 \ 10^6 \ kN/m^2$ $A_{AC} = 200 \ mm^2$ $A_{AB} = 1000 \ mm^2$ $A_{BC} = 1500 \ mm^2$

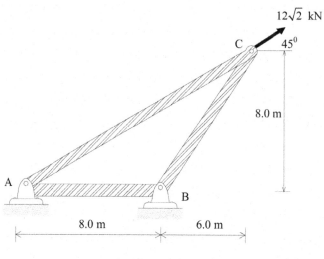

Fig. 15-4a

SOLUTION

The reactions, external loads, displacements, and internal forces are shown in Figs. 15-4b, 15-4c, and 15-4d.

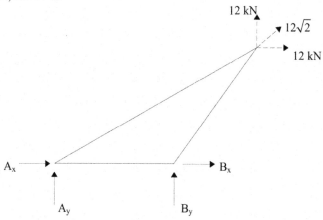

Fig. 15-4b. Structural Model

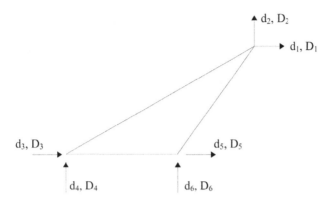

Fig. 15-4c. Degrees of Freedom

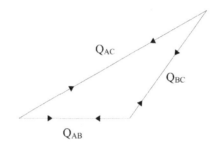

Fig. 15-4d. Internal Member Forces

The global member stiffness matrix [k], which is computed using Equation [15-9], is summarized in Figs. 15-4e,15-4f, and 15-4g.

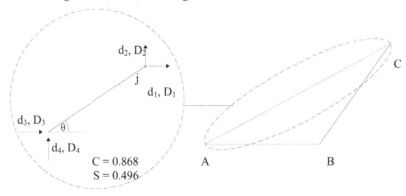

Fig. 15-4e

531

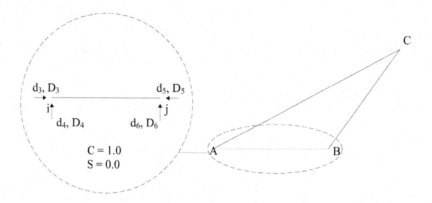

Fig. 15-4f

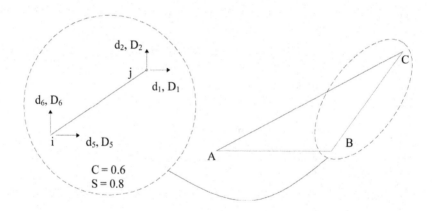

Fig. 15-4g

$$k_{AB} = \begin{array}{cccc} D_3 & D_4 & D_5 & D_6 \end{array}$$

$$k_{AB} = \begin{bmatrix} 2.5\times10^4 & 0 & -2.5\times10^4 & 0 \\ 0 & 0 & 0 & 0 \\ -2.5\times10^4 & 0 & 2.5\times10^4 & 0 \\ 0 & 0 & 0 & 0 \end{bmatrix} \begin{matrix} D_3 \\ D_4 \\ D_5 \\ D_6 \end{matrix}$$

$$\begin{array}{cccc} D_5 & D_6 & D_1 & D_2 \end{array}$$

$$k_{BC} = \begin{bmatrix} 1.08\times10^4 & 1.44\times10^4 & -1.08\times10^4 & -1.44\times10^4 \\ 1.44\times10^4 & 1.92\times10^4 & -1.44\times10^4 & -1.92\times10^4 \\ -1.08\times10^4 & -1.44\times10^4 & 1.08\times10^4 & 1.44\times10^4 \\ -1.44\times10^4 & -1.92\times10^4 & 1.44\times10^4 & 1.92\times10^4 \end{bmatrix} \begin{matrix} D_5 \\ D_6 \\ D_1 \\ D_2 \end{matrix}$$

$$\begin{array}{cccc} D_3 & D_4 & D_1 & D_2 \end{array}$$

$$k_{AC} = \begin{bmatrix} 1.869\times10^4 & 1.068\times10^4 & -1.869\times10^4 & -1.068\times10^4 \\ 1.068\times10^4 & 6.103\times10^3 & -1.068\times10^4 & -6.103\times10^4 \\ -1.869\times10^4 & -1.068\times10^4 & 1.869\times10^4 & 1.068\times10^4 \\ -1.068\times10^4 & -6.103\times10^3 & 1.068\times10^4 & 6.103\times10^3 \end{bmatrix} \begin{matrix} D_3 \\ D_4 \\ D_1 \\ D_2 \end{matrix}$$

The global structure stiffness matrix, which is obtained by superposing the global member stiffness of all members, is equal to

$$\begin{Bmatrix} 12 \\ 12 \\ R_3 \\ R_4 \\ R_5 \\ R_6 \end{Bmatrix} = \begin{bmatrix} 2.949\times10^4 & 2.508\times10^4 & -1.869\times10^4 & -1.068\times10^4 & -1.08\times10^4 & -1.44\times10^4 \\ 2.508\times10^4 & 2.53\times10^4 & -1.068\times10^4 & -6.103\times10^3 & -1.44\times10^4 & -1.92\times10^4 \\ -1.869\times10^4 & -1.068\times10^4 & 4.369\times10^4 & 1.068\times10^4 & -2.5\times10^4 & 0 \\ -1.068\times10^4 & -6.103\times10^3 & 1.068\times10^4 & 6.103\times10^3 & 0 & 0 \\ -1.08\times10^4 & -1.44\times10^4 & -2.5\times10^4 & 0 & 3.58\times10^4 & 1.44\times10^4 \\ -1.44\times10^4 & -1.92\times10^4 & 0 & 0 & 1.44\times10^4 & 1.92\times10^4 \end{bmatrix} \begin{Bmatrix} d_1 \\ d_2 \\ d_3 \\ d_4 \\ d_5 \\ d_6 \end{Bmatrix}$$

The nodes displacements, which are obtained using Equation [15-14], are equal to

$$[d_D] = \begin{bmatrix} d_1 \\ d_2 \end{bmatrix} = \begin{bmatrix} 2.949\times10^4 & 2.508\times10^4 \\ 2.508\times10^4 & 2.53\times10^4 \end{bmatrix}^{-1} \begin{bmatrix} 12 \\ 12 \end{bmatrix} = \begin{bmatrix} 2.25\times10^{-5} & \text{m} \rightarrow \\ 4.5\times10^{-4} & \text{m} \uparrow \end{bmatrix}$$

The support reactions, which are obtained using Equation [15-15], is equal to

$$[R] = \begin{bmatrix} -1.869 \times 10^4 & -1.068 \times 10^4 \\ -1.068 \times 10^4 & -6.103 \times 10^4 \\ -1.08 \times 10^4 & -1.44 \times 10^4 \\ -1.44 \times 10^4 & -1.92 \times 10^4 \end{bmatrix} \begin{bmatrix} 2.25 \times 10^{-5} \\ 4.5 \times 10^{-4} \end{bmatrix} = \begin{bmatrix} -5.25 \\ -3 \\ -6.75 \\ -9 \end{bmatrix} = \begin{bmatrix} 5.25\,\text{kN} \leftarrow \\ 3\,\text{kN} \downarrow \\ 6.75\,\text{kN} \leftarrow \\ 9\,\text{kN} \downarrow \end{bmatrix}$$

The bar forces are determined using the following equation:

$$\overline{D_{ij}} = \frac{AE}{L}[c \quad s]\begin{bmatrix} (u_j - u_i) \\ (v_j - v_i) \end{bmatrix} \qquad\qquad [15\text{-}17]$$

where u = horizontal displacements, and v = vertical displacements.

$$\overline{D_{AB}} = \frac{1000(200)}{8}[1 \quad 0]\begin{bmatrix} (d_5 - d_3) = 0.0 \\ (d_6 - d_4) = 0.0 \end{bmatrix} = 0.0$$

$$\overline{D_{AC}} = \frac{2000(200)}{16.12}[0.868 \quad 0.496]\begin{bmatrix} (d_1 - d_3) = 2.25 \times 10^{-5} \\ (d_2 - d_4) = 4.5 \times 10^{-4} \end{bmatrix} = 6.046 \ \text{kN}$$

$$\overline{D_{BC}} = \frac{1500(200)}{10}[0.6 \quad 0.8]\begin{bmatrix} (d_1 - d_5) = 2.25 \times 10^{-5} \\ (d_2 - d_6) = 4.5 \times 10^{-4} \end{bmatrix} = 11.249 \ \text{kN}$$

The bar forces, the external loads, and the reactions are shown in Fig. 15-4h.

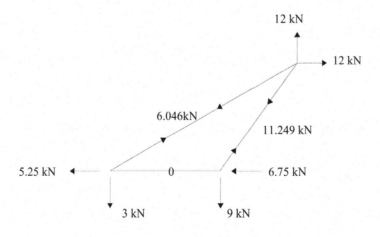

Fig.15-4h. Bar forces, Applied Loads, and Reactions

Example: 15-2

Determine for the beam shown in Fig 15-5a the following:
   a) The support reactions for the given loads.
   b) The support reactions if support B settles by 30 mm.
[$E = 200.10^6$ kN/m^2, $I = 200.10^6$ mm^4]

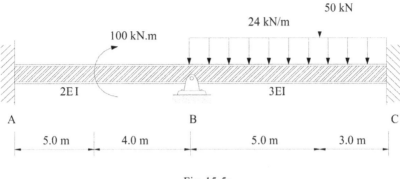

Fig. 15-5a

SOLUTION

Neglecting axial effects, the structure reactions, external loads, displacements, and fixed end forces are shown in Figs. 15-5b, 15-5c, 15-5d, and 15-5e.

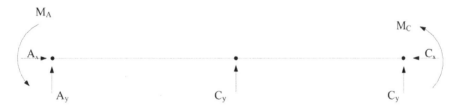

Fig. 15-5b. Structural Model

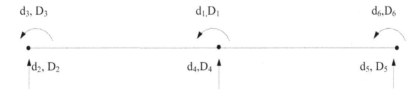

Fig. 15-5c. Degrees of Freedom

16.46   (96+16.46+15.82)   130.1kN

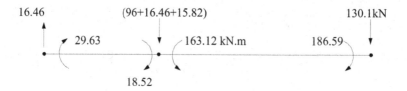

Fig. 15-5d. Fixed End Forces

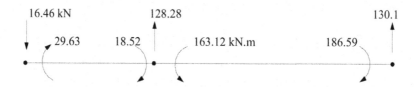

Fig. 15-5e. Fixed End Reactions

The member stiffness matrices [k$_{AB}$] and [k$_{BC}$], which are obtained using Equation [15-10], are equal to

$$
k_{AB} = \begin{array}{cccc} D_2 & D_3 & D_4 & D_1 \end{array}
$$

$$
k_{AB} = \begin{bmatrix} 1.317\times10^3 & 5.926\times10^3 & -1.317\times10^3 & 5.926\times10^3 \\ 5.926\times10^3 & 3.556\times10^4 & -5.926\times10^3 & 1.778\times10^4 \\ -1.317\times10^3 & -5.926\times10^3 & 1.317\times10^3 & -5.926\times10^3 \\ 5.926\times10^3 & 1.778\times10^4 & -5.926\times10^3 & 3.556\times10^4 \end{bmatrix} \begin{array}{c} D_2 \\ D_3 \\ D_4 \\ D_1 \end{array}
$$

$$
k_{BC} = \begin{array}{cccc} D_4 & D_1 & D_5 & D_6 \end{array}
$$

$$
k_{BC} = \begin{bmatrix} 2.812\times10^3 & 1.125\times10^4 & -2.812\times10^3 & 1.125\times10^4 \\ 1.125\times10^4 & 6\times10^4 & -1.125\times10^4 & 3\times10^4 \\ -2.812\times10^3 & -1.125\times10^4 & 2.812\times10^3 & -1.125\times10^4 \\ 1.125\times10^4 & 3\times10^4 & -1.125\times10^4 & 6\times10^4 \end{bmatrix} \begin{array}{c} D_4 \\ D_1 \\ D_5 \\ D_6 \end{array}
$$

The structure stiffness matrix [K], which is obtained by superposing the global stiffness matrices of AB and BC, is equal to

$$
\begin{bmatrix} -144.6 \\ R_2 \\ R_3 \\ R_4 \\ R_5 \\ R_6 \end{bmatrix} =
\begin{bmatrix}
9.556\times10^4 & 5.926\times10^3 & 1.778\times10^4 & 5.324\times10^3 & -1.125\times10^4 & 3\times10^4 \\
5.926\times10^3 & 1.317\times10^3 & 5.926\times10^3 & -1.317\times10^3 & 0 & 0 \\
1.778\times10^4 & 5.926\times10^3 & 3.556\times10^4 & -5.926\times10^3 & 0 & 0 \\
5.324\times10^3 & -1.317\times10^3 & -5.926\times10^3 & 4.129\times10^3 & -2.812\times10^3 & 1.125\times10^4 \\
-1.125\times10^4 & 0 & 0 & -2.812\times10^3 & 2.812\times10^3 & -1.125\times10^4 \\
3\times10^4 & 0 & 0 & 1.125\times10^4 & -1.125\times10^4 & 9.556\times10^4
\end{bmatrix}
\begin{bmatrix} d_1 \\ d_2 \\ d_3 \\ d_4 \\ d_5 \\ d_6 \end{bmatrix}
$$

Note that -(163.12 - 18.52) = -144.6
(-) equal and opposite

The displacement matrix [d_D], which is computed using Equation [15-14], is equal to

$$[d_D] = [d_1] = [9.556\times10^4]^{-1} [-144.6] = -1.513\times10^{-3} = 1.513\times10^{-3}\,\text{rad} \quad \curvearrowright$$

The unknown nodal force matrix [R], which is computed using Equation [15-15], is equal to

$$
[R] = \begin{bmatrix} R_2 \\ R_3 \\ R_4 \\ R_5 \\ R_6 \end{bmatrix} =
\begin{bmatrix} 5.926\times10^3 \\ 1.778\times10^4 \\ 5.324\times10^3 \\ -1.125\times10^4 \\ 3\times10^4 \end{bmatrix}
\begin{bmatrix} -1.513\times10^3 \end{bmatrix} =
\begin{bmatrix} -8.967 \\ -26.902 \\ -8.057 \\ 17.024 \\ -45.398 \end{bmatrix}
$$

The final structure forces due to span loading are computed using the following equation:

$$[Q] = [Q_0] + [R] \tag{15-18}$$

Where [Q_0] = Fixed End reactions shown in Fig. 15-5e.

$$
\begin{bmatrix} Q_2 \\ Q_3 \\ Q_4 \\ Q_5 \\ Q_6 \end{bmatrix} =
\begin{bmatrix} -16.46 \\ -29.63 \\ 128.28 \\ 130.118 \\ -186.59 \end{bmatrix} +
\begin{bmatrix} -8.967 \\ -26.902 \\ -8.057 \\ 17.024 \\ -45.398 \end{bmatrix} =
\begin{bmatrix} -25.427 \\ -56.532 \\ 120.223 \\ 147.142 \\ -231.988 \end{bmatrix}
$$

The structure reactions are shown in Fig. 15-5f.

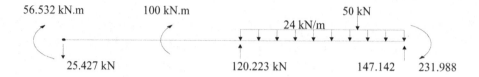

Fig. 15.5f. Loads and Reactions

b) To determine the support reactions due to settlement at support B, Equation [15-1] needs to be rearranged as follows.

$$
\begin{bmatrix} -144.6 \\ R_2 \\ R_3 \\ R_4 \\ R_5 \\ R_6 \end{bmatrix} =
\begin{bmatrix}
9.556\times10^4 & 5.926\times10^3 & 1.778\times10^4 & 5.324\times10^3 & -1.125\times10^4 & 3\times10^4 \\
5.926\times10^3 & 1.317\times10^3 & 5.926\times10^3 & -1.317\times10^3 & 0 & 0 \\
1.778\times10^4 & 5.926\times10^3 & 3.556\times10^4 & -5.926\times10^3 & 0 & 0 \\
5.324\times10^3 & -1.317\times10^3 & -5.926\times10^3 & 4.129\times10^3 & -2.812\times10^3 & 1.125\times10^4 \\
-1.125\times10^4 & 0 & 0 & -2.812\times10^3 & 2.812\times10^3 & -1.125\times10^4 \\
3\times10^4 & 0 & 0 & 1.125\times10^4 & -1.125\times10^4 & 9.556\times10^4
\end{bmatrix}
\begin{bmatrix} d_1 \\ d_2 = 0 \\ d_3 = 0 \\ d_4 = -0.03 \\ d_5 = 0 \\ d_6 = 0 \end{bmatrix}
$$

Solving for the displacement d_1

$-144.6 = 9.55\times10^4(d_1) + 5.926\times10^3 (0) + 1.778\times10^4 (0) + 5.324\times10^3 (-0.03) + (-1.125\times10^4)(0) + 3\times10^4 (0)$

$d_1 = 0.000158$ rad

The support reactions are determined using Equation [15-15] as follows.

$$
\begin{bmatrix} R_2 \\ R_3 \\ R_4 \\ R_5 \\ R_6 \end{bmatrix} =
\begin{bmatrix}
5.926\times10^4 & 1.317\times10^3 & 5.926\times10^3 & -1.317\times10^3 & 0 & 0 \\
1.778\times10^4 & 5.926\times10^3 & 3.556\times10^4 & -5.926\times10^3 & 0 & 0 \\
5.324\times10^4 & -1.317\times10^3 & -5.926\times10^3 & 4.129\times10^3 & -2.812\times10^3 & 1.125\times10^4 \\
-1.125\times10^4 & 0 & 0 & -2.812\times10^3 & 2.812\times10^3 & -1.125\times10^4 \\
3\times10^4 & 0 & 0 & 1.125\times10^4 & -1.125\times10^4 & 9.556\times10^4
\end{bmatrix}
\begin{bmatrix} 0.000158 \\ 0 \\ 0 \\ -0.03 \\ 0 \\ 0 \end{bmatrix}
$$

$$\begin{bmatrix} R_2 \\ R_3 \\ R_4 \\ R_5 \\ R_6 \end{bmatrix} = \begin{bmatrix} 40.444 \\ 180.591 \\ -123.039 \\ 82.595 \\ -332.752 \end{bmatrix}$$

The final structure forces due to span loading are computed using Equation [15-18] as follows.

$$\begin{bmatrix} Q_2 \\ Q_3 \\ Q_4 \\ Q_5 \\ Q_6 \end{bmatrix} = \begin{bmatrix} -16.46 \\ -29.63 \\ 128.28 \\ 130.118 \\ -186.59 \end{bmatrix} + \begin{bmatrix} 40.444 \\ 180.591 \\ -123.039 \\ 82.595 \\ -332.752 \end{bmatrix} = \begin{bmatrix} 23.984 \\ 150.961 \\ 5.241 \\ 212.713 \\ -519.342 \end{bmatrix}$$

The structure reactions are shown in Fig. 15-5g.

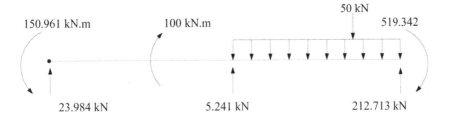

Fig. 15-5g. Applied Loads and Reactions

Example: 15-3

Determine the support reactions for the beam shown in Fig 15-6a.
[$E = 200.10^6$ kN/m^2, $I = 300.10^6$ mm^4, $K_{Spring} = 200$ kN/m]

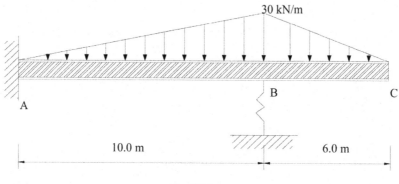

Fig. 15-6a

SOLUTION

Spring reactions, external loads, displacements, and fixed end forces are shown in Figs. 15-6b, 15-6c, and 15-6d.

Fig. 15-6b. Structural Model

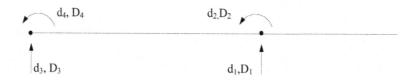

Fig. 15-6c. Degrees of Freedom

Fig. 15-6d. Fixed End Forces

Fig. 15-6e. Fixed End Reactions

Construct the member stiffness matrix [k], Equation [15-7]

$$k_{AB} = \begin{array}{cccc} D_3 & D_4 & D_1 & D_2 \\ \begin{bmatrix} 720 & 3.6\times10^3 & -720 & 3.6\times10^3 \\ 3.6\times10^3 & 2.4\times10^4 & -3.6\times10^3 & 1.2\times10^4 \\ -720 & -3.6\times10^3 & 720 & -3.6\times10^3 \\ 3.6\times10^3 & 1.2\times10^4 & -3.6\times10^3 & 2.4\times10^4 \end{bmatrix} & \begin{matrix} D_3 \\ D_4 \\ D_1 \\ D_2 \end{matrix} \end{array}$$

The structure stiffness matrix [K] is equal to:

$$\begin{bmatrix} -195 \\ -30 \\ \hline R_3 \\ R_4 \end{bmatrix} = \begin{bmatrix} 920 & -3.6\times10^3 & -720 & -3.6\times10^3 \\ -3.6\times10^3 & 2.4\times10^4 & 3.6\times10^3 & 1.2\times10^4 \\ \hline -720 & 3.6\times10^3 & 720 & 3.6\times10^3 \\ -3.6\times10^3 & 1.2\times10^4 & 3.6\times10^3 & 2.4\times10^4 \end{bmatrix} \begin{bmatrix} d_1 \\ d_2 \\ d_3 \\ d_4 \end{bmatrix}$$

541

It is worth noting that
$$K_{1,1} = K_{Spring} + K_{AB\,(1,1)}$$

$$= 200 + 720 = 920$$

The displacement matrix $[d_D]$, which is computed using Equation [15-14], is equal to

$$[d_D] = \begin{bmatrix} d_1 \\ d_2 \end{bmatrix} = \begin{bmatrix} 920 & -3.6 \times 10^3 \\ -3.6 \times 10^3 & 2.4 \times 10^4 \end{bmatrix}^{-1} \begin{bmatrix} -195 \\ -30 \end{bmatrix} = \begin{bmatrix} -0.525 \\ -0.08 \end{bmatrix}$$

The structure forces are determined using equation [15-15] as follows.

$$[Q] = [k_{AB}][d_D] + [Q_0]$$

where $[Q_0]$ = Fixed End Forces (Fig. 15-6e).

$$[Q] = \begin{bmatrix} 720 & 3.6 \times 10^3 & -720 & 3.6 \times 10^3 \\ 3.6 \times 10^3 & 2.4 \times 10^4 & -3.6 \times 10^3 & 1.2 \times 10^4 \\ -720 & -3.6 \times 10^3 & 720 & -3.6 \times 10^3 \\ 3.6 \times 10^3 & 1.2 \times 10^4 & -3.6 \times 10^3 & 2.4 \times 10^4 \end{bmatrix} \begin{bmatrix} d_3 = 0 \\ d_4 = 0 \\ d_1 = -0.525 \\ d_2 = -0.08 \end{bmatrix} + \begin{bmatrix} 45 \\ 100 \\ 195 \\ -150 \end{bmatrix}$$

$$Q = \begin{bmatrix} 135 \\ 1.03 \times 10^3 \\ 105 \\ -180 \end{bmatrix}$$

The structure reactions are shown in Fig. 15-6f.

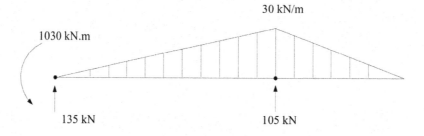

Fig. 15.6f. Applied Loads and Reactions

Example: 15-4

Determine the support reactions, displacements, and internal member forces for the frame shown in Fig. 15-7a. [E = 200.10^6 kN/m^2, I = 0.005208 m^4, A = 0.25 m^2]

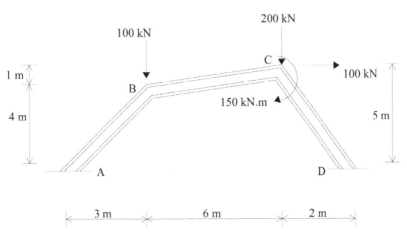

Fig. 15-7a

SOLUTION

The frame reactions, external loads, and displacements are shown in Fig. 15-7b and 15-7c.

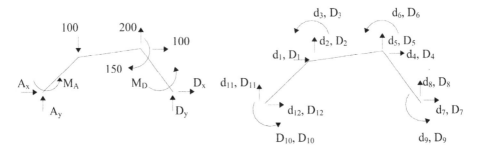

Fig. 15-7b. Structural Model   Fig. 15-7c. Degrees of Freedom

The construction of member stiffness matrix [k] is illustrated in Fig. 15-7d, 15-7e, and 15-7f.

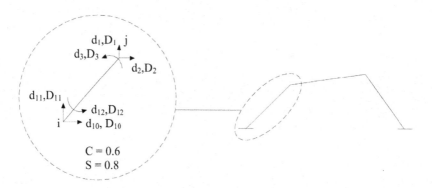

Fig. 15-7d

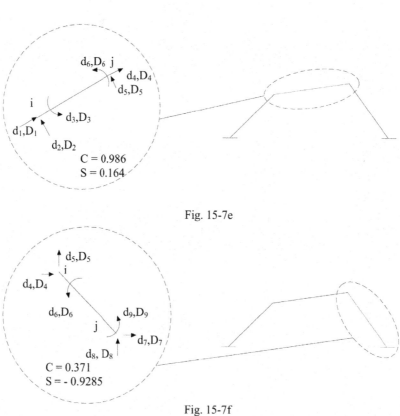

Fig. 15-7e

Fig. 15-7f

$$k_{AB} = \begin{bmatrix} 3.664 \times 10^3 & 4.752 \times 10^3 & -199.987 & -3.664 \times 10^3 & -4.752 \times 10^3 & -199.987 \\ 4.752 \times 10^3 & 6.436 \times 10^3 & 149.99 & -4.752 \times 10^3 & -6.436 \times 10^3 & 149.99 \\ -199.987 & 149.99 & 833.28 & 199.987 & -149.99 & 416.64 \\ -3.664 \times 10^3 & -4.752 \times 10^3 & 199.987 & 3.664 \times 10^3 & 4.752 \times 10^3 & 199.987 \\ -4.752 \times 10^3 & -6.436 \times 10^3 & -149.99 & 4.752 \times 10^3 & 6.436 \times 10^3 & -149.99 \\ -199.987 & 149.99 & 416.64 & 199.987 & -149.99 & 833.28 \end{bmatrix}$$

$$k_{BC} = \begin{bmatrix} 3.664 \times 10^3 & 4.752 \times 10^3 & 199.987 & -7.999 \times 10^3 & -1.324 \times 10^3 & -27.768 \\ 4.752 \times 10^3 & 6.436 \times 10^3 & -149.99 & -1.324 \times 10^3 & -276.196 & 166.61 \\ 199.987 & -149.99 & 833.28 & 27.768 & -166.61 & 342.476 \\ -7.999 \times 10^3 & -1.324 \times 10^3 & 27.768 & 1.35 \times 10^3 & -3.174 \times 10^3 & 200.09 \\ -1.324 \times 10^3 & -276.196 & -166.61 & -3.174 \times 10^3 & 8.015 \times 10^3 & 80.036 \\ -27.768 & 166.61 & 342.476 & 200.09 & 80.036 & 773.681 \end{bmatrix}$$

$$k_{CD} = \begin{bmatrix} 1.35 \times 10^3 & -3.174 \times 10^3 & 200.09 & -1.35 \times 10^3 & 3.174 \times 10^3 & 200.09 \\ -3.174 \times 10^3 & 8.015 \times 10^3 & 80.036 & 3.174 \times 10^3 & -8.015 \times 10^3 & 80.0361 \\ 200.09 & 80.036 & 773.6818 & -200.09 & -80.036 & 386.841 \\ -1.35 \times 10^3 & 3.174 \times 10^3 & -200.09 & 1.35 \times 10^3 & -3.174 \times 10^3 & -200.09 \\ 3.174 \times 10^3 & -8.015 \times 10^3 & -80.036 & -3.174 \times 10^3 & 8.015 \times 10^3 & -80.036 \\ 200.09 & 80.036 & 386.841 & -200.09 & -80.036 & 773.681 \end{bmatrix}$$

The structure stiffness matrix [K] is constructed as follows.

$$\begin{bmatrix} 0 \\ -100 \\ 0 \\ 100 \\ -200 \\ -150 \\ R_7 \\ R_8 \\ R_9 \\ R_{10} \\ R_{11} \\ R_{12} \end{bmatrix} = \begin{bmatrix} K_{DD} & K_{DR} \\ K_{RD} & K_{RR} \end{bmatrix} \begin{bmatrix} d_1 \\ d_2 \\ d_3 \\ d_4 \\ d_5 \\ d_6 \\ d_7 = 0 \\ d_8 = 0 \\ d_9 = 0 \\ d_{10} = 0 \\ d_{11} = 0 \\ d_{12} = 0 \end{bmatrix}$$

Where:

$$
K_{DD} = \begin{array}{c} \\ \\ \\ \\ \\ \\ \end{array}
\begin{bmatrix}
1.166\times10^4 & 6.076\times10^3 & 127.219 & -7.999\times10^3 & -1.324\times10^3 & -27.768 \\
6.076\times10^3 & 6.712\times10^3 & 16.62 & -1.324\times10^3 & -276.196 & 166.61 \\
127.219 & 16.62 & 1.518\times10^3 & 27.768 & -166.61 & 342.476 \\
-7.999\times10^3 & -1.324\times10^3 & 27.768 & 9.349\times10^3 & -1.85\times10^3 & 227.85 \\
-1.324\times10^3 & -276.196 & -166.61 & -1.85\times10^3 & 8.291\times10^3 & -86.574 \\
-27.768 & 166.61 & 342.476 & 227.858 & -86.574 & 1.459\times10^3
\end{bmatrix}
\begin{array}{l} d_1 \\ d_2 \\ d_3 \\ d_4 \\ d_5 \\ d_6 \end{array}
$$

with column headers $d_1 \quad d_2 \quad d_3 \quad d_4 \quad d_5 \quad d_6$

$$
K_{RD} =
\begin{bmatrix}
0 & 0 & 0 & -1.35\times10^3 & 3.174\times10^3 & -200.09 \\
0 & 0 & 0 & 3.174\times10^3 & -8.015\times10^3 & -80.036 \\
0 & 0 & 0 & 200.09 & 80.036 & 386.841 \\
-3.664\times10^3 & -4.752\times10^3 & -199.987 & 0 & 0 & 0 \\
-4.752\times10^3 & -6.436\times10^3 & 149.99 & 0 & 0 & 0 \\
199.987 & -149.99 & 416.64 & 0 & 0 & 0
\end{bmatrix}
\begin{array}{l} d_7 \\ d_8 \\ d_9 \\ d_{10} \\ d_{11} \\ d_{12} \end{array}
$$

with column headers $d_1 \quad d_2 \quad d_3 \quad d_4 \quad d_5 \quad d_6$

The displacement matrix $[d_D]$ is equal to

$$
[d_D] = \begin{bmatrix} d_1 \\ d_2 \\ d_3 \\ d_4 \\ d_5 \\ d_6 \end{bmatrix} = [K_{DD}]^{-1}[D] =
\begin{bmatrix}
0.331 \\
-0.254 \\
-7.452\times10^3 \\
0.276 \\
0.081 \\
-0.104
\end{bmatrix}
=
\begin{array}{l}
0.331 \text{ m} \rightarrow \\
0.254 \text{ m} \downarrow \\
7.452\times10^3 \text{ rad} \curvearrowleft \\
0.276 \text{ m} \rightarrow \\
0.081 \text{ m} \uparrow \\
0.104 \text{ rad} \curvearrowright
\end{array}
$$

The structure end force matrix $[R]$ is equal to

$$
[R] = \begin{bmatrix} R_7 \\ R_8 \\ R_9 \\ R_{10} \\ R_{11} \\ R_{12} \end{bmatrix} =
\begin{array}{l}
-96.064 \text{ kN} \leftarrow \\
238.855 \text{ kN} \uparrow \\
21.451 \text{ kN.m} \curvearrowleft \\
3.936 \text{ kN} \leftarrow \\
61.145 \text{ kN} \uparrow \\
101.142 \text{ kN.m} \curvearrowleft
\end{array}
$$

The displacements and reactions are shown in Fig. 15-7g.

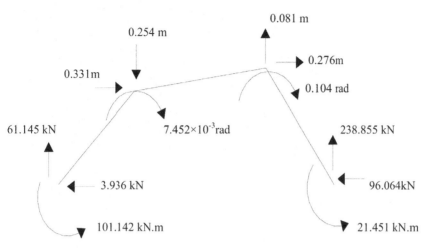

Fig.15-7g. Displacements and Reactions

What do you think of the displacements? Is the structure failing? Why ?

The member internal forces are determined as follows, Equations [15-11,13 and 16].

$$
[\overline{D}_{AB}] = \begin{bmatrix} 1\times10^4 & 0 & 0 & -1\times10^4 & 0 & 0 \\ 0 & 99.994 & 249.984 & 0 & -99.994 & 249.984 \\ 0 & 249.984 & 833.28 & 0 & -249.984 & 416.64 \\ -1\times10^4 & 0 & 0 & 1\times10^4 & 0 & 0 \\ 0 & -99.994 & -249.984 & 0 & 99.994 & -249.984 \\ 0 & 249.984 & 416.64 & 0 & -249.984 & 833.28 \end{bmatrix} \begin{bmatrix} 0.6 & 0.8 & 0 & 0 & 0 & 0 \\ -0.8 & 0.6 & 0 & 0 & 0 & 0 \\ 0 & 0 & 1 & 0 & 0 & 0 \\ 0 & 0 & 0 & 0.6 & 0.8 & 0 \\ 0 & 0 & 0 & -0.8 & 0.6 & 0 \\ 0 & 0 & 0 & 0 & 0 & 1 \end{bmatrix} \begin{bmatrix} 0 \\ 0 \\ 0 \\ 0.331 \\ -0.254 \\ -7.452\times10^{-3} \end{bmatrix}
$$

$$
[\overline{D}_{AB}] = \begin{bmatrix} 46.554 \\ 39.836 \\ 101.142 \\ -46.554 \\ -39.836 \\ 98.037 \end{bmatrix}
$$

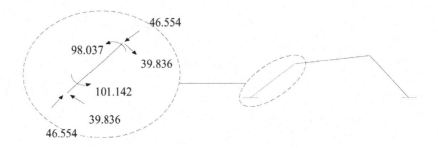

Fig. 15-7h. Internal Forces of Member AB

$$[\overline{D}_{BC}] = \begin{bmatrix} 8.22 \times 10^3 & 0 & 0 & -8.22 \times 10^3 & 0 & 0 \\ 0 & 55.537 & 168.908 & 0 & -55.537 & 168.908 \\ 0 & 168.908 & 684.952 & 0 & -168.908 & 342.476 \\ -8.22 \times 10^3 & 0 & 0 & 8.22 \times 10^3 & 0 & 0 \\ 0 & -55.537 & -168.908 & 0 & 55.537 & -168.908 \\ 0 & 168.908 & 342.476 & 0 & -168.908 & 684.952 \end{bmatrix} \begin{bmatrix} 0.986 & 0.164 & 0 & 0 & 0 & 0 \\ -0.164 & 0.986 & 0 & 0 & 0 & 0 \\ 0 & 0 & 1 & 0 & 0 & 0 \\ 0 & 0 & 0 & 0.986 & 0.164 & 0 \\ 0 & 0 & 0 & -0.164 & 0.986 & 0 \\ 0 & 0 & 0 & 0 & 0 & 1 \end{bmatrix} \begin{bmatrix} 0.331 \\ -0.254 \\ -7.452 \times 10^{-3} \\ 0.276 \\ 0.081 \\ -0.104 \end{bmatrix}$$

$$[\overline{D}_{BC}] = \begin{bmatrix} -10.27 \\ -37.679 \\ -98.037 \\ 10.27 \\ 37.679 \\ -131.158 \end{bmatrix}$$

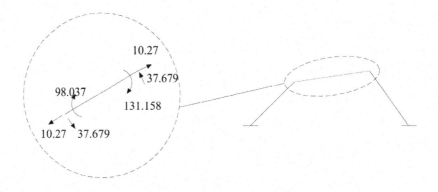

548

Fig. 15-7i. Internal Forces of Member BC

$$[\overline{D}_{CD}] = \begin{bmatrix} 9.285 \times 10^3 & 0 & 0 & -9.285 \times 10^3 & 0 & 0 \\ 0 & 80.036 & 215.503 & 0 & -80.036 & 215.503 \\ 0 & 215.503 & 773.681 & 0 & -215.503 & 386.841 \\ -9.285 \times 10^3 & 0 & 0 & 9.285 \times 10^3 & 0 & 0 \\ 0 & -80.036 & -215.503 & 0 & 80.036 & -215.503 \\ 0 & 215.503 & 386.841 & 0 & -215.503 & 773.681 \end{bmatrix} \begin{bmatrix} -0.371 & 0.928 & 0 & 0 & 0 & 0 \\ -0.928 & -0.371 & 0 & 0 & 0 & 0 \\ 0 & 0 & 1 & 0 & 0 & 0 \\ 0 & 0 & 0 & -0.371 & 0.928 & 0 \\ 0 & 0 & 0 & -0.928 & -0.371 & 0 \\ 0 & 0 & 0 & 0 & 0 & 1 \end{bmatrix} \begin{bmatrix} 0 \\ 0 \\ 0 \\ 0.276 \\ 0.081 \\ -0.104 \end{bmatrix}$$

$$[\overline{D}_{CD}] = \begin{bmatrix} 257.449 \\ 0.484 \\ 21.451 \\ -257.449 \\ -0.484 \\ -18.842 \end{bmatrix}$$

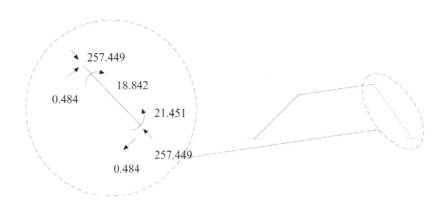

Fig. 15-7j. Internal Forces of Member CD

549

Example: 15-5

Determine the displacements and bar forces for the space truss shown in Fig 15-8a.
[E = 200.10^6 kN/m^2 and A = 200 mm^2 for all members]

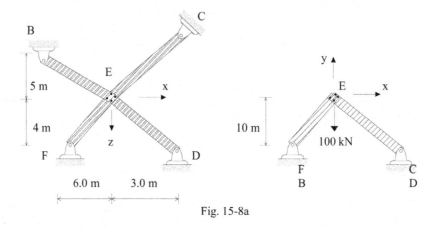

Fig. 15-8a

SOLUTION

Each element of the space truss has a total of six degrees of freedom three at each node as shown in Fig. 15-8b

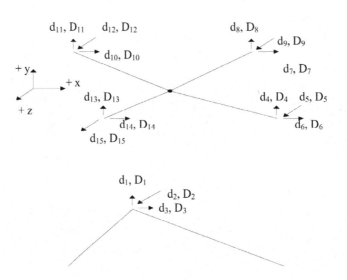

Fig. 15-8b, Degrees of Freedom

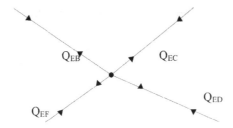

Fig. 15-8c. Internal Member Forces

The element stiffness matrix [k] for a space truss member is equal to

$$
[k] = \frac{AE}{L}
\begin{bmatrix}
C^2 & CS & ZC & -C^2 & -CS & -ZC \\
CS & S^2 & SZ & -CS & -S^2 & -SZ \\
ZC & ZS & Z^2 & -CZ & -SZ & -Z^2 \\
-C^2 & -CS & -CZ & C^2 & CS & CZ \\
-CS & -S^2 & -SZ & CS & S^2 & SZ \\
-ZC & -ZS & -Z^2 & CZ & SZ & Z^2
\end{bmatrix}
$$

Since node E is the only node with non-zero displacements (d_1, d_2, d_3), then the element stiffness matrix of each member will be reduced to a 3×3 matrix.

$$
[k_{FF}] = [k_{EF}] = \frac{AE}{L_{EF}}
\begin{bmatrix}
C^2 & CS & CZ \\
CS & S^2 & ZS \\
ZC & ZS & Z^2
\end{bmatrix}
=
\begin{bmatrix}
769.405 & 1.281 \times 10^3 & -511.883 \\
1.281 \times 10^3 & 2.134 \times 10^3 & -852.438 \\
-511.883 & -852.438 & 340.555
\end{bmatrix}
$$

Where:

$$
C = \frac{X_F - X_E}{\left(L_{EF} = \sqrt{(x_F - x_E)^2 + (y_F - y_E)^2 + (z_F - z_E)^2} \right)} = \frac{-6}{\sqrt{(6)^2 + (10)^2 + (4)^2}} = -0.487
$$

$$
S = \frac{y_F - y_E}{L_{EF}} = \frac{-10}{12.33} = -0.811
$$

$$
Z = \frac{z_F - z_E}{L_{EF}} = \frac{4}{12.33} = 0.324
$$

The values of C, S, and Z for the other members are summarized in Table 15-I

Table 15-I

member	L	C	S	Z	EA
EB	12.7	-0.472	-0.787	-0.394	40000
EC	11.576	0.259	-0.864	-0.432	40000
ED	11.18	0.268	-0.894	0.358	40000

The stiffness matrices of the other members are listed below.

$$[k_{EB}] = \begin{bmatrix} 701.682 & 1.17 \times 10^3 & 585.726 \\ 1.17 \times 10^3 & 1.951 \times 10^3 & 976.624 \\ 585.726 & 976.624 & 488.932 \end{bmatrix}$$

$$[k_{EC}] = \begin{bmatrix} 231.793 & -773.241 & -386.621 \\ -773.241 & 2.579 \times 10^3 & 1.29 \times 10^3 \\ -386.621 & 1.29 \times 10^3 & 644.865 \end{bmatrix}$$

$$[k_{ED}] = \begin{bmatrix} 256.973 & -857.216 & 343.27 \\ -857.216 & 2.86 \times 10^3 & -1.145 \times 10^3 \\ 343.27 & -1.145 \times 10^3 & 458.547 \end{bmatrix}$$

The structure stiffness matrix [K] is constructed as follows.

$$\begin{bmatrix} D_1 = 0 \\ D_2 = -10 \\ D_3 = 0 \end{bmatrix} = \begin{bmatrix} 1.96 \times 10^3 & 820.796 & 30.492 \\ 820.796 & 9.523 \times 10^3 & 268.828 \\ 30.492 & 268.828 & 1.933 \times 10^3 \end{bmatrix} \begin{bmatrix} d_1 \\ d_2 \\ d_3 \end{bmatrix}$$

The nodal displacements are equal to

$$\begin{bmatrix} d_1 \\ d_2 \\ d_3 \end{bmatrix} = \begin{bmatrix} 4.557 \times 10^{-4} \\ -1.093 \times 10^{-3} \\ 1.449 \times 10^{-4} \end{bmatrix} = \begin{bmatrix} 4.557 \times 10^{-4} \, m \rightarrow \\ 1.093 \times 10^{-3} \, m \downarrow \\ 1.449 \times 10^{-4} \, m \swarrow \end{bmatrix}$$

The bar forces [Q] are determined as follows.

$$[Q_{ij}] = \frac{AE}{L}[C \ S \ Z] \begin{bmatrix} u_j - u_i \\ v_j - v_i \\ z_j - z_i \end{bmatrix} \qquad [15\text{-}17]$$

where u = the displacement in the x-axis, v = the displacement in the y-axis, and z = the displacement in the z-axis

$$Q_{EF} = \frac{40000}{12.33}\begin{bmatrix} -0.487 & -0.811 & 0.324 \end{bmatrix}\begin{bmatrix} d_{13} - d_1 \\ d_{14} - d_2 \\ d_{15} - d_3 \end{bmatrix}$$

$Q_{EF} = -2.308 = 2.308$ kN   **Compression**

$Q_{EB} = -1.852 = 1.852$ kN   **Compression**
$Q_{EC} = -3.455 = 3.455$ kN   **Compression**
$Q_{ED} = -4.119 = 4.119$ kN   **Compression**

Summary

- Structural analysis computer software is generally programmed using the stiffness matrix. In matrix form, the equilibrium equation is

$$K\Delta = F$$

  where K = structure stiffness matrix, F = column vector of forces acting on the joints, and Δ = column vector of unknown joint displacements.

- The element k_{ij}, which is located in the ith row and jth column of K matrix, is called a stiffness coefficient. Coefficient k_{ij}, represents the joint force in the direction (or degree of freedom) of i due to a unit displacement in the direction of j. With this definition, the K matrix can be constructed by basic mechanics. For computer applications, however, it is more convenient to assemble the structure stiffness matrix from the member stiffness matrices.

- The equilibrium equations tied to the degrees of freedom that are allowed to move is used to calculate the unknown joint displacements.

- Once the unknown joint displacements are determined, the support reactions can be easily determined.

Problems

1-   Determine the reactions using the direct stiffness method. [EA=Const.]

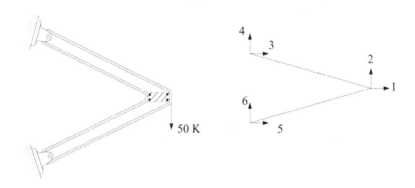

$$K = \frac{AE}{5} \begin{bmatrix} 0.3 & & & & & \\ 0.04 & 0.28 & & Sym & & \\ -0.09 & 0.12 & 0.09 & & & \\ 0.12 & -0.16 & -0.12 & 0.16 & & \\ -0.21 & -0.16 & 0 & 0 & 0.21 & \\ -0.16 & -0.12 & 0 & 0 & 0.16 & 0.12 \end{bmatrix} \qquad r = \begin{bmatrix} r1 \\ r2 \\ r3 \\ r4 \\ r5 \\ r6 \end{bmatrix}$$

2-   Determine the reactions using the direct stiffness method. [EI = Constant]

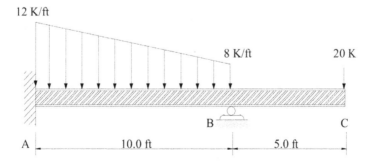

3-   Use the direct stiffness method to analyze the beam shown below. [EI= Constant]

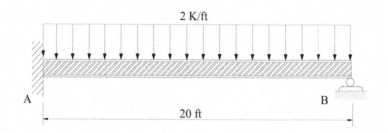

2 K/ft

A

B

20 ft

4-   Use the direct stiffness method to determine the reactions and the settlements at the support c.
[E = 200.10^6 kN/m, I =100.10^6 mm^4 , K$_{sp}$ = 200 kN/m]

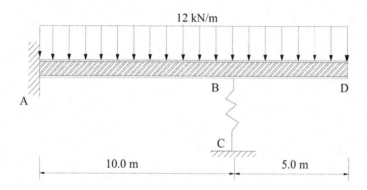

12 kN/m

A

B   D

C

10.0 m   5.0 m

5-   Determine support reactions, displacements and internal end forces for each member of the frame.

[E = 200.10^6 kN/m^2, I = 0.005208 m^4, A = 0.25 m^2]

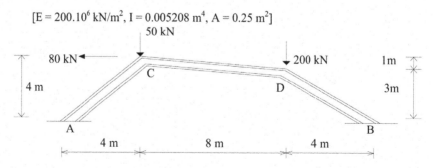

50 kN

80 kN

C

200 kN   1m

4 m

D   3m

A

B

4 m   8 m   4 m

6-   Determine the displacements and the bar forces of the space truss shown
below. [E = 200.10^6 kN/m^2 and A = 200 mm^2 for all members]

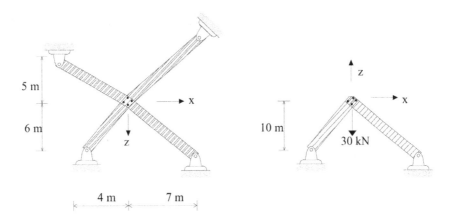

7-   Solve problem 11-9 using direct stiffness method.
8-   Solve problem 11-14 using direct stiffness method.
9-   Solve problem 11-16 using direct stiffness method.
10-   Solve problem 11-17 using direct stiffness method.
11-   Solve problem 11-18 using direct stiffness method.
12-   Solve problem 11-18 using direct stiffness method.
13-   Solve problem 13-7 using direct stiffness method.
14-   Solve problem 13-8 using direct stiffness method.

CHAPTER- 16
SELECTED TOPICS

JURUSALEM – Holy Land

Chapter 16. Selected Topics

16-1 INTRODUCTION TO REINFORCED CONCRETE DESIGN

Concrete, which is a non-homogenous manufactured stone, is a mixture of sand, gravel and crushed stone held together with a cement paste. Concrete is strong in resisting compressive forces, but weak in resisting tension forces. Reinforced concrete, which is a composite material made of steel and concrete, utilizes the steel in resisting tension forces and concrete in resisting compressive forces. Reinforced concrete is used to make structure elements such as slabs, beams, columns, walls and footings as shown in Figs. 16-1 and 16-2.

Reinforced concrete codes are usually used for the design of structural concrete elements. These codes state only the minimum performance standards and specifications that are necessary to provide safe and acceptable designs. There are a few common design methods that are used to design structural concrete elements such as the Ultimate Design Method (UDM), the Working Stress Method (WSM), and the Unified Design Method (UDM).

The general design procedure of structural reinforced concrete elements consists of the following steps:

1) Model the building into structural elements.
2) Compute the loads acting on the buildings structural elements.
3) Obtain the allowable bearing capacity of the soil for the footing design.
4) Perform the structure analysis to determine the design moments, shear and other forces.
5) Determine the size of the structural elements and its steel reinforcement.
6) Check the serviceability requirements of structural elements such as deflection, cracking, and vibration.
7) Prepare structural drawings and steel detailing of structural elements.

The Ultimate Design Method (UDM) will be used in the following examples.

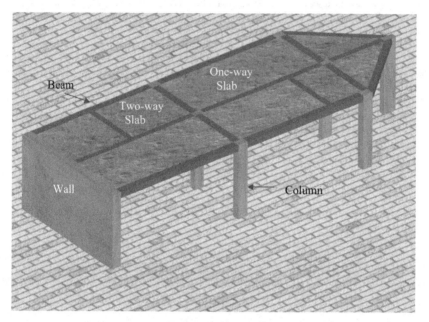

Fig. 16-1

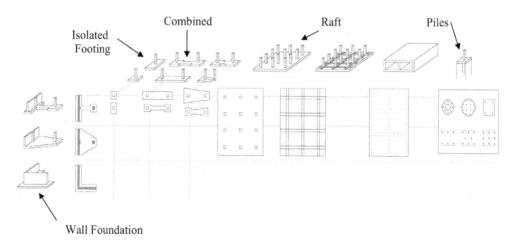

Fig. 16-2. Foundation Systems

Example: 16-1

A simply supported beam has a span length of 4.0 meter as shown in Fig. 16-3a.
1. Design the beam using ACI 318m-95 code.
2. Analyze the section using BS8110 and EC2.
Beam width (b) = 250 mm, concrete density = 25 kN/m³, F_y (steel reinforcement yield strength) = 400 MPa, and f'_c (compressive strength of concrete) = 30MPa

DL = 25 kN/m, LL = 15 kN/m

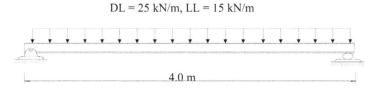

Fig. 16-3a

SOLUTION
(1) The structural element is modeled as shown in Fig. 16-3a.
(2) Compute the factored loads on the beam
   $W_u = 1.4DL + 1.7LL$
   $= 1.4(25) + 1.7(15) = 60.5$ kN/m
(3) Perform the structural analysis and draw the shear and moment diagrams
   as shown in Figs. 16-3b, 16-3c, 16-3d, and 16-3e.

60.5 kN/m

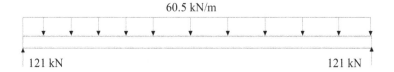

121 kN   121 kN

Fig. 16-3b. Applied Loads and Reactions

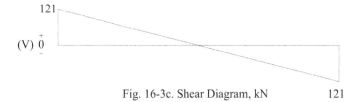

Fig. 16-3c. Shear Diagram, kN

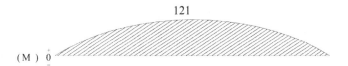

Fig. 16.3d. Moment Diagram, kN.m

(4) The beam size and reinforcement are determined using the following ACI-code equation:

$$M_u = \phi \rho F_y \, bd^2 \left(1 - 0.59 \rho \frac{F_y}{f'c} \right) \qquad [16\text{-}1]$$

Where M_u = External ultimate moment (design moment), ϕ = bending strength reduction factor ($\phi = 0.9$), ρ = Ratio of the tensile steel area (A_s) to the effective concrete area ($\rho = \frac{A_s}{bd}$). The ratio ρ must satisfy the following equation:

$$\left(\rho_{min} = \frac{1.4}{F_y} \right) \le \rho \le \left(\rho_{max} = 0.75 \, \rho_b \right)$$

Where the balanced reinforcement ratio ρ_b is given by the following equation:

$$\rho_b = \left(\frac{0.85 \beta f'_c}{F_y} \right) \left(\frac{600}{600 + F_y} \right)$$

The parameter β is given by the following equation:

$$\beta = 0.85 - \left(\frac{f'c - 30}{7} \right) (0.05) \ge 0.65$$

It is a common practice to assume a value of ρ equal to $\frac{1}{2} \rho_b$. Applying equation [16-1] and making use of Table A-1:

$M_u = 121$ kN.m

$\rho = \dfrac{1}{2} \rho_b = 0.0162$ $\qquad\qquad$ ($F_y = 400$ MPa and $f'_c = 30$ MPa)

$b = 250$ mm

$$121 \times 10^6 = 0.9(0.0162)(400)bd^2 \left(1 - 0.59 \rho \frac{400}{30} \right)$$

$$bd^2 = 23777848.5 \text{ mm}^3 \Rightarrow \left[\overset{b}{250} \times \overset{d}{308} \right]$$

Use b = 250 mm, d= 330 mm, d_1 (distance from the external tension fiber to the centeroid of the tensile steel) $\approx$ 70 mm
Thus, h = d + d_1 = 400 mm

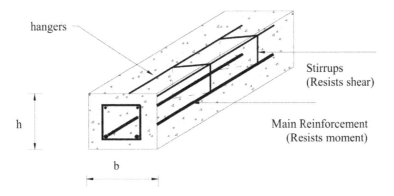

Fig. 16-3e. Beam Size and Steel Reinforcement Types

The final beam dimensions are 250 mm × 400 mm

Check h_{min} (Table A-2) $= \dfrac{L\,(4000mm)}{16} = 250$ mm < 400 OK

The beam depth can also be determined by using Table A-3 of Appendix A.

$\rho = 0.0162$

$\dfrac{M_u}{\phi b d^2} = 5.657$ MPa   (From Table A-3)

$bd^2 = \dfrac{121 \times 10^6}{0.9(5.657)} = 23766032$ mm³

Once the beam width (b) and depth (d) are determined, a bar selection is made to obtain the required area of steel reinforcement A_s as shown in Table 16-1.

Table 16-1

b (mm)	h (mm)	$d = h-d_1$	ρ	As = ρ bd (mm²)	Number of bars
250	400	330	0.0162	1336.5	$2\phi30$ (1413>136.5) $3\phi25$(1472>136.5)

The selected bars have to be spaced according the code as shown in Fig. 16-4. The minimum spacing S_{min} is equal to the following:

bar diameter (d_b) + the largest of $\left(25\ mm,\ d_b,\ \dfrac{4}{3}\ (maximum\ aggregate\ size) \right)$

$S_{min} = 25 + 25 = 50 < S_{available} = \dfrac{250 - 2(70)}{2} = 55$ Ok

$3\,\phi\,25$ ($A_s = 1472 < A_{smax} = 0.75\,\rho_b\,bd = 2013\ mm^2$)
and
$$S_{available} < S_{max} = 500\ mm$$

The beam width is checked as to satisfy the following code equation:

$b_{min} = 2d_1 + 4d_b = 2(70) + 4(25)$
$b_{min} = 240\ mm < b = 250\ mm$ \qquad Ok.

The final design is as follows.
Beam size of 250 mm x 400 mm with 3 $\phi\,25$ ($A_s = 1472\ mm^2$)

The check of the minimum steel reinforcement is as follows.

$$A_s = 1472\ mm^2 > As_{min} = \left(\frac{1.4}{F_y}\right)bd = 289\ mm^2 \qquad\qquad Ok$$

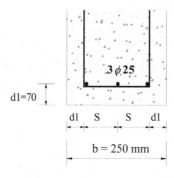

$d1=70$

$3\,\phi\,25$

dl   S   S   dl

b = 250 mm

Fig. 16-3f. Steel Reinforcement Detailing

(5) Nominal Bending Moment
M_n is calculated to determine the ultimate moment capacity of the section M_c based on the size and reinforcement of the beam as shown in Fig. 16-3g.

566

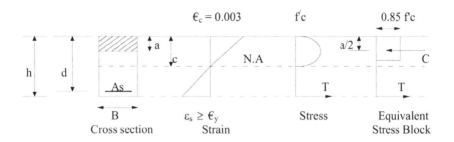

Fig. 16-3.g

In Fig. 16-3.g, the concrete compressive force **C** is given by the following equation:

$$C = 0.85 \, f_c' \, b \, a$$

On the other, the steel reinforcement tensile force **T** is given by the following equation:

$$T = A_s \, F_y$$

The ultimate moment capacity M_c of the section is equal to the following equation:

$$M_c = \phi \, T \left(d - \frac{a}{2}\right) = \phi \, A_s \, F_y \left(d - \frac{a}{2}\right) \qquad [16\text{-}2]$$

where a = depth of the equivalent rectangular concrete stress block (a = β c =

$a = \dfrac{A_s \, F_y}{0.85 f_c' b}$); $\beta = 0.85 - \left(\dfrac{f'c - 30}{7}\right)(0.05) \geq 0.65$; c = distance from the external

compression fiber to the neutral axis; ε_c = strain of the top fiber of the compressive

concrete; ε_y = yield strain of the tensile reinforcement ($\varepsilon_y = \dfrac{F_y}{E_s}$); E_s = modulus of

elasticity of steel.

Therefore,

$$M_c = 0.9 \, (1472)(400)\left(300 - \frac{92.36}{2}\right)\left(\frac{1}{10^6}\right)$$

$$M_c = 150.4 \text{ kN.m}$$

Check $M_c > M_{ut}$

where $M_{ut} = M_u + M_{u \, beam}$ (Moment due to the beam own weight)

$$M_{u \, beam} = \frac{W_{ub} \, (L)^2}{8} = \frac{1.4(250 \times 400)(\gamma_c = 25)(4)^2}{(10^6)8}$$

$$M_{u \, beam} = 7 \text{ kN.m}$$

$M_c = 150 > 121 + 7 = 128$ kN.m   Ok.

(6) Design for shear

The beam has to be checked for shear adequacy. For the section to be adequate in shear, it has to satisfy the following equation:

$$V_u \leq \phi(5V_c)$$

Where V_u = shear force due to factored loads, V_c = concrete shear strength of concrete $(V_c = \frac{1}{6}\sqrt{f_c'}bd)$, and ϕ = strength reduction factor ($\phi = 0.85$ for shear)

Substituting

$$121 \text{ kN} < 0.85(5)\left(\frac{75311.9}{1000}\right) = 320 \text{ kN} \qquad \text{Ok}$$

Stirrups are required because $\frac{1}{2}\phi V_c = 37.66\ kN \prec V_u = 121\ kN$

Therefore, stirrups are required with a spacing S given by the following equation:

$$S = \frac{(Av = 2A_b)(F_y)(d)}{V_s}$$

Where A_v = shear reinforcement cross-sectional area; A_b = cross-sectional area of an individual bar, and V_s = shear reinforcement strength $(V_s = \frac{V_u}{\phi} - V_c)$

Select stirrups with a diameter of $\phi 8$ => $A_v = 100$ mm^2

$$S = \frac{100(400)(330)}{67(1000)} = 197 \text{ mm}$$

The stirrup spacing S has to be checked against the maximum allowable spacing set by the code as follows.

$$S_{max} \leq \frac{d}{2} \text{ or } 600mm \quad \text{if} \qquad V_s < \frac{1}{3}\sqrt{f_c'}bd$$

$$S_{max} \leq \frac{d}{4} \qquad\qquad \text{if} \qquad V_s > \frac{1}{3}\sqrt{f_c'}bd$$

$$V_s = 67 \text{ kN} < 150.6 \text{ kN} => S_{max} = \frac{d}{2} = \frac{330}{2} = 165 \text{ mm} < 600$$

$$S_{max} = 165 \text{ mm}$$

Thus the spacing $S = 197$ mm $> S_{max} = 165$ mm

Use $\phi 8 @ 160mm$ as shown Fig. 16-3h

Note

Generally the most economical design is obtained by varying the stirrup spacing over the length of the beam because the shear force decreases towards the centre of the span of the beam. However, the maximum spacing allowed by the code was used for this beam.

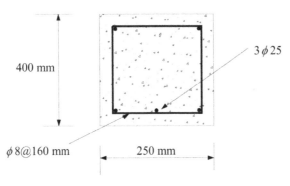

$3 \phi 25$

400 mm

$\phi 8@160$ mm

250 mm

Fig. 16-3h

(7) Section Analysis using British code BS8110:

$$a = \frac{0.95 F_y A_s}{0.45 f_c' b}$$

$$= \frac{0.95(400)(1472)}{0.45(30)(250)} = 165.74 \text{ mm}$$

$$M_c = 0.95 \ F_y \ A_s \left(d - \frac{a}{2} \right)$$

$$= 0.95 \ (400)(1472) \left(330 - \frac{165.7}{2} \right) \times 10^{-6}$$

$$= 138 \text{ kN.m}$$

$M_c = 138 \text{ kN.m} > M_u = 128 \text{ kN.m}$ \hspace{2cm} Ok.

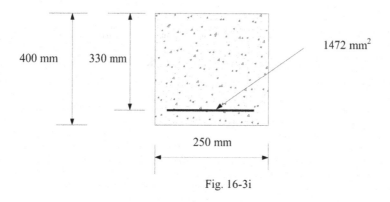

Fig. 16-3i

(8) Section Analysis using Euro Code2:

$$a = \frac{0.87 F_y A_s}{0.567 f_c^{'} b}$$

$$= \frac{0.87(400)(1472)}{0.567(30)(250)} = 120.5 \text{ mm}$$

$$M_c = 0.87 \, F_y \, A_s \left(d - \frac{a}{2} \right)$$

$$= 0.87 \, (400)(1472) \left(330 - \frac{120.5}{2} \right) \times 10^{-6}$$

$$= 138 \text{ kN.m}$$

$$M_c = 138 \text{ kN.m} > M_u = 128 \text{ kN.m} \qquad \text{Ok.}$$

Therefore the design is adequate by all three codes as shown in Table 16-2

Table 16-2

Code	Depth of stress block (a)	Moment Capacity (M_c)
ACI	92.36 mm	150.4 kN.m
BS8110	165.74 mm	138 kN.m
EC2	120.5 mm	138 kN.m

Example: 16-2

Design a short square tied column loaded axially with P_u = 2000 kN using ACI 318M-95 code. Analyze the column using British code BS8110. f'c = 30MPa and F_y = 400 MPa,

SOLUTION

(1) Column size

$$P_u = \phi 0.8\left[0.85f'c\left(A_g - A_{st}\right)+F_y A_{st}\right] \qquad \text{(ACI-Eq.10-1)}$$

Where ϕ = strength reduction factor (ϕ=0.7 for tied column), A_g = gross or total area of cross section, A_{st} = total steel sectional area, and ρ = steel reinforcement ration ($\rho = \dfrac{A_s}{b\ h}$)

$$2000(1000) = 0.7(0.8)[0.85(30)(A_g - A_{st}) + 400A_{st}]$$

setting ρ = 0.02 => A_{st} = 0.02A_g

$$2000(1000) = 0.7(0.8)[0.85(30)(A_g - 0.02A_g) + 400(0.02A_g)]$$

A_g = 108258 mm^2, Use a 330 mm × 330 mm square tied column
(A_g = 108900 mm^2)

(2) Reinforcement

Because a large column cross section is selected, the reinforcement steel is computed using ACI-Equation 10-2

$$2000(1000) = 0.7(0.8)[0.85(30)(108900 - A_{st}) + 400 A_{st}]$$
A_{st} = 1153 mm^2 => Use 6 ϕ 16 (1194 mm^2)

(3) Ties

Using ϕ 10 ties with a spacing, which is less than or equal to the smallest of
  a) 16× (d_b = 16mm) = 256 mm
  b) 48× (d_{bt} = 10) = 480 mm
  c) Least column dimension = 330 mm

Use ϕ 10 ties @ 250 mm

(4) Cross section detailing and code requirements

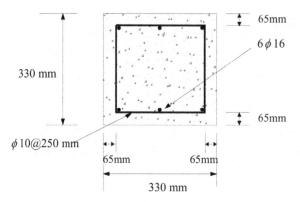

Fig. 16-4a

Checking code requirements:

- Bar spacing $= \dfrac{(b-2d_1)}{N_B-1} = \dfrac{330-2(65)}{3-1} = 100mm$

  $S_{max} = 150mm > S = 100mm > S_{min} = 25 + (d_b = 16) = 41$ mm Ok

- $\rho_{min} = 0.01 < \rho = \dfrac{A_{st}}{A_g} = \dfrac{1194}{330(330)} = 0.11 < \rho_{max} = 0.08$ Ok

- Total Number of bars = 6 > 4 Ok.

(5) Section analysis using BS 8110

$$P_u = 0.4f'c\,A_g + A_{st}\left(0.8F_y - 0.4f'c\right)$$
$$P_u = [0.4(30)(330)^2 + 1194(0.8\times400 - 0.7\times30)] \times 10^{-3}$$
$$P_u = 1674 \text{ kN} < 2000 \text{ kN}$$

The column, which is inadequate based on the BS8110, needs increasing the steel area or enlarging the cross section of the column. If $6\,\phi\,22$ ($A_s = 2322$ mm^2) bars are to be used

$$P_u = [0.4(30)(330)^2 + 2322(0.8\times400 - 0.4\times30)] \times 10^{-3}$$

$$P_u = 2022 \text{ kN} > 2000 \text{ kN} \quad \text{Ok.}$$

Example: 16-3

Design a wall footing to support a 200-mm-thick reinforced wall subjected to a dead load (DL = 30 kN/m) and live load (LL = 60 kN/m). The top of the footing (D_f) is to be 1.0 m below the final grade. The soil density, the concrete density, the allowable soil pressure q_a are estimated at 15 kN/m^3, 25 kN/m^3, and 200 kN/m^2, respectively. F_y = 400 MPa, f'_c = 30MPa

SOLUTION

Assume a 300-mm-thick footing (h) based on the minimum requirements set by the ACI code as shown in Fig. 16-5a

(1) The effective soil pressure = q_e

$q_e = q_a$ – footing weight – soil weight

$q_e = 200 - \dfrac{300}{1000}(25) - 1(15) = 177.5$ kN/m

(2) Footing Area

Footing area = Footing width (B) × wall width (b)

The required footing Area = $\dfrac{DL + LL}{q_e} = 0.5$ m^2 Fig. 16-5a

B = Footing width = $\dfrac{0.5\,\text{m}^2}{(b = 1\text{m})} = 0.5$ m

Use 1.0 m width section for a convenient and practical design
=> Use B = 1.0 m

(3) The ultimate soil pressure

$q_u = \dfrac{1.4(DL) + 1.7(LL)}{B(b)} = \dfrac{1.4(30) + 1.7(60)}{1(1)} = 144$ kN/m^2

(4) The required depth based on one-way shear at a distance d from the wall as shown in Fig. 16-5b.

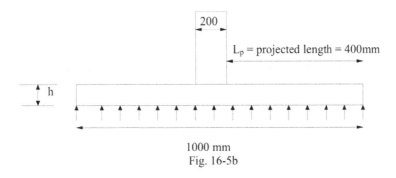

L_p = projected length = 400mm

1000 mm
Fig. 16-5b

$V_u = q_u (L_p - d) (b) = 144 (0.4-(300-75))$ (1)

$\quad = 25.2$ kN

$$d = \frac{V_u}{\phi\left(0.1667\sqrt{f'c}\right)(b)} = \frac{25.2(1000)}{0.85\left(0.1667\sqrt{30}\right)(1000)}$$

$d = 32.6$ mm $<$ actual depth $= h - $cover$ = 225$mm   Ok.

Note
If the required (d) was larger than the actual depth, steps 1 and 2 must be repeated.

Use h = footing thickness = 300mm

(5) Footing Reinforcement

The Moment at the face of the wall is equal to,

$$M_u = q_u \frac{\left(L_p\right)^2}{2}(b) = (144)\frac{(0.4)^2}{2}(1) = 11.52 \text{ kN.m}$$

$$\frac{M_u}{\phi bd^2} = \frac{11.52 \times 10^6}{0.9(1000)(225)^2} = 0.253$$

Using Table 16-3I, $\rho = 0.0035$

$$\rho_{min} = \frac{1.4}{F_y} = 0.0035$$

Use $\rho = 0.0035$

$A_s = \rho\, bd = 0.0035(1000)(225) = 788$ mm^2

Use $\phi\, 14 @ 200$ mm (As $= 924$ mm$^2 > 788$ mm^2)   Ok

Spacing 200 mm $< 3(h) = 900$ mm

$\qquad\qquad\qquad\quad < 450$ mm   Ok

(6) Developmental Length (Ld)

The Developmental length (Ld) is the required length for the steel bar to develop its own strength.

$$Ld = \text{larger of} \begin{cases} 0.019(A_b)(F_y)\left(\sqrt{f'c}\right) = 213 \text{ mm} \\ \\ 0.058\,(d_b)\,(F_y) = 324.8 \text{ mm} \end{cases}$$

Ld = 324.8 mm

where d_b = bar diameter (d_b = 14 mm), and A_b = area of one bar (A_b = 153.94 mm^2)

Check

Ld (required) < La (available length)

324.8 < lp – Cover = 400 – 75 = 325 mm   Ok.

Note

If Ld was larger than the available length (La), select smaller bars or increase the footing width and repeat steps 3 to 7.

(7) Shrinkage and Temperature

As = 0.002(b)(h) = 0.002(1000)(300) = 600 mm^2

Use ϕ 12 @ 200 mm => As = 678 mm^2 > 600 mm^2 Ok.

Note

For practical use most municipalities require larger bar sizes ($\geq$ 16 mm) and higher footing depths with 200 mm as a maximum steel bar spacing.

(8) Detailing

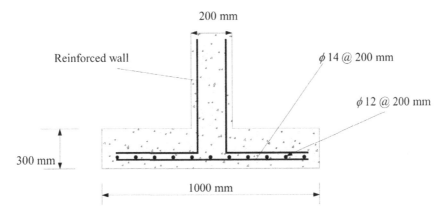

Fig. 16-5c

16.2   Plastic Analysis and Design of Structures

Plastic design considers the overall ultimate capacity of structures. The design of structural elements based on their behavior in the plastic region depends mainly on the resistance to bending actions caused by the loads. Let us consider the steel stress-strain diagram shown in Fig. 16-6a.

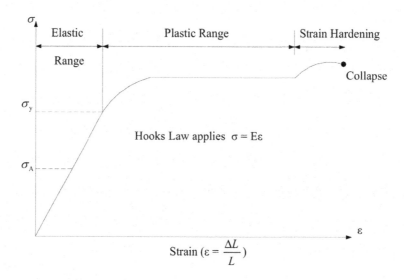

σ_y = yield stress
σ_a = allowable stress
E = Modulus Elasticity

Fig. 16-6a.   Stress–Strain Diagram

Plastic Hinge

A plastic hinge is a point of maximum moment where the structural element can not sustain any additional moment. The structural element will fail and collapse if one plastic hinge develops as shown in Fig. 16-6b.

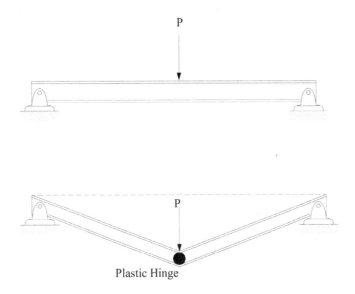

Plastic Hinge

Fig. 16-6b

Plastic Moment.

Let us consider the bending of a beam with a rectangular cross section as shown in Fig. 16-6c. The bending moment that produces a plastic hinge in a structural element is called the plastic moment (M_p). Its magnitude is the product of the plastic modulus (Z_p) of the structural element and its yield stress of the structural element material (σ_y) as follows

$$\mathbf{M_p = Z_p \ \sigma_y}$$

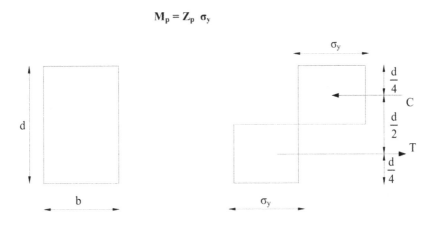

Fig. 16-6c.

In Fig. 16-6c, the total compressive and tensile forces (C) and (T) are equal and are given by the following equation:

$$C = T = \text{stress} * \text{area} = \sigma_y \left(\frac{d\,b}{2}\right)$$

Plastic Moment = $M = M_p$ = Force × lever arm

$$= \left(\sigma_y \frac{d}{2} b\right)\left(\frac{d}{2}\right)$$

$$= \frac{\sigma_y\, b\, d^2}{4}$$

Hence $Z_p = \dfrac{M_p}{\sigma_y} = \dfrac{b\, d^2}{4}$

Beam Analysis by Virtual work.

The plastic moment can be determined by introducing a small virtual displacement in a structure that make forces and moments perform work (energy). This internal work (w_i) done by the moment is the product of the moment and the angle of its rotation while the external work (w_e) done by the force is the product of the force and the distance traveled by the force. Let us consider the beam shown in Fig. 16-6d and the principle of conservation of energy

External Work (W_e) = Internal work (W_i)

$$P(\Delta) = M_p\,(\theta) + M_p\,(2\,\theta) + M_p\,(0)$$

$$P\,\frac{L}{2}\,(\theta) = M_p\,(3\theta)$$

$$P = \text{Collapse load} = \frac{6M_p}{L}$$

and

$$M_p = \text{Plastic Moment} = \frac{P\,L}{6}$$

The following examples will employ the virtual work method

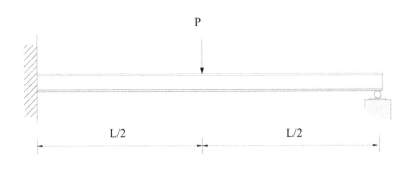

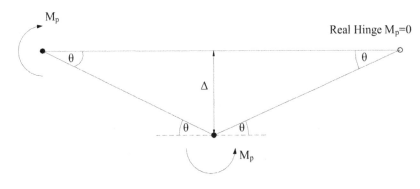

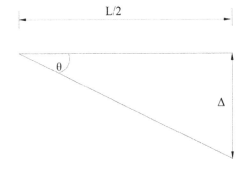

$$\frac{\Delta}{\left(\dfrac{L}{2}\right)} = \tan \theta$$

$$\Delta = \left(\frac{L}{2}\right)\tan \theta$$

$$\theta = \tan \theta \quad \left(\text{Small Angle Theory}\right)$$

$$\therefore \ \Delta = \left(\frac{L}{2}\right)\theta$$

Fig. 16-6d.

Example: 16-4

Select the most economical W-section for the fixed end beam shown in 16-7a using the AISC manual, the British manual, and the Canadian manual. [$\sigma_y = 250$MPa]

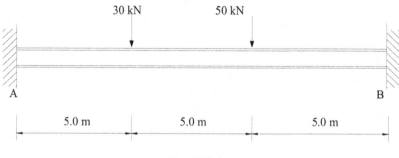

Fig. 16-7a

SOLUTION

The beam has two failure patterns due to plastic hinge formation under the loads. These failure patterns, which are called failure mechanisms, are shown in Figs. 16-7b and 16-7c. A load factor of 1.7 is used to factor the working loads to compensate for load uncertainties and to provide structural safety against collapse.

1st Failure Mechanism

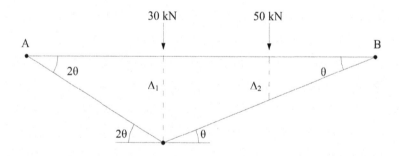

Fig. 16-7b. First Failure Mechanism

Set the rotation (θ_B) at B equal to θ, then $\Delta_1 = 10\,\theta$ and $\Delta_2 = 5\,\theta$. Using the value of Δ_1, the rotation at A is equal to $\dfrac{\Delta_1}{5} = \dfrac{10\theta}{5} = 2\theta$

From the principle of energy conservation, we have $W_i = W_E$

$M_p (2\theta) + M_p (2\theta + \theta) + M_p (\theta) = 1.7 [30(10\ \theta) + 50(5\ \theta)]$

$$M_p = 155.8 \text{ kN.m}$$

2nd Failure Mechanism

Set $\theta_A = \theta$, then $\Delta_1 = 10\ 0$; $\Delta_2 = 5\ 0$; $\theta_B = 2\ \theta$

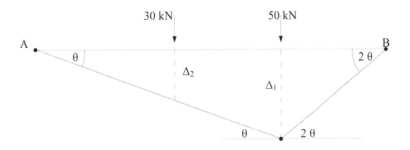

30 kN   50 kN

A   B

θ $2\ \theta$

Δ_2 Δ_1

θ $2\ \theta$

Fig. 16-7c. Second Failure Mechanism

From the principle of energy conservation, we have

$$W_i = W_E$$

$M_p (6\theta) = 1.7 [30(5\ \theta) + 50(10\ \theta)]$

$$M_p = 184.17 \text{ kN.m}$$

Note that 2nd Mechanism shown in Fig. 16-7c controls the design of the beam. Thus,

$M_p = M_{max} = 184.17$ kN.m

$Z_p = \dfrac{M_p}{\sigma_y} = \dfrac{184.17}{250 \times 10^6} = 736680 \text{ mm}^3$

$Z_p = 736.7 \text{ cm}^3$ about the major axis

The most economical W-section will be selected based on the Z_p value

Manual	Section	Z_x	Condition
AISC	W 360 × 44.8	764.8 > 736.7	Ok
BRITISH	W 356 × 171 ×45	773.7 > 736.7	Ok
CANADIAN	W 360 ×45	778 > 736.7	Ok

Example: 16-5

Select the most economical W-section for the continuous beam shown in Fig. 16-8a using the AISC and the British manuals. [$\sigma_y = 250MPa$]

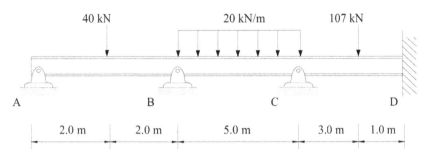

Fig. 16-8a

SOLUTION

The continuous beam has plastic hinge forming at the supports where moment is largest. Thus each span will be analyzed separately and the plastic moment will be determined for each span based on its failure mechanism as shown in Fig. 16-8b, 16-8c, and 16-8d.

Span AB

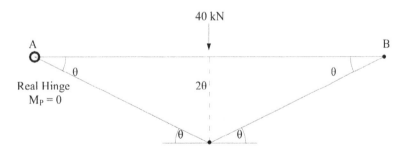

Fig. 16-8b. Span AB

From the principles of energy conservation, we have $W_i = W_E$

$$M_p (\theta+2\theta) = 1.7 [40(2\theta)]$$

$$\mathbf{M_p = 45.3 \ kN.m}$$

Span BC

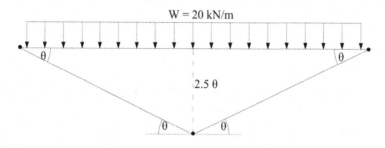

W = 20 kN/m

θ   θ

2.5 θ

θ   θ

Fig. 16-8c. Span BC

From the principles of energy conservation, we have

$$W_i = W_E$$

$$M_p (\theta + 2\theta + \theta) = 1.7 \,[(wL)(2.5\,\theta)(0.5)]$$

$$\mathbf{M_p = 53.125 \ kN.m}$$

Span CD

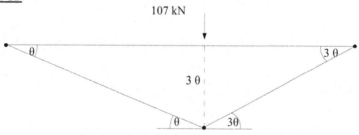

107 kN

θ   3 θ

3 θ

θ   3θ

Fig. 16-8d. Span CD

$$M_p (\theta + 4\theta + 3\,\theta) = 1.7 \,[107(3\,\theta)]$$

$$\mathbf{M_p = 68 \ kN.m}$$

Thus the maximum $M_p = 68$ kN.m

$$Z_p = \frac{M_p}{\sigma_y} = 272 \text{ cm}^3 \text{ about major axis}$$

Manual	Section	Z_x	Condition
AISC	W 310 × 21	287 > 272	Ok
BRITISH	W 305 × 102 × 25	337.5 > 272	Ok

Example: 16-6

Select the most economical W-section for the frame shown in Fig. 16-9a using the
AISC and the British manuals. [σ_y = 250MPa]

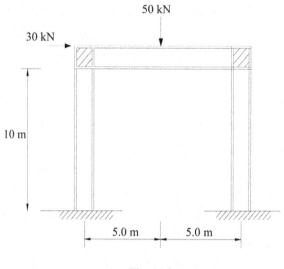

Fig. 16-9a

SOLUTION

The following three failure mechanisms are considered: Beam mechanism, column
mechanism and combined mechanism as shown in Figs. 16-9b and 16-9c.

Beam Mechanism

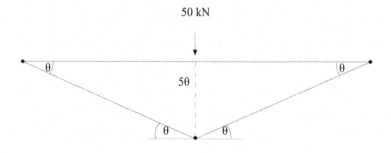

Fig. 16-9b. Beam Mechanism

From the principle of energy conservation, we have $W_i = W_E$

$$M_p (\theta + 2\theta + \theta) = 1.7 [50(5\theta)]$$

$$\mathbf{M_p = 106.25 \ kN.m}$$

Column Mechanism

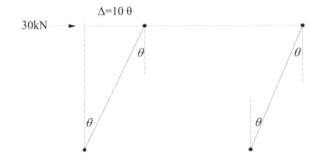

Fig. 16-9c. Column Mechanism

From the principle of energy conservation, we have $W_i = W_E$

$$M_p (\theta + \theta + \theta + \theta) = 1.7 [30(10 \ \theta)]$$

$$\mathbf{M_p = 127.5 \ kN.m}$$

Combined Mechanism

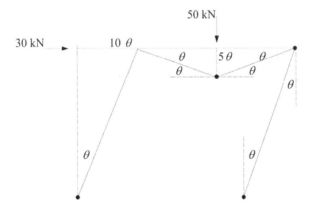

Fig. 16-9d. Combined Mechanism

$$M_p \ (\theta + 2\theta + 2\theta + \theta) = 1.7 \ [30(10 \ \theta) + 50(5 \ \theta)]$$

$M_p = 155.8$ kN.m

Thus, the maximum $M_p = 155.8$ kN.m

$$Z_p = \frac{M_p}{\sigma_y} = 623.2 \ \text{cm}^3 \ \text{about the major axis}$$

Manual	Section
AISC	W 410 × 38.8
BRITISH	UB 356 × 127 ×39

16-3 IRREGULAR STRUCTURAL MEMBERS

In this section we will analyze irregular structural members such as: 1) non prismatic section, 2) curved beams, 3) sandwich beams, 4) beams with side openings, 5) fixed arches, and 6) concrete beams with irregular cross section.

NON PRISMATIC BEAMS

Non prismatic beams are beams having a variable flexural rigidity (moment of inertia) as shown in Fig. 16-10a

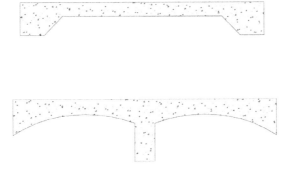

Fig. 16-10a

CURVED BEAM IN PLANE

The curved beam, shown in Fig. 16-10b, is subjected to bending moment (M_θ), torsional moment (T_θ), and shear force (V) as shown in Fig. 16-10c.

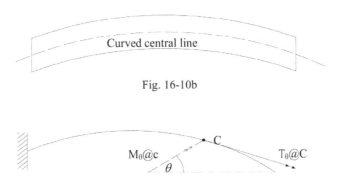

Curved central line

Fig. 16-10b

$M_\theta @c$ θ C $T_\theta @C$

Fig. 16-10c

Chapter 16. Selected Topics

SANDWICH BEAMS

Sandwich beams are structural elements made of thin, stiff, and sheets of metallic or
fiber composite materials separated by a thick layer of low density material known as
the core (Fig. 16-10d)

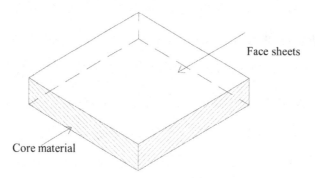

Fig. 16-10d

BEAMS WITH SIDE OPENINGS

The irregularity of these beams comes from the presence of side openings with any
that can be placed any where as shown in Fig. 16-10e.

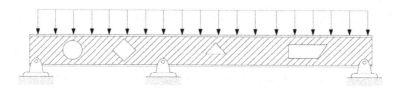

Fig. 16-10e

FIXED ARCHES

Fixed arches structures, which are used for long spans, are statically indeterminate to
the third degree as shown in Fig. 16-10f.

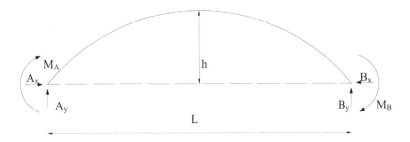

Fig. 16-10f Fixed Arches

CONCRETE BEAMS

Concrete beams with irregular cross section are analyzed based on their shape to
determine their strength as shown in Fig. 16-10g.

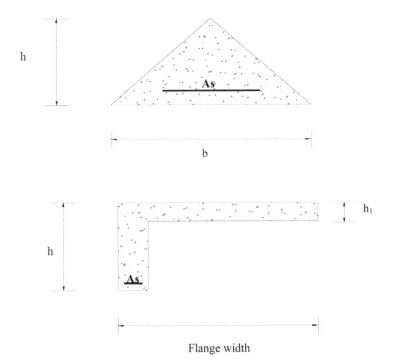

Flange width

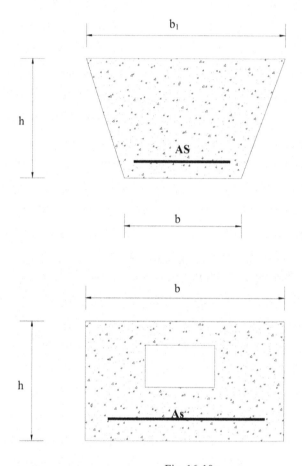

Fig. 16-10g

16-4 PLATES AND SHELLS

Plates are flat rigid elements whose thickness (t) is small compared to their length (L) and width (W) as shown in Fig. 16-10i. There are two types of plates: 1) thin plates $\left(\dfrac{t}{W} \leq \dfrac{1}{20}\right)$, and 2) thick plates $\left(\dfrac{t}{W} > \dfrac{1}{20}\right)$.

Concrete slabs are plates which can be one way slabs $\left(\dfrac{L}{W} \geq 2\right)$ or two-way slabs $\left(\dfrac{L}{W} < 2\right)$ as shown in Fig. 16-10i. A number of analytical and theoretical methods have been developed for the calculation of plate deflection. The method to be used in this section is a simplified method that uses the concept of the stiffness method to

calculate the deflection of a plate with specified boundary cond Plates and Shells
uniformly distributed load.

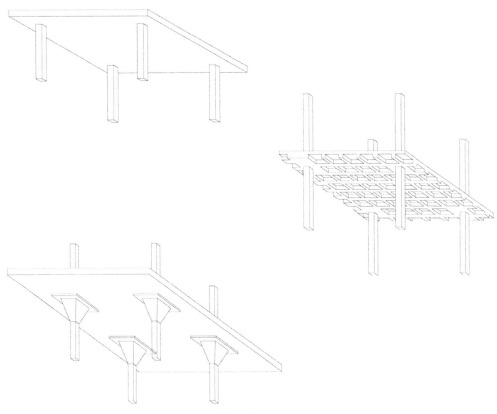

Fig. 16-10i

Shells are thin-walled elements in which the thickness is rather small compared to the
other shell dimensions. Shell is a solid continuum lying between two closely spaced
curved surfaces. Apart from roofing purposes, shells have been used in a wide range
of structures such as liquid containers, bunkers, silos, marine structures, etc., Fig. 16-
10j. Shell roofs are structurally superior since the whole cross-section is uniformly
stressed due to normal forces with minimal and negligible bending effects.   Because
of that, the thickness of shells is usually very small in the range of 80 mm to 160 mm.
Shell roofs, which are architecturally expressive, are generally used for hangers,
sports auditoriums, exhibition halls, and a variety of other large span structures. Shells
are analyzed by the membrane theory, the beam theory, the arch theory, and the finite
element methods.

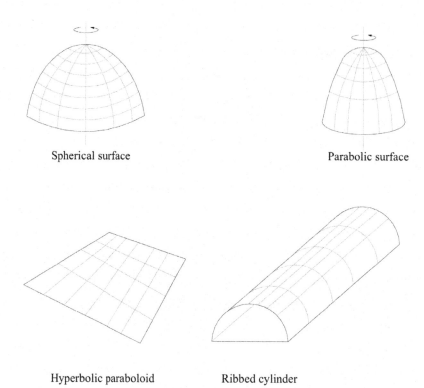

Spherical surface   Parabolic surface

Hyperbolic paraboloid   Ribbed cylinder

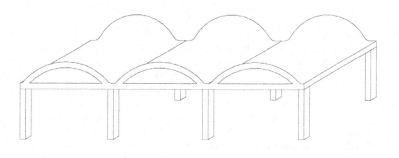

Multiple cylindrical shells

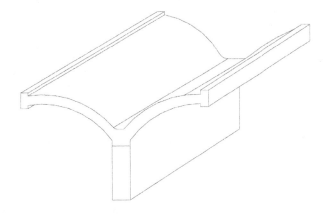

Butterfly shell

Fig. 16-10j

16-5 COMPUTER PROGRAMS

The computer programs available in the market can be classified in two categories, as shown in Fig. 16-11a. The first one includes the analysis and design programs such as STAAD, SAP, etc. These programs are mostly for professional use in the analysis and design of structures. The second category consists of programs developed using mathematical programming tools such as MathCAD, Mathematica, Matlab, etc. These programs can be used in teaching and learning in the academic environment. The instructor can use these programs to show the students a step by step solution of engineering problem that he had solved in his way. The student on the other hand can use these programs to solve and understand any engineering problem he is interested in by having the program perform the same steps as the manual solution with greater accuracy and a much shorter time. Theses programs will save students' and instructor time and will help students understand the theoretical meaning through different numerical answers resulting from changing the input data. For example in calculating the deflection at mid-span of a simple beam, the student will notice by using the program repeatedly with different input data that the deflection will decrease if the moment of inertia is increased provided that the other parameters are held constants, Fig. 16-19. In this section MathCAD program will be used as a teaching and learning tool for the design and the analysis of structural elements. For a better understanding of Figs. 16-20 and 16-21, one should refer to examples 16-12 and 16-13.

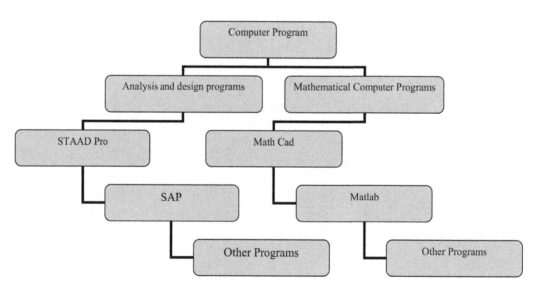

Fig. 16-11a

Example: 16-7

Determine the final moments at the supports of the one-meter-thick beam shown in Fig. 16-12a using the moment distribution method and the stiffness method.
[E = Constant]

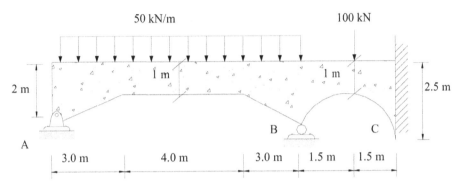

Fig. 16-12a

SOLUTION

Since the beam spans are non-prismatic, special factors for fixed end moments, stiffness and carry over factors, which are tabulated in the Handbook of Frame constants published by the Portland Cement Association (PCA), are used herein.

Span AB (Straight Haunches)

[PCA Hand Book of Frame Constants, Table 39 P. 18]

$$a_A = a_B = \frac{3}{10} = 0.3$$

$$r_A = r_B = \frac{2-1}{1} = 1$$

$$C_{AB} = C_{BA} = 0.705$$
$$K_{AB} = K_{BA} = 10.85$$

$$\overline{K}_{AB} = \frac{K_{AB}\, E\, I_C}{L} = \frac{(10.85)(E)\left(\frac{1}{12}\right)(1)(1)^3}{10} = 0.09$$

$$\overline{K}_{BA} = \overline{K}_{AB} = 0.09$$

UNIFORM LOAD

$$FEM_{AB} = -0.1034(50)(10)^2 = -517 \text{ kN.m}$$

$FEM_{BA} = 517$ kN.m

Span BC (Parabolic Hunches)

[PCA Hand Book of Frame Constant, Table15 p.9]

$$a_B = a_A = \frac{1.5}{3} = 0.5$$

$$r_B = \frac{2-1}{1} = 1, \qquad r_C = \frac{2.5-1}{1} = 1.5$$

$C_{BC} = 0.781, \qquad C_{CB} = 0.664$

$K_{BC} = 13.12, \qquad K_{CB} = 15.47$

$$\overline{K}_{BC} = \frac{(13.12)(E)\left(\dfrac{1}{12}\right)(1)(1)^3}{3} = 0.364$$

$$\overline{K}_{CB} = \frac{(15.47)(E)\left(\dfrac{1}{12}\right)(1)(1)^3}{3} = 0.43$$

CONCENTRATED LOAD

$$b = \frac{1.5}{3} = 0.5$$

$FEM_{BC} = -0.1456(100)(3) = -43.68$ kN.m

$FEM_{CB} = 0.1939(100)(3) = 58.17$ kN.m

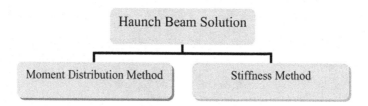

Moment Distribution Method

Once the fixed end moments, the stiffness factors and the carry over factors are determined for the spans AB and BC, the distribution factors (DF_{BA} and DF_{BC}) are computed to determine the final moments and reactions as shown in Fig. 16-12b, 16-12c, 16-12d.

Pinned End Span $\longrightarrow$ $\hat{k}_{BA} = \overline{k}_{BA}\left(1 - C_{AB}C_{BA}\right)$
$$= 0.09(1-(0.705)(0.705)) = 0.045$$

Thus

$$DF_{BA} = \frac{\hat{k}_{BA}}{\hat{k}_{BA} + \overline{k}_{BC}} = \frac{0.045}{(0.045) + (0.364)} = 0.11$$

$$DF_{BC} = \frac{\overline{k}_{BC}}{\hat{k}_{BA} + \overline{k}_{BC}} = \frac{0.364}{(0.045) + (0.364)} = 0.89$$

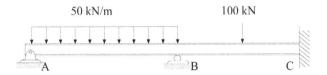

Fig. 16-12b Beam model

Moment Distribution Table

No.	DF	1		0.11	0.89		0	
1.	FEM	-517		517	-43.68		58.17	
	Balance	517	C_{AB}					
	COM		0.705	364.485				
	Balance			-92.16	-745.65	C_{BC}		
						0.781	-582.35	
	Σ	0		789	-789		-524	Final Moments

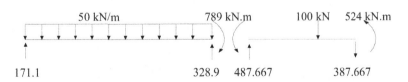

Fig. 16-12c. Member Free Body Diagram

Stiffness Method

The beam reactions, the external loads, the displacements, and the internal forces are shown in Fig. 16-12d, 16-12e, 16-12f, and 16-12g.

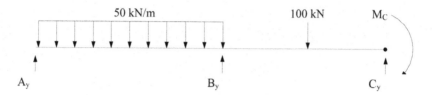

Fig. 16-12d. Structural Model

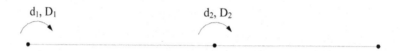

Fig. 16-12e. Degrees of Freedom

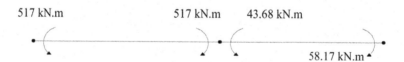

Fig. 16-12f. Fixed End Forces

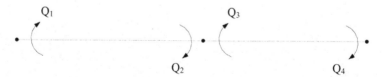

Fig. 16-12g. Member Internal Forces

The displacement transformation matrix [a] is equal to

$$[S] = \begin{array}{c} \begin{array}{cc} d_1 & d_2 \end{array} \\ \begin{bmatrix} 1 & 0 \\ 0 & 1 \\ 0 & 1 \\ 0 & 0 \end{bmatrix} \begin{bmatrix} d_1 \\ d_2 \end{bmatrix} \end{array}$$

The element stiffness matrix [k] for the non-prismatic element AB is equal to

$$[k] = \begin{bmatrix} \overline{k}_{AB} & \overline{k}_{AB}C_{AB} \\ \overline{k}_{BA}C_{BA} & \overline{k}_{BA} \end{bmatrix}$$

The element stiffness matrix [k] for the two span beam is equal to

$$[k] = \begin{bmatrix} 0.09 & 0.09(0.705) & 0 & 0 \\ 0.09(0.705) & 0.09 & 0 & 0 \\ 0 & 0 & 0.364 & 0.364(0.781) \\ 0 & 0 & 0.43(0.664) & 0.43 \end{bmatrix}$$

Constructing the structure stiffness matrix [K] Equation [14-11], The displacement matrix [d] is equal to

$$[d] = [K]^{-1}[D] = [d] = \begin{bmatrix} 7.188 \times 10^3 \\ -2.047 \times 10^3 \end{bmatrix}$$

Where $[D] = \begin{bmatrix} -(-517) = 517^{kN.m} @d_1 \\ -(517 - 43.68) = -473.32^{kN.m} @d_2 \end{bmatrix}$

The structure internal force matrix [Q] is equal to

$$[Q] = \begin{bmatrix} 517 \\ 271.818 \\ -745.138 \\ -584.483 \end{bmatrix}$$

The final structure internal force matrix $\left[\overline{Q}\right]$ due to span loading is equal to

$$\left[\overline{Q}\right] = [Q] + [Q_0]$$

wher $[Q_0]$ = Fixed End Moments shown inFig. 16-12f.

Thus

$$\left[\overline{Q}\right] = \begin{bmatrix} 517 \\ 271.818 \\ -745.138 \\ -584.483 \end{bmatrix} + \begin{bmatrix} -517 \\ 517 \\ -43.68 \\ 58.17 \end{bmatrix} = \begin{bmatrix} 0 \\ 788.818 \\ -788.818 \\ -526.313 \end{bmatrix}$$

Example: 16-8

The circular beam with a radius of 4.0 meters and equidistant supports, which is shown in Fig. 16-13a is subjected to uniformly distributed load of 50 kN/m over its entire span. Determine the support moment (M_s), the mid-span moment (M_m), and the maximum torsional moment T_{max}.

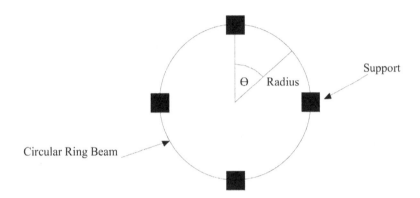

Fig. 16-13a

SOLUTION

The central angle Θ is given by

$$\theta = \frac{2\pi}{n} = \frac{360}{4} = 90^0 = \frac{\pi}{2}$$

Where n = number of columns or supports

The circular ring beam has no torsional moments at the supports due to symmetry. Thus, the support reactions are equal to

$$R_s = \frac{wR\theta}{2} = (50)(4)\left(\frac{\pi}{2}\right)$$
$$= 314.16 \text{ kN}$$

The moment coefficients m_1, m_2, and m_3 listed in Table 16-8I are used to compute M_s and M_c as follows.

$$M_s = m_1 \ w \ R^2 \ \theta = 0.137(50)(4^2)\left(\frac{\pi}{2}\right) = 172.16 \text{ kN.m}$$

$$M_c = m_2 \ w \ R^2 \ \theta = 0.07(50)(4^2)\left(\frac{\pi}{2}\right) = 88 \text{ kN.m}$$

$$T_{max} = m_3 \ w \ R^2 \ \theta = 0.021(50)(4^2) \left(\frac{\pi}{2} \right) = 26.4 \ kN$$

Table 16-8I Bending Moments Coefficients

# of supports	θ	m_1	m_2	m_3
4	90	0.137	0.07	0.021
5	72	0.108	0.054	0.0148
6	60	0.89	0.045	0.009
8	45	0.066	0.03	0.005
10	36	0.054	0.023	0.003
12	30	0.045	0.017	0.002

For the special case of a semi-circular beam fixed at both ends, Fig. 16-13b.

$$R_A = R_B = R_S = \left(\frac{\pi}{8} \right) wr^2$$

$$M_A = M_B = M_S = -\frac{w \ r^3}{3}$$

$$T_A = T_B = T_{max} = -0.11 \ wr^3$$

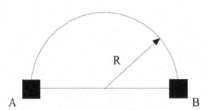

Fig. 16-13b, Semi-Circular Beam

For more details on the subject see Reference [12].

Example: 16-9

Determine the deflection at A for the cantilever sandwich beam shown in Fig. 16-14a.
[E_{Fs} = $E_{Face\ sheet}$ = 180000 MPa, E_C = E_{Core} = 240000 MPa]

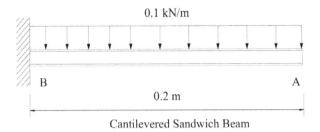

0.1 kN/m

B   A

0.2 m

Cantilevered Sandwich Beam

0.01m

0.2 m

0.01m

Face Sheet

Core

0.5 m

Beam Cross-Section

Fig. 16-13a

SOLUTION

The moment Inertia of the section, Fig. 16-13b

$$I = \frac{B\,t\,h^2}{2} = \frac{0.5\,(0.01)\,(0.21)^2}{2} = 0.00011 \quad m^4$$

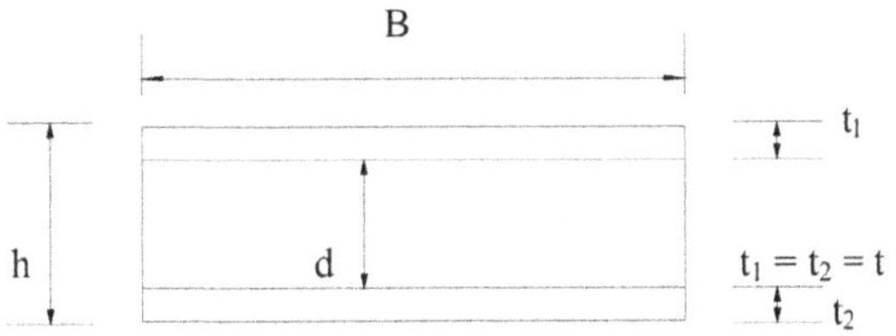

B

h

d

t_1

$t_1 = t_2 = t$

t_2

Fig. 16-13b

Chapter 16. Selected Topics

The deflection in a sandwich beam is made up of the bending deflection (Δ_B) and the shear deflection (Δ_S). Therefore the total deflection (Δ) is equal to

$$\Delta = \Delta_B + \Delta_S$$

With

$$\Delta_B = \frac{w\,L^4}{8E_{fs}I} = \frac{0.1(0.2)^4}{8(180)(0.00011)} = 0.001\text{m}$$

$$\Delta_S = \frac{0.6w\,L^2}{E_C\,B\,d} = \frac{0.6(0.1)(0.2)^2}{240(0.5)(0.2)} = 0.0001 \text{ m}$$

Thus the total deflection of the sandwich beam

$$\Delta = 0.001 + 0.0001 = 0.0011 \text{ m} \downarrow$$

For more details on the subject see Reference [15].

Example: 16-10

Determine the Deflection at A for the cantilever beam with one opening shown in Fig. 16-14a. [E = 27.4 $\times 10^6$ kN/m^2]

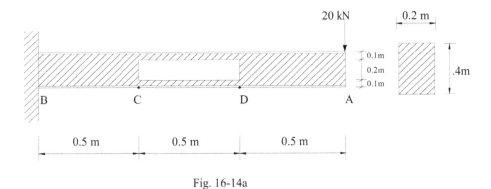

Fig. 16-14a

SOLUTION

Because of the presence of the hole, the moment of inertia and the moments are changing at C and D. This requires a new model for the beam as shown in Fig. 16-14b.

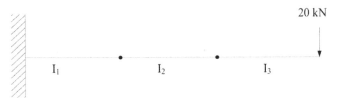

Fig. 16-14b

$$I_1 = \frac{(0.2)(0.4)^3}{12} = 0.00107 \text{ m}^4$$

$$I_2 = \frac{(0.2)(0.4)^3}{12} - \frac{(0.2)(0.2)^3}{12} = 0.00093 \text{ m}^4$$

The beam reactions, degrees of freedom, external loads, displacements, and internal forces are shown in Figs. 16-14c, 16-14d, and 16-14e.

M_B 20 kN

B_y

Fig. 16-14c, Structural Model

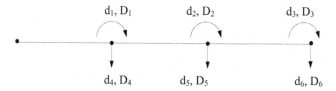

d_1, D_1 d_2, D_2 d_3, D_3

d_4, D_4 d_5, D_5 d_6, D_6

Fig. 16-14d. Degrees of Freedom

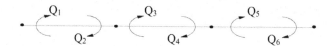

Q_1 Q_3 Q_5

Q_2 Q_4 Q_6

Fig. 16-14d. Member Internal Forces

The displacement transformation matrix [a] is equal to

$$[S] = \begin{bmatrix} 0 & 0 & 0 & -2 & 0 & 0 \\ 1 & 0 & 0 & -2 & 0 & 0 \\ 1 & 0 & 0 & 2 & -2 & 0 \\ 0 & 1 & 0 & 2 & -2 & 0 \\ 0 & 1 & 0 & 0 & 2 & -2 \\ 0 & 0 & 1 & 0 & 2 & -2 \end{bmatrix} \begin{bmatrix} d_1 \\ d_2 \\ d_3 \\ d_4 \\ d_5 \\ d_6 \end{bmatrix}$$

The element stiffness matrix [k] is equal to

$$[k] = \begin{bmatrix} 2.345 \times 10^5 & 1.173 \times 10^5 & 0 & 0 & 0 & 0 \\ 1.173 \times 10^5 & 2.345 \times 10^5 & 0 & 0 & 0 & 0 \\ 0 & 0 & 2.039 \times 10^5 & 1.019 \times 10^5 & 0 & 0 \\ 0 & 0 & 1.019 \times 10^5 & 2.039 \times 10^5 & 0 & 0 \\ 0 & 0 & 0 & 0 & 2.345 \times 10^5 & 1.173 \times 10^5 \\ 0 & 0 & 0 & 0 & 1.173 \times 10^5 & 2.345 \times 10^5 \end{bmatrix}$$

The structure stiffness matrix [K] is given by

$$[K] = \begin{bmatrix} 4.384 \times 10^5 & 1.019 \times 10^5 & 0 & -9.206 \times 10^4 & -6.116 \times 10^5 & 0 \\ 1.019 \times 10^5 & 4.384 \times 10^5 & 1.173 \times 10^5 & 6.116 \times 10^5 & 9.206 \times 10^4 & -7.036 \times 10^5 \\ 0 & 1.173 \times 10^5 & 2.345 \times 10^5 & 0 & 7.036 \times 10^5 & -7.036 \times 10^5 \\ -9.206 \times 10^5 & 6.116 \times 10^5 & 0 & 5.261 \times 10^6 & -2.446 \times 10^6 & 0 \\ -6.116 \times 10^5 & 9.206 \times 10^5 & 7.036 \times 10^5 & -2.446 \times 10^6 & 5.261 \times 10^6 & -2.815 \times 10^6 \\ 0 & -7.036 \times 10^5 & -7.036 \times 10^5 & 0 & -2.815 \times 10^6 & 2.815 \times 10^6 \end{bmatrix}$$

The displacement matrix [d] is determined as follows.

$$[d] = [k]^{-1}[D] = \begin{bmatrix} 2.132 \times 10^{-3} \\ 3.603 \times 10^{-3} \\ 4.03 \times 10^{-3} \\ 5.685 \times 10^{-4} \\ 2.043 \times 10^{-3} \\ 3.987 \times 10^{-3} \end{bmatrix}$$

where

$$[D] = \begin{bmatrix} 0 \\ 0 \\ 0 \\ 0 \\ 0 \\ 20 @ d_6 \end{bmatrix}$$

The displacement @ A = d_6 = 0.003987 m $\downarrow$

Note

The displacement at A without the beam opening is equal to

$$\delta = \frac{pL^3}{3EI} = \frac{20(1.5)^3}{3(27.4 \times 10^6)(0.00107)} = \textbf{0.00077 m} \downarrow$$

Example: 16-11

Determine the reactions at the fixed ends and the moment at the crown for the fixed parabolic arch shown in Fig. 16-15a. [$I_x = I_c \sec \theta$]

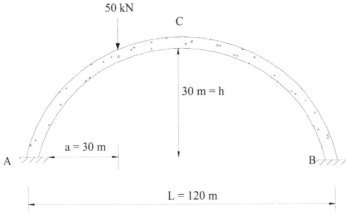

50 kN

C

30 m = h

a = 30 m

A

B

L = 120 m

Fig. 16-15a

SOLUTION

  For a parabolic arch with a concentrated load and $I_x = I_c \sec \theta$, Fig. 16-15b, we have

$$A_x = \frac{15}{4} P \frac{L}{h} k^2 \left(1 - k^2\right)$$

$$k = \frac{a}{L} = \frac{30}{120} = \frac{1}{4}$$

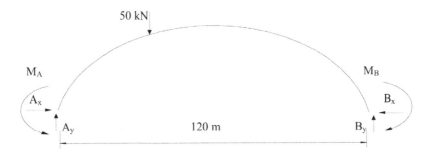

50 kN

M_A M_B

A_x B_x

A_y 120 m B_y

Fig. 16-15b

$$A_x = \frac{15}{4}(50)\left(\frac{120}{3}\right)\left(\frac{1}{4}\right)^2\left(1-\frac{1}{4}\right)^2 = 26.367 \text{ kN} \rightarrow$$

$$\sum F_x = 0 \rightarrow + \quad B_x = 26.367 \text{ kN} \leftarrow$$

$$A_y = P (1-k)^2 (1+2k)$$

$$= 50 \left(1-\frac{1}{4}\right)^2\left(1+2\left(\frac{1}{4}\right)\right)$$

$$= 42.188 \text{ kN} \uparrow$$

$$\sum F_y = 0 \uparrow + \quad B_y = 50 - 42.188 = 7.8 \text{ kN} \uparrow$$

$$M_A = \frac{PL}{2} k (1-k)^2 (2-5k)$$

$$M_A = \frac{50(120)}{2}\left(\frac{1}{4}\right)\left(1-\frac{1}{4}\right)^2\left(2-5\left(\frac{1}{4}\right)\right) = 316.4 \text{ kN.m} \quad \circlearrowleft$$

$$\sum M_B = 0 \quad +\circlearrowright \qquad M_B = 246 \text{ kN.m} \quad \circlearrowleft$$

OR

$$M_B = \frac{PL}{2} \overline{k} \left(1-\overline{k}\right)^2 \left(2-5\overline{k}\right)$$

$$\overline{k} = \frac{L-a}{L} = \frac{90}{120} = \frac{3}{4}$$

$$M_B = \frac{50(120)}{2}\left(\frac{3}{4}\right)\left(1-\frac{3}{4}\right)^2\left(2-5\left(\frac{3}{4}\right)\right) = -246 \text{ kN.m} = 246 \text{ kN.m} \quad \circlearrowleft$$

The bending moment at the crown M_C, Fig. 16-15c, is determined as follows.

$$\sum M_C = 0 \circlearrowleft + \qquad M_c = 76.13 \text{ kN.m} \qquad \circlearrowright$$

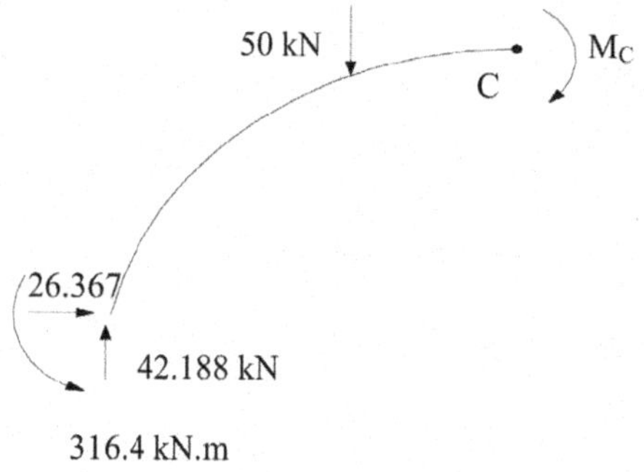

Fig. 16-15c

For more details on the subject of fixed arches see Reference [9].

Example: 16-12

Determine the design strength (M_c) of the concrete beam shown in Fig. 16-16a.
[$f'c$ = 30 MPa, F_y = 400 MPa, E = 200GPa]

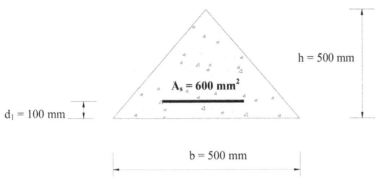

$$h = 500 \text{ mm}$$

$$A_s = 600 \text{ mm}^2$$

$$d_1 = 100 \text{ mm}$$

$$b = 500 \text{ mm}$$

Fig. 16-16a

SOLUTION

The first step is to assume that the stress f_s in the tension reinforcement equals the yield strength F_y. Based on this assumption, the tensile force becomes

$$T = A_s \times F_y$$
$$= 600(400)$$
$$\mathbf{T = 240000 \text{ N}}$$

The second step is to compute the compressive force C that satisfies the equilibrium condition (C = T) as shown in Fig. 16-16b.

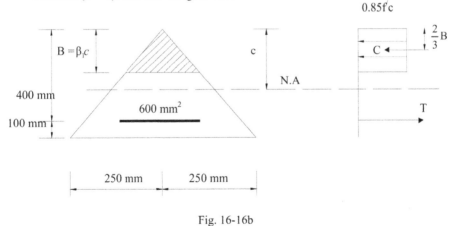

Fig. 16-16b

Because of the beam cross-section shape, the width of the compression zone B is equal to the depth of the compression zone B. Therefore, the compressive force C is equal to

$$C = 0.85 \text{ f'c } A_c = 0.85(30)\left(\frac{B^2}{2}\right)$$

$$C = 12.75 \text{ B}^2$$

Where A_c = compression zone area

Making use of equilibrium, T = C
$$240000 = 12.75 \text{ B}^2 \quad \Rightarrow \quad B = 137.2 \text{ mm}$$

The third step is to check the validity of the assumption that f_s is equal to F_y. The assumption is true if the tensile reinforcement strain ϵ_s is larger than the yield strain ϵ_y.

$$\epsilon_s = \frac{d-c}{c}(0.003) = \frac{400-161.412}{161.412}(0.003)$$

Where $c = \dfrac{B}{\beta_1} = \dfrac{137.2}{0.85} = 161.412$ mm

Thus $\epsilon_s = 0.00443$ and $\epsilon_y = \dfrac{Fy}{E} = \dfrac{400}{200000} = 0.002$

Finally $\epsilon_s = 0.00445 > 0.002 \Rightarrow f_s = F_y$ and the assumption is true.

The fourth step is to compute the beam design strength

$$M_c = \varphi M_u = \varphi \text{ As Fy}\left(d - \frac{2}{3}B\right)\left(10^{-6}\right)$$

$$= 0.9(600)(400)\left(400 - \frac{2}{3}(137.2)\right)\left(10^{-6}\right)$$

$$= \textbf{66.64 kN.m}$$

Example: 16-13

Determine the deflection at point c of the fixed plate shown in Fig. 16-17a.
[F = Flexural rigidity = 93.34]

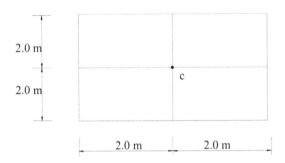

2.0 m

2.0 m

2.0 m | 2.0 m

Fig. 16-17a

SOLUTION

The structural model for a plate with fixed boundaries is shown in Fig. 16-17b.

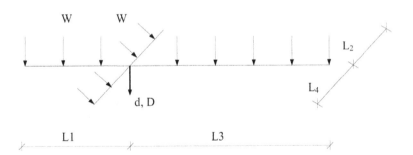

W   W

L_2

L_4

d, D

L1   L3

Fig. 16-17b. Applied Loads and Degrees of Freedom

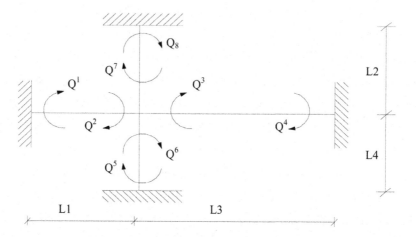

Fig. 16-17c. Member Internal Forces

The displacement transformation matrix [a] is equal to

$$[S] = \begin{bmatrix} -\dfrac{1}{L_1} \\[6pt] -\dfrac{1}{L_1} \\[6pt] -\dfrac{1}{L_3} \\[6pt] -\dfrac{1}{L_3} \\[6pt] -\dfrac{1}{L_4} \\[6pt] -\dfrac{1}{L_4} \\[6pt] -\dfrac{1}{L_2} \\[6pt] -\dfrac{1}{L_2} \end{bmatrix} [d]$$

The element stiffness matrix [k] is equal to the following:

$$[k] = \frac{F}{L} \begin{bmatrix} 4 & 2 & 0 & 0 & 0 & 0 & 0 & 0 \\ 2 & 4 & 0 & 0 & 0 & 0 & 0 & 0 \\ 0 & 0 & 4 & 2 & 0 & 0 & 0 & 0 \\ 0 & 0 & 2 & 4 & 0 & 0 & 0 & 0 \\ 0 & 0 & 0 & 0 & 4 & 2 & 0 & 0 \\ 0 & 0 & 0 & 0 & 2 & 4 & 0 & 0 \\ 0 & 0 & 0 & 0 & 0 & 0 & 4 & 2 \\ 0 & 0 & 0 & 0 & 0 & 0 & 2 & 4 \end{bmatrix}$$

The structure stiffness matrix [K] is determined as follows.

$[K] = [a]^T[k][a]$

The displacement matrix [d] is determined as follows.

$[d] = [k]^{-1}[D]$

$$[d] = \frac{L_1^2 L_2^2 L_3^2 L_4^2 w \left(L_1 + L_2 + L_3 + L_4\right)}{\left[24\left(F\left[L_3^2 L_4^2 L_2^2 + L_1^2 L_4^2 L_2^2 + L_1^2 L_3^2 L_2^2 + L_1^2 L_3^2 L_4^2\right]\right)\right]}$$

$[d] = [0.043] = 0.043 \text{ m} \downarrow$

Where $L_1 = L_2 = L_3 = L_4 = 2.0$ m and $[D] = \dfrac{w}{2}\left(L_1 + L_2 + L_3 + L_4\right)$

For more details on the deflection calculation method of plates of other boundaries and different shapes see Reference [19].

Example: 16-14

For a loading of 6 kN/m^2, determine the member stresses for the spherical dome with a base diameter of 12 m, a thickness of 100 mm, and a rise of 3.0 m shown in Fig. 16-18.

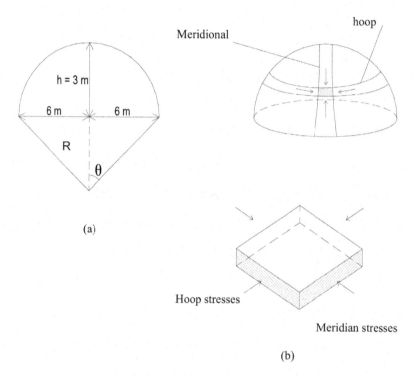

(a)

(b)

Fig. 16-18.

SOLUTION

The radius R of a spherical dome is given by the following equation:

$$R = \frac{r^2 + h^2}{2r} = \frac{6^2 + 3^2}{2(3)} = 7.5 \; m$$

Where r = base radius, h = dome rise

$$\sin\theta = \frac{r}{R} = \frac{6}{7.5} = 0.8$$

Thus θ = 53.13,   cos (θ) = 0.6

Once θ is determined the meridional stresses and hoop stresses are computed,

The meridional stresses are maximum at $\theta = 53.13^0$ => $f_m = \dfrac{meridional\ force\ (F_m)}{thickness\ (t)}$

Where $F_m = \dfrac{Rw}{1 + \cos\theta} = \dfrac{7.5(6)}{1 + 0.6} = 28.125kN$

Thus

$$f_m = \dfrac{28.125}{0.1} = 281.25 \text{ kN/m}^2$$

The hoop stresses are maximum at $\theta = 0$ => $f_h = \dfrac{Rw}{t}\left[\cos\theta - \dfrac{1}{1 + \cos\theta}\right]$

$$= \dfrac{7.5(6)}{0.1}\left[1 - \dfrac{1}{1 + 1}\right] = 225 \text{ kN/m}^2$$

For more details on this subject see Reference [4]

Simple Beam

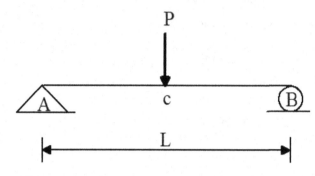

Beam Data

$P \equiv 100$	kN		" Concentrated Load "
$L \equiv 5$	m		" Span Length in meters "
$E \equiv 200 \cdot 10^6$	$\dfrac{kN}{m^2}$		" Modulus of Elasticity "
$I \equiv 400 \cdot 10^{-6}$	m^4		" Moment of Inertia "

Beam Analysis

$$\delta c := \frac{P \cdot L^3}{48 \cdot E \cdot I}$$

$\delta c = 3.255 \times 10^{-3}$ m   " Central Deflection "

$$Mc := \frac{P \cdot L}{4}$$

$Mc = 125$ kN·m   " Central Moment "

Fig. 16 - 14a

Fig. 16-19

Design Strength of Triangular Section with Equal Base (b) and Hight (h)

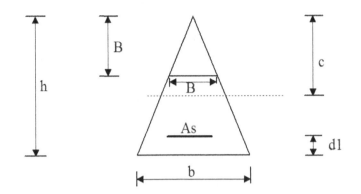

Design Data

fc ≡ 30	MPa	" Compressive Srength of Concrete "
fy ≡ 400	MPa	" Yield Strength of Steel Reinforcement "
E ≡ 200000	MPa	" Modulus of Elasticity "
d1 ≡ 100	mm	" Distance from Extreme tension Fiber to Centriod of Tension Steel "
f'c ≡ 30	MPa	" Compressive Srength of Concrete
As ≡ 600	mm^2	" Area of Tension Steel "
h ≡ 500	mm	" Total Depth of the Beam "
b ≡ 500	mm	" Beam Width "

Design Calculation (fs = fy   Assumed)

$d := h - d1$	$d = 400$	mm	" Depth to tension Steel "
$T := As \cdot fy$	$T = 2.4 \times 10^5$	N	"Tension Force "
$B := \sqrt{\dfrac{T \cdot 2}{0.85 f'c}}$	$B = 137.199$	mm	" Depth of Compression Area "
$Mc := 0.9 \cdot As \cdot fy \left(d - \dfrac{2}{3} \cdot B \right) \cdot 10^{-6}$	$Mc = 66.643$	kN·m	" Design Strength of the Beam "

Fig. 16-20

Simplified Method for Plate Deflection Calculation

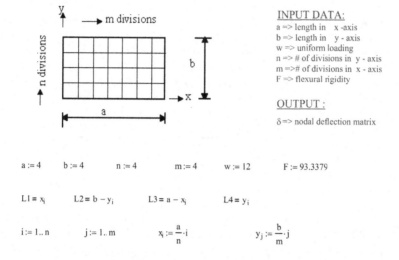

INPUT DATA:

a => length in x -axis
b => length in y - axis
w => uniform loading
n => # of divisions in y - axis
m =># of divisions in x - axis
F => flexural rigidity

OUTPUT :

δ => nodal deflection matrix

a := 4   b := 4   n := 4   m := 4   w := 12   F := 93.3379

$L1 = x_i$ $L2 = b - y_i$ $L3 = a - x_i$ $L4 = y_i$

$i := 1 .. n$ $j := 1 .. m$ $x_i := \dfrac{a}{n} \cdot i$ $y_j := \dfrac{b}{m} \cdot j$

Fixed Supported Boundaries:

$$d_{i,j} := \frac{\left[(x_i)^2 \cdot (b - y_j)^2 \cdot (a - x_i)^2 \cdot (y_j)^2 \right] \cdot w \cdot \left[(x_i + y_j) + (a - x_i) + (b - y_j) \right]}{24 \left[F \cdot \left[(a - x_i)^2 \cdot (y_j)^2 \cdot (b - y_j)^2 + (x_i)^2 \cdot (y_j)^2 \cdot (b - y_j)^2 + (a - x_i)^2 \cdot (x_i)^2 \cdot (b - y_j)^2 + (a - x_i)^2 \cdot (x_i)^2 \cdot (y_j)^2 \right] \right]}$$

$$d = \begin{pmatrix} 0 & 0 & 0 & 0 & 0 \\ 0 & 0.019 & 0.027 & 0.019 & 0 \\ 0 & 0.027 & 0.043 & 0.027 & 0 \\ 0 & 0.019 & 0.027 & 0.019 & 0 \\ 0 & 0 & 0 & 0 & 0 \end{pmatrix}$$

Fig. 16-21

APPENDICES

OBJECTIVE QUESTIONS

1. Define the following terms: load, force, shear, moment, deflection, center of gravity, moment of inertia, radius of gyration, section modulus, support reaction, and equilibrium.
2. What is the difference between a center of gravity and a centroid.
3. State and explain the principle of virtual work.
4. How will you apply the principle of virtual work to analyze trusses, beams, and frames?
5. What the methods that are used to determine truss bar forces ?
6. What is the principle of superposition?
7. What is Poison's ratio?
8. Explain briefly the relationship between a shear force and a bending moment at a section.
9. What is a spring? What are its uses?
10. What is a column?
11. Explain the failure of long and short columns.
12. What is the relationship between the slope deflection and the radius of curvature.
13. What is the area moment method?
14. What is the conjugate beam method?
15. What is a simple beam, a fixed beam, a cantilever beam, a continuous beam, a curved beam, a cable, and an over-hang beam?
16. What is the three moment method?
17. Define the force method and explain its uses.
18. Define the slope deflection method and explain its uses.
19. Define the moment distribution method and explain its uses.
20. State the differences between symmetrical and unsymmetrical frames.
21. What is the stiffness factor, the distribution factor, the carry-over factor? What is their importance in the moment distribution method?
22. What is meant by side-sway in frames?
23. Define boundary conditions and their uses.
24. What is a statically indeterminate structure?
25. What is a redundant?
26. Explain internal and external statical indeterminacy.
27. What is the relationship between the force and the flexibility methods?
28. What is the relationship between the slope deflection and the stiffness methods?
29. Why a carry-over factor for a prismatic member is different from that of a non prismatic member?
30. What is the difference between plane and space structures?
31. What is the difference between a three-hinged arch, a two-hinged arch, and a fixed arch?
32. What are the internal forces at any section of an arch?
33. What is the difference between a curved beam in plan and an arch?

34. What is the maximum moment in a simply-supported beam? Is it positive or negative moment?
35. How do you relate steel placement in a beam to its bending moment?
36. What are stirrups ? Explain their uses.
37. What are the functions of beams, columns, and footings?
38. What is the difference between plates and shells?

GLOSSARY

Absolute flexural stiffness: The moment that produces a rotation of 1 radian at a pin-supported end of a beam whose far end is fixed.

Abutment: A wall or a vertical element that transfers applied loads to the foundation.

Beam: A member that supports loads that are acting transverse to the member's axis.

Bending moment: Algebraic sum of the moments of all of the external forces to one side or the other of a particular section in a member, the moments being taken about an axis passing through the centroid of the section.

Beam-column: Columns subjected to both axial forces and moments.

Box beam: A hollow beam of rectangular shape. By eliminating material at the center of the member, the weight is reduced, while the bending stiffness is not significantly affected.

Braced frame: A frame that has resistance to lateral loads supplied by some type of auxiliary bracing.

Building code: A set of provisions that controls design and construction in a given region. Its provisions establish the minimum architectural, structural, mechanical, and electrical design requirements for buildings and other structures.

Cantilever: A projecting or overhanging beam.

Castigliano's theorem: Energy methods for computing deformations and for analyzing statically indeterminate structures.

Column: A structural member whose primary function is to support compressive loads.

Concrete: A mixture of sand, gravel, crushed rock, or other aggregates held together in a rocklike mass with a paste of cement and water.

Conjugate beam: An imaginary beam that has the same length as a real beam being analyzed, and that has a set of boundary and internal continuity conditions such that the slopes and deflections in the real beam equal the shear and moment in the fictitious beam when it is loaded with the M/EI diagram.

Dead load: Loads of constant magnitude that remain in one position. Examples are weights of walls, floors, roofs, fixtures, structural frames, and so on.

Elastic behavior: A linear relationship between stress and strain. When the external forces are removed, an elastic member will return to its original length.

Factored load: Loads obtained by multiplying design loads by load factors.

Free-body diagram: A sketch of a structure or a part of a structure showing all forces and dimensions required for an analysis.

Fixed-end moment: The moments at the ends of loaded members when the member joints are clamped to prevent rotation.

Flat plate: Solid concrete floor or slabs of uniform depths that transfer loads directly to supporting columns without the presence of beams or capital or drop beams.

Girder: A large beam that supports one or more secondary beams.

Gravity load: see dead load.

Indeterminate structure: A structure whose reactions and internal forces cannot be determined by using statics equations only.

Inertia forces: Dynamic forces acting on a moving body and defined as the product of the body mass and its acceleration..

Influence line: A diagram whose ordinates show the magnitude and character of some function of a structure (shear, moment, etc.) as a unit load moves across the structure.

Kinetic energy: Energy gained by a moving body and defined as the product of the body mass and the square of its velocity.

Principle of least work: The internal work accomplished by each member or each portion of a structure subjected to a set of external loads is the least possible amount necessary to maintain equilibrium in supporting the loads.

Live loads: Loads that change position and magnitude. They move or are moved . Examples are trucks, people, furniture, and so on.

Modulus of elasticity: A measure of a material's stiffness. It is defined as the ratio of stresses over strain.

Modulus of rupture: The flexural tensile strength of concrete.

Moment distribution: A successive correction or iteration method of analysis whereby fixed end and/or sidesway moments are balanced by a series of corrections.

One-way slab: A slab that mainly deflects along its short direction .

Plane frame: A frame that for purposes of analysis and design is assumed to lie in a single (or two dimensional) plane.

Plastic deformation: Permanent deformation occurring in a member after reaching its yield strength.

Principle of superposition: If a structure is linearly elastic, the forces acting on the structure may be separated in any convenient fashion and the structure analyzed for the separate cases. The final results can be obtained by adding together the individual parts.

Qualitative influence line: A sketch of an influence line in which no numerical values are given.

Quantitative influence line: An influence line that shows numerical values.

Reinforced concrete: A combination of concrete and steel reinforcing wherein the steel provides the tensile strength lacking in the concrete. The steel reinforcing also can be used to help the concrete resist compressive forces.

Section modulus: A property of the cross-sectional area that measures a member's capacity to carry moment.

Seismic load: Loads produced by earthquake ground motions .

Service loads: The actual loads that are assumed to be applied to a structure when it is in service (also called working loads).

Serviceability: Pertains to the performance of structure under service loads and is concerned with such items as deflections, vibrations, cracking, and slipping.

Shear: The algebraic summation of the external forces in a member to one side or the other of a particular section that are perpendicular to the axis of the member.

Sidesway: The lateral movement of a structure caused by unsymmetrical loads and/or by an unsymmetrical arrangement of the members of the structure.

Slope deflection: A classical method of analyzing statically indeterminate structure in which the moments at the ends of the members are expressed in terms of the rotations (or slopes) and deflections of the joints.

Space truss: A three-dimensional truss.

Statically determinate structure: Structures for which the equations of equilibrium are sufficient to compute all of the external reactions and internal forces.

Statically indeterminate structure: Structures for which the equations of equilibrium are insufficient for computing the external reactions and internal forces.

Steel: An alloy consisting almost entirely of iron (usually over 98 percent). It also contains small quantities of carbon, silicon, manganese, sulfur, phosphorus, and other elements.

Stirrups: Vertical reinforcement added to reinforced concrete beams to increase their shear capacity.

Strain: The ratio of change in length divided by the original length.

Stress: Force per unit area.

Strength design: A method of design where the estimated dead and live loads are multiplied by certain load or safety factored.

Structural analysis: The computation of the forces and deformations of structures under load.

Superposition principle: See principle of superposition.

Three-moment theorem: A classical theorem that presents the relationship between the moments in the different supports of a continuous beam.

Truss: A structure formed by a group of members arranged in the shape of one or more triangles.

Unbraced frame: A frame whose resistance to lateral forces is provided by its members and their connections.

Unstable equilibrium: A support situation whereby a structure is stable under one arrangement of loads but is not stable under other load arrangements.

Virtual displacement: A fictitious displacement imposed on a structure.

Virtual work: The work performed by a set of real forces during a virtual displacement.

Working loads: See Service loads.

Yield stress: The stress at which there is a decided increase in the elongation or strain in a member without a corresponding increase in stress.

Zero bars: A bar of a truss that remains unstressed under a particular loading condition.

Table A-1   Steel Reinforcement Ratios

F_y(MPa)	300			400		
F'c(MPa)	**20**	**30**	**40**	**20**	**30**	**40**
ρ_b	0.0321	0.0482	0.0582	0.0217	0.0325	0.0393
$0.35\rho_b$	0.1123	0.0168	0.0204	0.0076	0.0114	0.0137
$0.5\rho_b$	0.0160	0.0241	0.0291	0.1085	0.0163	0.0196
$0.75\rho_b$	0.0241	0.0361	0.0436	0.0163	0.0244	0.0295

Table A-2   Minimum Thickness (ACI)

	Simply supported	One end continuous	Both ends continuous	Cantilever
Solid one-way slabs	$\dfrac{L}{20}$	$\dfrac{L}{24}$	$\dfrac{L}{28}$	$\dfrac{L}{10}$
Beams or ribbed one-way slabs	$\dfrac{L}{16}$	$\dfrac{L}{18.5}$	$\dfrac{L}{21}$	$\dfrac{L}{8}$

Table A-3   Nominal Flexural Strength of Rectangular Sections.

F_y(MPa)	300			350			400		
F'c(MPa)	20	30	40	20	30	40	20	30	40
ρ_b	0.0321	0.0482	0.0582	0.0261	0.0391	0.0472	0.0217	0.0325	0.0393
P					q_n	(MPa)			
0.0005	0.15	0.15	0.15	0.17	0.17	0.20	0.20	0.20	0.20
0.0010	0.30	0.30	0.30	0.35	0.35	0.35	0.40	0.40	0.40
0.0015	0.44	0.45	0.45	0.52	0.52	0.52	0.59	0.59	0.59
0.0020	0.59	0.59	0.59	0.69	0.69	0.69	0.78	0.79	0.79
0.0025	0.73	0.74	0.74	0.85	0.86	0.86	0.97	0.98	0.99
0.0030	0.88	0.88	0.89	1.02	1.03	1.03	**1.16**	**1.17**	**1.18**
0.0035	1.02	1.03	1.03	**1.18**	**1.20**	**1.20**	1.34	1.36	1.37
0.0040	1.16	1.17	1.18	1.34	1.36	1.37	1.52	1.55	1.56
0.0045	**1.30**	**1.31**	**1.32**	1.50	1.53	1.54	1.7	1.74	1.75
0.0050	1.43	1.46	1.47	1.66	1.69	1.70	1.88	1.92	1.94
0.0055	1.57	1.60	1.61	1.82	1.85	1.87	2.06	2.11	2.13
0.0060	1.70	1.74	1.75	1.97	2.01	2.04	2.23	2.29	2.32
0.0065	1.84	1.86	1.89	2.12	2.17	2.20	2.40	2.47	2.50
0.0070	1.97	2.01	2.04	2.27	2.33	2.36	2.57	2.65	2.68
0.0075	2.10	2.15	2.18	2.42	2.49	2.52	2.74	2.82	2.87
0.0080	2.23	2.29	2.32	2.57	2.65	2.68	2.90	3.00	3.05
0.0085	2.36	2.42	2.45	2.71	2.80	2.84	3.06	3.17	3.23
0.0090	2.49	2.56	2.59	2.86	2.96	3.00	3.22	3.35	3.41
0.0095	2.61	2.69	2.73	3.00	3.11	3.16	3.38	3.52	3.59
0.0100	2.74	2.82	2.87	3.14	3.26	3.32	3.53	3.69	3.76
0.0105	2.86	2.96	3.00	3.28	3.41	3.48	3.68	3.85	3.94
0.0110	2.98	3.09	3.14	3.41	3.56	3.63	3.83	4.02	4.12
0.0115	3.10	3.22	3.27	3.55	3.71	3.79	3.98	4.19	4.29
0.0120	3.22	3.35	3.41	3.68	3.85	3.94	4.12	4.35	4.46
0.0125	3.34	3.47	3.54	3.81	4.00	4.09	4.26	4.51	4.63
0.0130	3.45	3.60	3.68	3.94	4.14	4.25	4.40	4.67	4.80
0.0135	3.57	3.73	3.81	4.07	4.29	4.40	4.54	4.83	4.97
0.0140	3.68	3.85	3.94	4.19	4.43	4.55	4.68	4.99	5.14
0.0145	3.79	3.98	4.07	4.32	4.57	4.70	4.81	5.14	5.31
0.0150	3.90	4.10	4.20	4.44	4.71	4.84	4.94	5.29	5.47
0.0155	4.01	4.23	4.33	4.56	4.85	4.99	5.07	5.45	5.63
0.0160	4.12	4.35	4.46	4.68	4.99	5.14	5.20	5.60	5.80
0.0165	4.23	4.47	4.59	4.79	5.12	5.28	5.32	5.75	5.96
0.0170	4.34	4.59	4.72	4.91	5.26	5.43	5.44	5.89	6.12
0.0175	4.44	4.71	4.84	5.02	5.39	5.57	5.56	6.04	6.28
0.0180	4.54	4.83	4.97	5.13	5.52	5.72	5.68	6.18	6.44
0.0185	4.64	4.95	5.10	5.24	5.65	5.86	5.79	6.33	6.59
0.0190	4.74	5.06	5.22	5.35	5.78	6.00	5.90	6.47	6.75
0.0195	4.84	5.18	5.35	5.45	5.91	6.14	6.01	6.61	6.91
0.0200	4.94	5.29	5.47	5.56	6.04	6.28	6.12	6.75	7.06
0.0205	5.04	5.41	5.59	5.66	6.17	6.42	6.22	6.88	7.21
0.0210	5.13	5.52	5.72	5.76	6.29	6.56	6.32	7.02	7.36
0.0215	5.23	5.63	5.84	5.86	6.41	6.69	6.42	7.15	7.51
0.0220	5.32	5.75	5.96	5.96	6.54	6.83	6.52	7.28	7.66
0.0225	5.41	5.86	6.08	6.05	6.66	6.96	6.62	7.41	7.82
0.0230	5.50	5.97	6.20	6.14	6.78	7.10	6.71	7.54	7.96
0.0235	5.59	6.08	6.32	6.24	6.90	7.23	6.80	7.67	8.10

632

0.0240	**5.68**	6.18	6.44	6.32	7.02	7.36	6.89	**7.79**	8.24
0.0245	5.76	6.29	6.56	6.41	7.13	7.49	6.98	7.92	8.39
0.0250	5.85	6.40	6.67	6.50	7.25	7.62	7.06	8.04	8.53
0.0255	5.93	6.50	6.79	6.58	7.36	7.75	7.14	8.16	8.67
0.0260	6.01	6.61	6.91	6.66	7.48	7.88	7.22	8.28	8.81
0.0265	6.09	6.71	7.02	6.74	7.59	8.01	7.30	8.40	8.95
0.0270	6.17	6.81	7.14	6.82	7.70	8.14	7.37	8.51	9.08
0.0275	6.25	6.92	7.25	6.90	7.81	8.26	7.44	8.63	9.22
0.0280	6.32	7.02	7.36	6.98	7.92	8.39	7.51	8.74	9.36
0.0285	6.40	7.12	7.47	7.05	8.02	8.51	7.58	8.85	9.49
0.0290	6.47	7.22	7.59	7.12	**8.13**	8.63	7.64	8.96	9.62
0.0295	6.55	7.31	7.70	7.19	8.23	8.76	7.70	9.07	**9.75**
0.0300	6.62	7.41	7.81	7.26	8.34	8.88	7.76	9.18	9.88
0.0305	6.69	7.51	7.92	7.32	8.44	9.00	7.82	9.28	10.01
0.0310	6.76	7.60	8.03	7.39	8.54	9.12	7.88	9.39	10.14
0.0315	6.82	7.70	8.14	7.45	8.64	9.24	7.93	9.49	10.27
0.0320	6.89	7.79	8.24	7.51	8.74	9.36	7.98	9.59	10.39
0.0325	6.95	7.89	8.35	7.57	8.84	9.47	8.03	9.69	10.51
0.0330	7.02	7.98	8.46	7.63	8.93	9.59	8.03	9.78	10.64
0.0335	7.08	8.07	8.56	7.68	9.03	9.70	8.12	9.88	10.76
0.0340	7.14	8.16	8.67	7.74	9.12	9.82	8.16	9.97	10.88
0.0345	7.20	8.25	8.77	7.79	9.22	9.93	8.20	10.07	11.00
0.0350	7.26	8.34	8.88	7.84	9.31	**10.04**	8.24	10.16	11.12
0.0355	7.31	8.43	8.98	7.88	9.40	10.15	8.27	10.25	11.23
0.0360	7.37	**8.51**	9.08	7.93	9.49	10.27	8.30	10.33	11.35
0.0365	7.42	8.62	9.19	7.97	9.57	10.37	8.33	10.42	11.47
0.0370	7.48	8.68	9.29	8.02	9.66	10.48	8.36	10.51	11.58
0.0375	7.53	8.77	9.39	8.06	9.75	10.59	8.38	10.59	11.69
0.0380	7.58	8.85	9.49	8.10	9.83	10.70	8.40	10.67	11.80
0.0385	7.63	8.93	9.59	8.13	9.91	10.80	8.42	10.75	11.91
0.0390	7.67	9.02	9.69	8.17	10.00	10.91	8.44	10.83	112.02
0.0395	7.72	9.10	9.78	8.20	10.08	11.01	8.46	10.91	12.13
0.0400	7.76	9.18	9.88	8.24	10.16	11.12	8.47	10.98	12.24
0.0405	7.81	9.26	9.98	8.27	10.24	11.22	8.48	11.05	12.34
0.0410	7.85	9.33	10.08	8.29	10.31	11.32	8.49	11.13	12.44
0.0415	7.89	9.41	10.17	8.32	10.39	11.42	8.50	11.20	12.55
0.0420	7.93	9.49	10.27	8.34	10.46	11.52	8.50	11.27	12.65
0.0425	7.97	9.56	10.36	8.37	10.54	11.62	8.50	11.33	12.75
0.0430	8.01	9.64	10.45	8.39	10.61	11.72	8.50	11.40	12.85
0.0435	8.04	9.71	**10.55**	8.41	10.68	11.82	8.50	11.46	12.95
0.0440	8.08	9.78	10.64	8.42	10.75	11.91	8.49	11.53	13.04
0.0445	8.11	9.86	10.73	8.44	10.82	12.01	8.48	11.59	13.14
0.0450	8.14	9.93	10.82	8.45	10.89	12.10	8.47	11.65	13.24

Table A-4 . Reinforcing Steel Bars' Cross-Sectional Area A_s (mm^2)

Diameter	1	2	3	4	5	6	7	8	9
6	28.29	56.57	84.86	113.14	141.43	169.71	198.00	226.29	254.57
8	50.29	100.57	150.86	201.14	251.43	301.71	352.00	402.29	452.57
10	78.57	157.14	235.71	314.29	392.86	471.43	550.00	628.57	707.14
12	113.14	226.29	339.43	452.57	565.71	687.86	792.00	905.14	1018.3
14	154.00	308.00	462.00	616.00	770.00	924.00	1078.0	1232.0	1386.0
16	201.14	402.29	603.43	804.57	1005.71	1206.86	1408.0	1609.1	1810.3
18	254.57	509.14	763.71	1018.29	1272.86	1527.43	1782.0	2036.5	2291.1
20	314.29	628.57	942.86	1257.14	1571.43	1885.71	2200.0	2514.2	2828.5
22	380.29	760.57	1140.88	1521.14	1901.43	2261.71	2662.0	3042.3	3422.5
24	452.57	905.14	1357.71	1810.29	2262.86	2715.43	3168.0	3620.5	4073.1
26	531.14	1062.29	1593.43	2124.57	2655.71	3186.86	3718.0	4249.1	4780.3
28	616.00	1232.00	1848.00	2464.00	3080.00	3696.00	4312.0	4928.0	5544.0
30	707.14	1414.29	2121.43	2828.57	3535.71	4242.86	4950.0	5657.1	6364.3
32	804.57	1609.14	2413.71	3218.29	4022.86	4827.43	5632.0	6436.5	7241.1
34	908.29	1816.57	2724.86	3633.14	4541.43	5449.71	6358.0	7266.3	8174.5
36	1018.29	2036.57	3054.86	4073.14	5091.43	6109.71	7128.0	8146.3	9164.5

ANSWERS TO SELECTED PROBLEMS

CHAPTER 2

2.1 $Ay = 65$ K $\uparrow$, $M_A = 475$ K.ft $\curvearrowright$, $Ax = 0.0$

2.2 $Ay = 44$ K $\uparrow$, $M_A = 312$ K.ft $\curvearrowright$, $Ax = 0.0$

2.3 $Ay = 168$ K $\uparrow$, $M_A = 720$ K.ft $\curvearrowright$, $Ax = 0.0$

2.4 $Ay = 16.25$ K $\uparrow$, $By = 31.25$ K $\uparrow$, $Ax = 0.0$

2.5 $Ay = 5$ K $\downarrow$, $By = 5$ K $\uparrow$, $Ax = 7.5$ K $\leftarrow$, $Bx = 2.5$ K $\leftarrow$

2.6 $Ay = 6.96$ K $\uparrow$, $R_B = 3.8$ K $\nearrow$, $Ax = 2.28$ K $\leftarrow$

2.7 $Ay = 63$ K $\uparrow$, $By = 39$ K $\uparrow$, $Ax = 34.75$ K $\rightarrow$, $B_x = 74.75$ K $\leftarrow$

2.8 $Ay = 30$ kN $\uparrow$, $Cy = 140$ kN $\uparrow$, $Ax = 80$ kN $\leftarrow$

2.9 $Ay = 17.33$ kN $\uparrow$, $By = 18.667$ kN $\uparrow$, $Ax = 13$ kN $\rightarrow$, $Bx = 11$ kN $\rightarrow$

2.10 $Ay = 4.96$ kN $\uparrow$, $M_A = 50$ kN.m $\curvearrowright$, $Ax = 5$ kN $\leftarrow$, $Bx = 0.0$, $By = 7.04$ kN $\uparrow$

2.11 $Ay = 86.26$ kN $\uparrow$, $Tc = 72.9$ kN $\uparrow$, $Ax = 1.68$ kN $\leftarrow$

2.12 $Ay = 24.793$ kN $\uparrow$, $M_A = 497.92$ kN.m $\curvearrowright$, $Ax = 25$ kN $\leftarrow$, $Bx = 0.0$, $By = 35.208$ $\uparrow$

2.13 $Ay = 237$ kN $\uparrow$, $Cy = 32$ kN $\uparrow$, $Ax = 255.38$ kN $\leftarrow$, $Bx = 175.38$ kN $\rightarrow$

2.14 $Ay = 222.67$ kN $\uparrow$, $Tc = 303.406$ kN $\uparrow$, $Ax = 134.76$ kN $\leftarrow$, $R_B = 248.46$ kN

CHAPTER 3

3.2 $F_{BC} = 37.72$ kN $\underline{C}$, $F_{FG} = 6.67$ kN $\underline{T}$, $F_{CG} = 0.0$

3.3 $F_{AB} = 26.667$ kN $\underline{C}$, $F_{AE} = 66.667$ kN $\underline{C}$, $F_{EC} = 33.338$ kN $\underline{T}$, $F_{BE} = 0.0$, $F_{DE} = 0.0$

3.5 $F_{CD} = 70$ kN $\underline{C}$, $F_{IK} = 20.62$ kN $\underline{T}$, $F_{BA} = F_{CJ} = F_{EK} = 0.0$

3.6 $Ay = 62$ kN $\uparrow$, $By = 52$ kN $\uparrow$, $Ax = 48$ kN $\rightarrow$

3.7 $F_{BC} = 9.321$ kN $\underline{C}$, $F_{IC} = 11.783$ kN $\underline{C}$, $F_{IH} = 0.0$

3.8 $F_{HG} = 30$ kN $\underline{C}$, $F_{DE} = F_{FE} = 0.0$

3.9 $F_{AB} = 71.43$ kN $\underline{T}$, $F_{BC} = 80.82$ kN $\underline{C}$, $Ay = 57.14$ kN $\downarrow$, $Cy = 57.14$ kN $\uparrow$, $Ax = 42.86$ kN $\leftarrow$, $Cx = 57.14$ kN $\leftarrow$

3.10 $F_{BH} = 33.33$ k $\underline{C}$, $Ay = 40$ k $\uparrow$, $Cy = 150$ k $\uparrow$, $Cx = 60$ k $\rightarrow$, $Mc = 1125$ k-ft $\curvearrowright$

3.11 $Ay = 40$ kN $\uparrow$, $Ey = 70$ kN $\uparrow$, $M_E = 405$ kN.m $\curvearrowright$

CHAPTER 4

4.1 $M = 281.25$ kN.m $\curvearrowright$ at $x = 12.5$m from A.

4.5 $M = 12.64$ kN.m $\curvearrowright$ at $x = 3.33$ m from A.

4.6 $M = 16.81$ kN.m $\curvearrowright$ at $x = 3.06$ m from A.

4.8 $M = 23.08$ kN.m $\curvearrowright$ at $x = 9.06$m from A.

4.10   Vc = 17.67 kN ↓ , Mc = 73.258 kN.m ↻

4.11   M = 259.2 kN.m ↻ at 7.8m from A.

4.12   Vc = 8.4 k↓ , Mc = 26 k-ft ↻

4.14   Vc = 82.167 k ↓ , Mc = 96.501 k-ft ↻

4.16   Vc = 81.43 k ↓ , Mc = 222.84 k-ft ↻

4.17 V_F = 60 kN ↑ , M_F = 160 kN.m ↻

4.18   Vc = 40 kN↑ ,   Mc = 200 kN.m ↻

4.20   Vc = 7 kN↓ , Mc = 21 kN.m ↻

4.21   Vc = 3 kN↓ , Mc = 29 kN.m ↻ , V_D = 33 kN↓ , M_D = 25 kN.m ↻

4.22 V_D = 2 kN↓ , M_D = 35 kN.m ↻ , V_{ED} = 12 kN↓ , M_E = 60 kN.m ↻

4.23 V_C = 11.5 kN ↓ , Mc = 0.0 , V_{CD} = 36.5 kN↓ , M_D = 120 kN.m ↻

4.24 V_B = 25 kN ↓ , M_B = 225 kN.m ↻ , V_C = 25 kN↓ , M_C = 150 kN.m ↻

4.25 V_D = 20 kN ↓ , M_D = 0.0 , V_B = 120 kN↓ , M_B = 350 kN.m ↻

CHAPTER 5

5.1 δ_B = 0.3125 m ↓

5.3 δ_B = 0.333 m ↓

5.5 δ_B = 0.054 m ↓

5.6 δ_B = 5.586 in ↓ , θ_B = 0.0621 rad ↻

5.8 δ_B = 0.063 m ↓ , δ_C = 0.1845 m ↓ .

5.9 δ_B = 0.137 m ↓ , θ_B = 0.019 rad ↻

5.12 δ_C = 3.2 in ↓ , θ_B = 0.013 rad ↻

5.14 δ_B = 0.072 m ↓ , δ_C = 0.1872 m ↓

5.15 δ_D = 0.0336 m ↓ , θ_B = 0.0036 rad ↻

5.17 δ_B = 0.056 m ↓ , θ_B = 0.007 rad ↻

5.20 δ_B = 0.0156 m ↓ , θ_B = 0.0026 rad ↻

5.23 δ_B = 0.621 in ↓ , δ_D = 2.9 in ↓

5.25 δ_A = 0.01 m ↓ , θ_E = 0.00208 rad ↻ δ_C = 0.0

5.27 δ_B = 0.0256 m↑ , θ_D = 0.02108 rad ↻ , δ_D = 0.0964 m ↓

5.29 δ_B = 0.104 m↓ , θ_B = 0.0

5.30 δ_B = 0.015625 m↓ , θ_B = 0.003125 rad ↻

5.31 δ_B = 0.03125 m↓ , θ_D = 0.00625 rad ↻ , δ_D = 0.1875 m↑ .

5.33 $\delta_B = 0.021$ m $\uparrow$

5.35 $\delta_D = 0.000241$ m $\uparrow$, $\theta_D = 0.000125$ rad $\curvearrowright$

5.39 $\delta_C = 0.0059$ m $\uparrow$, $\theta_C = 0.005$ rad $\curvearrowright$

5.40 $\delta_C = 0.0888$ m $\downarrow$

5.42 $\delta_{V@B} = 0.00073$ in $\downarrow$, $\delta_{H@B} = 0.00015$ in $\rightarrow$

5.43 $\delta_{V@B} = 0.1333$ mm $\downarrow$, $\delta_{H@B} = 0.295$ mm $\rightarrow$

5.46 $\delta_{V@B} = 0.092$ in $\downarrow$, $\delta_{H@C} = 0.052$ in $\rightarrow$

5.48 $\delta_{V@C} = 0.0052$ m $\downarrow$

5.52 $\delta_{V@C} = 0.026$ m $\downarrow$

5.53 $\delta_B = 0.79167$ m $\downarrow$, $\theta_B = 0.1167$ rad $\curvearrowright$

CHAPTER 6

6.2 $R_B = 77.305$ kN $\uparrow$, $R_E = 76.702$ kN $\uparrow$, $M_D = 147$ kN.m $\curvearrowright$

6.3 $R_B = 100.4$ kN $\uparrow$, $R_D = 4.6$ kN $\uparrow$, $Mc = 17.5$ kN.m $\curvearrowleft$

6.4 $Mc = 81.25$ kN.m $\curvearrowright$, $M_D = 68.125$ kN.m $\curvearrowright$

6.5 $R_A = 43.75$ kN $\uparrow$, $M_D = 48.75$ kN.m $\curvearrowright$

6.8 $M_D = 120.0$ kN.m $\curvearrowleft$, $M_E = 37.5$ kN.m $\curvearrowleft$

CHAPTER 7

7.4 $Ay = 17.5$ kN $\uparrow$, $By = 42.5$ kN $\uparrow$, $Ax = 35$ kN $\rightarrow$, $Bx = 45$ kN $\rightarrow$

7.6 $Ay = 57.27$ kN $\uparrow$, $Dy = 72.73$ kN $\uparrow$, $Ax = 218.19$ kN $\leftarrow$, $Dx = 218.19$ kN $\rightarrow$, $S = 2.625$ m

7.7 $L_1 = 55.848$ m, $d_c = 160.68$ mm, $L_c = 101.141$ m

CHAPTER 9

9.2 $M_B = 30$ kN.m $\curvearrowright$, M at 2.367 m from A = 9.965 kN.m $\curvearrowright$

9.4 $Ay = 26.6$ kN $\uparrow$, $By = 73.4$ kN $\uparrow$, $M_A = 633$ kN.m $\curvearrowleft$

9.5 $Ay = 32.4$ kN $\uparrow$, $By = 39.6$ kN $\uparrow$, $M_A = 100.8$ kN.m $\curvearrowleft$

9.7 M at 3.125m from A = 19.6 kN.m $\curvearrowright$

9.8 $Ay = 457.55$ kN $\uparrow$, $By = 118.45$ kN $\uparrow$, $M_A = 2745$ kN.m $\curvearrowleft$

9.10 $Ay = 50.78$ kN $\uparrow$, $By = 24.22$ kN $\uparrow$, $M_A = 174.36$ kN.m $\curvearrowleft$

9.11   Ay = 30 kN ↑, Tc = 20 kN, M_A = 100 kN.m ↶

9.12   Ay = 51.14 kN ↑, By = 2.86 kN ↑, M_A = 90.84 kN.m ↶

9.13   By = 10.775 kN ↑, Ay = 44.6125 kN ↑, Cy = 44.6125 kN ↑

9.15 M_B = 129 kN.m ↷

9.17 F_{BD} = 25.58 kN **T**

9.19   Ay = 4.38 kN ↑, By = 6.24 kN ↑, Cy = 0.62 kN ↓

9.22 F_{BD} = 50 kN **T**

9.23   Mc = 160.032 kN.m ↶, M under the load = 346.655 kN.m ↷

CHAPTER 10

10.2   Ay = 288.4 kN ↑, By = 223.6 kN ↑, M_A = 494.6 kN.m ↶

10.3   Ay = 40.5 kN ↑, By = 34.5 kN ↑, M_A = 66 kN.m ↶

10.4 M_B = 254.5 kN.m ↶

10.7 M_B = 12 kN.m ↷, Mc = 12 kN.m ↶

CHAPTER 11

11.1   Ay = 23.82 kN ↑, By = 5.06 kN ↓, Cy = 17.64 kN ↑, Dy = 5.6 kN ↑,
        M_A = 66.02 kN.m ↷

11.4   Ay = 27.45 kN↑, By = 148.37 kN↑, Cy = 59.16 kN ↑

11.6 M_B = ± 1492.46 kN.m, Mc = ± 1509.4 kN.m

11.7 M_A = 68.75 kN.m ↶, M_B = ± 72.6 kN.m, Mc = 71.7 kN.m ↷

11.8 M_B = ± 57.44 kN.m, Mc = ± 300 kN.m

11.10 M_D = 7.45 kN.m ↷

11.13   Ay = 51.58 kN↑, Ax = 26.16 kN ←, M_A = 20.53 kN.m ↶, Cy = 28.42 kN↑
        Cx = 23.84 kN←

11.14   Ay = 53.5 kN ↑, Dy = 96 kN ↑, Cy = 42.5 kN ↑, Dx = 19.14 kN←,
        M_A = 78.65 kN.m ↶, Mc = 49.55 kN.m ↷

CHAPTER 13

13.1   Ay = 144.2 kN ↑, By = 111.8 kN ↑, M_A = 247.335 kN.m ↶

13.2   Ay = 336.813 kN ↑, By = 101.627 kN ↑, Cy = 61.56 kN ↑,
        M_A = 1029.38 kN.m ↶

13.3   Ay = 27.45 kN↑, By = 148.39 kN↑, Cy = 59.16 kN↑

13.6   Mc = 160.032 kN.m ↷, M under the load = 346.655 kN.m ↷

CHAPTER 14

14.1a $Ay = 27.73$ kN $\uparrow$, $By = 4.89$ kN $\downarrow$, $Cy = 16.16$ kN $\uparrow$

14.1c $Ay = 16.83$ kN $\uparrow$, $By = 16.94$ kN $\uparrow$, $T_C = 12.22$ kN

14.1e $Ay = 1.09$ kN $\uparrow$, $By = 33.82$ kN $\uparrow$, $Cy = 1.09$ kN $\uparrow$,

      $M_A = 63.46$ kN.m $\curvearrowleft$

14.1h $Ay = 21.71$ kN $\uparrow$, $By = 0.91$ kN $\downarrow$, $Cy = 8.64$ kN $\uparrow$, $Dy = 4.56$ kN $\uparrow$,

14.1j $Ay = 170.33$ kN $\uparrow$, $By = 153.67$ kN $\uparrow$, $M_B = 318.33$ kN.m $\curvearrowright$,

      $M_A = 393.33$ kN.m $\curvearrowleft$

14.1k $Ay = 65.8$ kN $\uparrow$, $By = 42.2$ kN $\uparrow$, $M_A = 106.18$ kN.m $\curvearrowleft$

14.1p $Ay = 61.146$ kN $\downarrow$, $Ax = 46$ kN $\rightarrow$, $Dy = 58.85$ kN $\downarrow$, $Dx = 96$ kN $\leftarrow$

      $M_A = 20.17$ kN.m $\curvearrowright$, $M_D = 37.332$ kN.m $\curvearrowright$

14.1s $Ay = 303.58$ kN $\uparrow$, $Ax = 235.5$ kN $\rightarrow$, $Ey = 303.58$ kN $\uparrow$, $Ex = 235.5$ kN $\leftarrow$

      $M_A = 340.76$ kN.m $\curvearrowright$, $M_E = 340.76$ kN.m $\curvearrowleft$

14.3 $Ay = 67.18$ kN $\uparrow$, $Ax = 47.72$ kN $\rightarrow$, $Dy = 112.82$ kN $\uparrow$, $Dx = 7.72$ kN $\leftarrow$

      $M_A = 48$ kN.m $\curvearrowleft$, $M_D = 99$ kN.m $\curvearrowleft$, $M_C = 326.9$ kN.m

14.5 $Ay = 58.073$ kN $\uparrow$, $Ax = 0$, $Tc = 61.93$ kN, $M_A = 213$ kN.m $\curvearrowleft$

CHAPTER 15

15.1 $Cx = 24.029$ kN $\rightarrow$, $Cy = 17.961$ kN $\uparrow$, $Ax = 24.029$ kN $\leftarrow$, Ay 32.039 kN $\uparrow$

15.2 $Ay = 51$ kN $\uparrow$, $By = 69$ kN $\uparrow$, $M_A = 76.67$ kN.m $\curvearrowleft$

15-4 $Ay = 81.923$ kN $\uparrow$, $By = 98.077$ kN $\uparrow$, $M_A = 369.231$ kN.m $\curvearrowleft$,

      $\delta_c = 0.49$ m $\downarrow$

CONVERSION FACTOR

To convert	To	Multiply By
1- Length		
Inch	Millimeter	25.4
Foot	Millimeter	304.8
Yard	Meter	0.9144
Meter	Foot	3.281
Meter	Inch	39.37
2- Area		
Square inch	Square millimeter	645
Square foot	Square meter	0.0929
Square yard	Square meter	0.836
Square meter	Square foot	10.76
3- Volume		
Cubic inch	Cubic millimeter	16390
Cubic foot	Cubic meter	0.02832
Cubic yard	Cubic meter	0.765
Cubic foot	Liter	28.3
Cubic meter	Cubic foot	35.31
Cubic meter	Cubic yard	1.308
4- mass		
Ounce	Gram	28.35
Pound (Ib)	Kilogram	0.454
Pound	Gallon	0.12
Short ton (2000 Ib)	Kilogram	907
Long ton (2240 Ib)	Kilogram	1016
Kilogram	Pound (Ib)	2.205
Slug	Kilogram	14.59
5- Density		
Pound/cubic foot	Kilogram/cubic meter	16.02
Kilogram/cubic meter	Pound/cubic foot	0.06243
6- Force		
Pound (Ib)	Newton (N)	4.448
Kip (1000 Ib)	KiloNewton (kN)	4.448
Newton (N)	Pound	0.2248
KiloNewton (kN)	Kip (K)	0.225
7- Moments		
Foot . Kip	KiloNewton . meter	1.356
Inch . Kip	KiloNewton . meter	0.113
Inch . Kip	Kilogram force . meter	11.52
KiloNewton . meter	Foot . Kip	0.7375

REFERENCES

1- R .E. Sennett, *Matrix analysis of structure*, Prentice Hall, 1994

2- Mosley, Bungey, and Hulse, *Reinforced Concrete Design*, Macmillan Press,1999.

3- R. C. Hibbeler, *Structural Analysis*, Prentice Hall, 2002.

4- D. L. Schodoek, *Structures*, Prentice Hall, 2004.

5- Pandit & Gupta, *Structural Analysis - A Matrix Approach* , Tata McGraw Hill, 1981.

6- J.F. Fleming, *Analysis of Structural Systems*, Prentice Hall,1997.

7- Y-Y, Hsieh, *Elementary Theory of Structures*, Prentice Hall, 1988.

8- McGuire, Gallagher, and Ziemian, *Matrix Structural analysis*, Wiley, 2000.

9- Sujit and Subrata, *Fundamentals of structural Analysis*, S. Chand, 2003.

10- West and Geschwindner, *Fundamentals of structural Analysis*, Wiley,2002.

11- Leet and Uang, *Fundamentals of structural Analysis*, McGraw Hill, 2002.

12- S. S. Bhavikatti, *Structural Analysis*, Vikas & UBS, 1999.

13- F. Arbabi, *Structural Analysis and Behavior*, McGraw Hill, 1991.

14- S. D. Rajan, *Introduction to Structural Analysis and Design*, Wiley, 2001.

15- Vazirani and Ratwani, *Analysis of Structures*, Khanna, 1993.

16- SR. l. Sack, *Matrix Structural Analysis*, Waveland Press, 1989.

17- J. C. McCormac, *Design of Reinforced Concrete*, Wiley, 2001.

18- M. N. Hassoun, *Structural Concrete Theory and Design*, Prentice Hall, 2002.

19- M. S. Al – Ansari, Simplified Method for Plate Deflection Calculation, Engineering Journal Of The University of Qatar, Vol. 10, pp51-67.

20- S. Timoshenko, *Theory of Plates and Shells*, McGraw Hill, 1959.

21- R.S.Khurmi, *Theory of Structures*, S & C Pub., 1992.

22- S. Handa, *Gate (Civil Engineering)*, Satya Prakashan, 2001.

23- J. Nelson & J. McCormak, *Structural Analysis: Using Classical and Matrix Method*, Wiley, 2003.

INDEX

Printed in the United States
By Bookmasters